NUMBER-CRUNCHING

NUMBER

CRUNCHING

Taming Unruly Computational Problems from Mathematical Physics to Science Fiction

A collection of challenging problems in mathematical
physics that roar like lions when attacked analytically,
but which purr like kittens when confronted by a high-speed
electronic computer and its powerful scientific software
(plus some speculations for the future from science fiction)

PAUL J. NAHIN

Princeton University Press
Princeton and Oxford

Copyright © 2011 by Princeton University Press

Published by Princeton University Press, 41 William Street, Princeton, New Jersey 08540

In the United Kingdom: Princeton University Press, 6 Oxford Street, Woodstock, Oxfordshire OX20 1TW

All Rights Reserved

Library of Congress Cataloging-in-Publication Data

Nahin, Paul J.

Number-crunching : taming unruly computational problems from mathematical physics to science / Paul J. Nahin.

p. cm.

"A collection of challenging problems in mathematical physics that roar like lions when attacked analytically, but which purr like kittens when confronted by a high-speed electronic computer and its powerful scientific software (plus some speculations for the future from science fiction)."

Includes bibliographical references and index.

ISBN 978-0-691-14425-2 (hardcover : alk. paper)

1. Mathematical physics—Data processing. 2. Mathematical physics—Problems, exercises, etc.

I. Title.

QC20.7.E4N34 2011

530.150285—dc22

2010048685

British Library Cataloging-in-Publication Data is available

This book has been composed in ITC New Baskerville
Printed on acid-free paper. ∞
press.princeton.edu
Printed in the United States of America

10 9 8 7 6 5 4 3 2 1

FRONTISPIECE CAPTION: Two 1950s teenagers explore the mysteries of chemistry under the watchful gaze of their high school science instructor, Paul Norris (1895–1989). (To see what happened next, turn to the start of the Solutions section.) The mystery of the Bunsen burner and the flaming "something" (who can remember *what* it was after fifty-five years!) wasn't deep enough, however; both boys knew the real action in the future would be in physics, math, and electronics. Both became electrical engineers, the author (on the left) in industry and academia, and his boyhood friend John Baker as a member of the technical staff at the Naval Research Laboratory and the Lawrence Livermore National Laboratory. Photographs reproduced from Baker's copy of *Gusher*, the Brea-Olinda Union High School (Brea, California) yearbook for 1956.

For Patricia Ann

Sunsets are red and clear skies are blue,

Both are explained by physics,

and mathematics, too.

Physics and math—each is wonderful, it's true,

But if I had to choose,

and could choose only one,

I'd choose you.

SHOULD YOU READ THIS BOOK?

In his 1981 autobiographical book *Psalm Sunday*, the late science fiction writer Kurt Vonnegut tells us what his position on algebra was:

Question: If Farmer A can plant 300 potatoes an hour, and Farmer B can plant potatoes fifty percent faster, and Farmer C can plant potatoes one third as fast as farmer B, and 10,000 potatoes are to be planted to an acre, how many nine-hour days will it take Farmers A, B, and C, working simultaneously, to plant 25 acres?

Answer: I think I'll blow my brains out.

If your reaction was somewhat less extreme, and instead you picked up a pen and started to work out the answer, then this is a book for you! And just so you don't go away with the wrong impression of Vonnegut, he also writes elsewhere in the same book: "I believe that the best minds of my generation were probably musicians and physicists and mathematicians." Kurt may have stunk at math, but you just have to admire somebody who has that kind of penetrating insight.

CONTENTS

INTRODUCTION

I didn't become a mathematician because mathematics was so full of beautiful and difficult problems that one might waste one's power in pursuing them without finding the central problem.
   —Albert Einstein

Mathematics is trivial, but I can't do my work without it.
   —Richard Feynman

These two views, from two of the most famous physicists of the last century, on the relationship between physics and mathematics, are quite different. Both won the Nobel Prize (Einstein in 1921 and Feynman in 1965), and their words, however different, deserve some thought. Einstein's are ones without sting, carefully crafted to perhaps even bring a surge of pride to the practitioners of the rejected mathematics. Feynman's, on the other hand, are just what we have come to expect from Feynman—brash, outrageous, almost over-the-top. Feynman's comment really goes too far, in fact, and I think it was uttered as a joke, simply to get attention.[1] Physicist Feynman was also a highly skilled mathematician, and not for a moment do I believe he really thought mathematics to be "trivial." (Mathematicians shouldn't take such "Feynman put-downs" too seriously; I'm an electrical engineer and while Feynman had some jibes for EEs, too, I have never let them influence my appreciation for his genius.[2])

The mathematician Peter Lax, in his 2007 Gibbs Lecture to the American Mathematical Society, gave a good, concise summary of the interplay between mathematics and physics.[3] His talk opened with these words: "Mathematics and physics are different enterprises: physics is looking for laws of nature, mathematics is trying to invent the structures and prove the theorems of mathematics. *Of course these structures are not invented out of thin air but are linked, among other things, to physics*" (my emphasis). A few years earlier a physicist, in his own Gibbs Lecture, gave a specific illustration of this connection:[4] "The influence of general relativity in twentieth-century mathematics has been clear enough. Learning that Riemannian geometry is so central in physics gave a big boost to its growth as a mathematical subject; it developed into one of the most fruitful branches of mathematics, with applications in many other areas." The influence can flow in the other direction, too: it was the *mathematician* Kurt Gödel (Einstein's friend at the Institute for Advanced Study at Princeton) who in 1949 discovered that the equations of general relativity allow for the possibility of time travel into the past, a totally unexpected result—previously thought to be a childish science fiction fantasy—that quite literally left the *physicist* Einstein nearly speechless.[5] I have also long marveled at the intimate connections between physics and mathematics, and this book is the result of that fascination. Indeed, *Number-Crunching* is a continuation of my 2009 book, *Mrs. Perkins's Electric Quilt*, inspired by the same fascination.

As you read through the chapters of this book you'll see how important I find the role of the modern high-speed electronic computer to the study of both mathematics and physics. The relatively new subject of computer science can legitimately be viewed, in fact, as a bridge connecting those two much older subjects (whose pure practitioners have sometimes appeared to be at extreme odds). As one pioneer in the development of electronic computers once bluntly wrote,

> The relationship ... between physics and mathematics may best be described as an unsuccessful marriage, with no possibility of divorce. Physicists internalize whatever mathematics they require, and eventually claim priority for whatever mathematical theory they become acquainted with. Mathematicians see to it that every physical theory, sooner or later, is freed from all shackles of reality to fly in the thin air of pure reason.[6]

The author of that rather harsh assessment had a background that included all three subjects: he held a PhD in physics, made important contributions to applied mathematics, and was the guiding spirit behind one of the earliest electronic computing machines (the famous MANIAC-I at the Los Alamos Scientific Laboratory), and so I think his words are worth consideration. I personally don't believe, however, that the schism between mathematics and physics is nearly as wide as he asserted, and whatever their actual divide may have been when he wrote, the development of the electronic computer has certainly narrowed rather than widened it. With electronic computers and the development of powerful software by computer scientists to "power" these machines, mathematicians can now perform physics "experiments" via simulation, and physicists can perform mathematical feats that once defeated even such genius as that of Isaac Newton. In this book you'll see examples of both such uses of computers.

The laws of physics are quite compact. In their natural language—mathematics—these laws are easy to write down on paper. All of the known fundamental laws can in fact be completely written out on just a couple of pages. And for many analysts it is that compactness that is the real puzzle, not the laws themselves, as much of the observed behavior of the real world appears far too complex to be explainable by a mere few lines of symbols. Einstein famously gave life to this puzzle in a 1936 essay when he declared, "The fact that it [the world] is comprehensible is a miracle." Feynman included a particularly poetic statement of the wonderful entanglement of mathematics and physics in his *Lectures on Physics* when he ended one presentation with these words: "The next great era of awakening of human intellect may well produce a method of understanding the *qualitative* content of equations. Today we cannot. Today we cannot see that the water flow equations contain such things as the . . . structure of turbulence. . . . Today we cannot see whether Schrödinger's equation [the probability wave equation of quantum mechanics] contains frogs, musical composers, or morality—or whether it does not. We cannot say whether something beyond it like God is needed, or not. And so we can all hold strong opinions either way."[7]

The ancients had an effective way of avoiding a need for mathematical physics; everything they found perplexing was "explained" (in the form of a good story usually involving revenge, envy, sex, and death—see any good book on mythology) by just saying "the gods make it happen." And, of course, it *is* a lot easier to make up good stories around the campfire at night than it is to discover good physics, which is why the god myths are thousands of years old and the origins of good physics are considerably more recent.

The theme here remains the same as in *Mrs. Perkins*. The theme of *Mrs. Perkins* and of *Number-Crunching* is to show by example how the fundamental laws of mathematical physics, combined with the tremendous computational power of modern computers and their software, can explain extremely complicated behaviors, behaviors often so counterintuitive as to bring gasps of astonishment even from experienced, professional analysts. If you have studied the first two years of college physics and math and understand the basic ideas behind the writing of computer codes in a high-level language (I use MATLAB), then you should be able to read this book. Each chapter ends with at least one challenge problem (solutions are provided at the end of the book) to let you enjoy your own gasps of astonishment. (Here's a quick example of what I mean by a surprise: a straight, precisely one-mile-long stretch of continuous railroad track is laid during a cold night, with the two ends firmly fixed in the ground. The next morning the hot sun causes the track to expand by exactly one foot, and so the track buckles upward. Now, quickly, off the top of your head, is the midpoint of the track raised above the ground by (1) several inches, (2) several feet, or (3) several yards? The answer is at the end of this Introduction.)

As a second, far more serious example of both the sort of analyses in this book and the minimum level of the mathematics and physics you should feel comfortable with to read them, let me now describe a situation that at first glance may seem to present an impossible task. To set the stage for this problem, consider first the following recent extreme weather event that I personally experienced. In December 2008 a huge ice storm devastated the northeastern United States, and my home state of New Hampshire in particular. As the storm developed, vast numbers of trees, covered in thick layers of ice, collapsed, and as they fell they tore down nearby power lines that hadn't already had their wires snapped by the weight of the heavy ice that coated them. The damage was enormous, with lines down everywhere, resulting in widespread power outages that shut down hundreds of thousands of home heating systems and so drove large numbers of people from their freezing houses and into hotels and community shelters.

It took up to two weeks to restore electrical power in some areas of the state, but the repair crews could at least *see* the lines that were down.

Just imagine how much more difficult it would have been if the repair crews couldn't have seen the damaged wires. Now, that might seem to be an artificial complication, but it really isn't. It occurs in a natural way, for example, in even more extreme fashion, if there is a break in an undersea communication cable. Locating such "invisible" breaks became a problem right from the beginning of the laying of such cables in the middle of the nineteenth century. One could, of course, imagine a repair ship sailing along the known path of the submerged cable, periodically grabbling for the cable, and pulling it up to the surface until the break (or *fault*) is found. Considering that a cable might be hundreds of miles long, in hundreds or even in thousands of feet of water, however, should make it clear that that could easily be a lengthy and most expensive approach. Can we do better in finding the location of a fault, even if we can't see the cable? Yes. What I'll next show you is a technique called the *Blavier method*, invented by the French telegraph engineer Edouard Ernst Blavier (1826–1887).

I should, however, start with some words on how a fault might occur in an undersea cable. You won't be surprised, I think, if I begin by saying that an ice storm generally is not the cause! Much more likely is that shifting water currents would move the cable back and forth across a rough sea bottom and thereby cause its outer shielding to be abraded. Another, perhaps less obvious cause of a cable fault was a fish bite! As one observer noted in an 1881 letter to the English trade journal *The Electrician*, a cable laid in 1874 was soon after found to have suffered "at least four indubitable fish-bite faults ... where the iron sheathing had been forcibly crushed up and distorted from the core as if by the powerful jaws of some marine animal." That writer presented convincing evidence pointing to *Plagyodus ferox* ("one of the most formidable of deep-sea fishes") as the culprit. In any case, before the cable broke completely it would first have developed a so-called *electrical leakage fault* from its still physically intact internal wires to the surrounding water.

Well, no matter the origin of the fault; we'll mathematically model it and the cable as shown in Figure I.1. Let's assume the cable has a fixed, known electrical resistance per unit length, and that we write the total resistance of the cable from one end to the other as a ohms, the resistance from the left end of the cable to the fault as x ohms, the resistance from the fault to the right end of the cable as $a - x$ ohms,

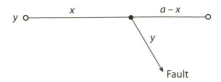

FIGURE I.1. Locating the fault

and the resistance from the fault to the sea water (which we'll take as equivalent to a good earth ground) as y ohms. If we can determine the value of x, then we can use the resistance per unit length of the cable to calculate how far the fault is from either end.

To start, let's write

$$b = x + y.$$

We can measure the value of b from the left end of the cable by open circuiting the right end, applying a battery of known voltage between the left end and the earth, and measuring the resulting current. Ohm's law then gives us b. Next, we'll ground the right end of the cable, which puts the $a - x$ portion of the cable in parallel with the fault resistance. If we call c the resulting resistance now seen from the left end of the cable, then

$$c = x + \frac{y(a - x)}{y + a - x}.$$

We now have two equations in two unknowns, x and y, so we can eliminate the fault resistance y and solve for x. Since $y = b - x$ from the first equation, substituting this into the second equation gives, with just a little algebra, the quadratic equation

$$x^2 - 2cx + c(a + b) - ab = 0,$$

which is easily solved (with the famous quadratic formula) to give us two *real* values for x:

$$x = c \pm \sqrt{(a - c)(b - c)}.$$

(Can you see why both $(a - c) > 0$ *and* $(b - c) > 0$, and so we have the square root of a positive quantity, and thus x is *guaranteed* to be real?)

The fault is located somewhere in *one* place, of course, not two, and so we can't have two values for x. So, which sign do we use in front of the square root, the plus or the minus? The answer is the minus sign, because (by definition) $x < a$ (and so $a - x > 0$), and the fault resistance must obviously be positive ($y > 0$). This says

$$x = c - \frac{y(a - x)}{y + a - x} < c.$$

Thus, the unique solution for x is

$$x = c - \sqrt{(a - c)(b - c)}.$$

Some years ago, while researching a biography of the English mathematical electrophysicist Oliver Heaviside (1850–1925), I discovered one of his notebooks in which Heaviside had recorded an application of this formula that he had made while working as a telegraph operator in Denmark.[8] Observing that an electrical fault had occurred somewhere along the 360-mile undersea cable between Sondervig in Denmark (the right end) and Newbiggin-by-the-Sea in England (the left end), the opened and shorted resistance measurements gave the values of $b = 1,040$ ohms and $c = 970$ ohms, respectively. Knowing that the cable had a resistance of 6 ohms per mile, he had $a = (6)(360) = 2,160$ ohms. Therefore,

$$x = 970 - \sqrt{(2,160 - 970)(1,040 - 970)} = 682 \, \text{ohms}.$$

Thus, the fault was located $682/6 = 113\frac{2}{3}$ miles from the left end (the English end). Heaviside was clearly quite pleased with this pretty calculation: his notebook entry (dated January 16, 1871) recording his work ends with the happy words, "All over. Dined roast beef, apple tart and rabbit pie with claret and enjoyed ourselves." As you work your way through this book I hope you'll find it isn't all just numbers, theorems, and calculus calculations. It is permissible, indeed *required*, that you have fun, too. If a serious fellow like Heaviside can enjoy solving a quadratic equation, so can you.

Let me give you an example of what I mean by having fun. More than thirty years ago (1978) the publisher of *Penthouse* started a new magazine called *Omni*. *Omni* lasted not quite two decades, ceasing publication in 1995, but while it existed *Omni* was a bright star in a part of the literary sky that is all too often ignored. It was devoted to the reporting of cutting-edge scientific breakthroughs, the use of beautiful science-based art, and the publication of short science fiction stories. Other than *Playboy*, *Omni* was the only glossy print magazine publishing first-rate science fiction *and* paying *Playboy*-scale money to fiction writers. One of *Omni*'s regular features was its interview essay, similar to a regular feature in *Playboy*. Rather than interviewing rock music stars, Hollywood celebrities, or politicians, however, *Omni*'s interviews focused on visionary thinkers and world-class scientists; Feynman's friend, the mathematical physicist Freeman Dyson of the Institute for Advanced Study in Princeton, was the subject of the very first interview. A few months later, Feynman himself was the featured personality in the magazine.

Another regular feature in *Omni* was the one-page humor essay that always appeared as the final page of each issue, and so was appropriately called "Last Word." In 1982 I wrote the following "Last Word," which I've reproduced below with no editing other than adding some elaborative endnotes for which there was no room in the magazine.[9]

Just about everyone who learns to read eventually comes across the story of the monkeys and the typewriters. The idea is elegantly simple. Merely put a bunch of monkeys in front of an equal number of typewriters (and a *lot* of paper) and let them bang away on the keys. They'll produce mostly gibberish, of course, but the immutable laws of probability predict they will also reproduce verbatim *everything* that has ever been written.[10] Like this issue of *Omni*, this essay included.

And they'll produce everything else that will and even could be written. All the books of the future will be included in the monkeys' output: masterpieces of science fiction, fascinating volumes of "history" describing and analyzing events yet to occur and a lot that won't, and treatises on marvelous scientific breakthroughs.

This cosmic collection of written words is called the Universal Library, a term coined by Kurd Lasswitz, a German philosopher and mathematics professor. According to Willy Ley, the original concept of such a collection can be traced back to the medieval Spanish philosopher and mystic Ramon Lully.[11]

Whatever you want to call it, this incredible storehouse of knowledge would lead to tantalizing revelations. If a time machine could ever be built, somewhere in the megatons of paper spewing from the typewriters will be contained the directions on how to put one together. So would the recipe for a faster-than-light spaceship drive, a cure for cancer, and the secret of longevity.

But the rub, of course, is that there would be a multitudinous quantity of chaff with the wheat. How could we search through this literary equivalent of the Augean stables to find such gems? Until now, we couldn't, and the idea has remained just a vague speculation, good for a quick chuckle, but soon discarded so we can attend to more practical and more pressing matters.

No more. Using high-speed digital computers, coupled with the latest innovations in high-resolution imaging, instant developing, and reusable microfilm, we could begin right now realizing the ancient dream of the Universal Library. Better yet, we can do something more impressive: put together the Universal Photo Album (TUPA).

To understand TUPA, it is necessary to grasp just a few elementary concepts. First, any black-and-white photograph is nothing more than a composite of small flecks of brightness levels varying from black through various shades of gray to white.[12] Second, a computer can easily be programmed to generate sets of flecks in a fine mesh grid. The programmer can also instruct the fleck-generating machine never to repeat itself. The computer would toil, like a team of monkeys, to produce all possible combinations and permutations of brightness levels. Instead of generating a mountain of typed gibberish, it would create a mountain of visual gibberish, countless random images flying out from this tireless image factory.

Then, to examine the enormous number of images, we could take advantage of recent advances in machine vision, artificial intelligence, and work in pattern recognition. Special computers would be used to process the first batch of photos, rejecting the ones that are clearly nonsensical and referring the interesting ones to humans for interpretation. (Since this would require just about every piece of film ever made, we could save on film costs by recycling the rejected photos and using the film over again.)

What would we get from all of this? Simply all possible black-and-white photographs: all that have ever been taken, all now being taken, and all that could ever be taken. We get TUPA.

What a treasure it would be! In TUPA would be included pictures of every human who has ever lived—Moses, Jesus, Henry VIII, you, me—as well as every human who will ever live. There will be images of every creature in the universe, some of them more incredible than we could imagine. You'll see in TUPA all the vacation photos you will ever take, including the ones where you left the lens cap on.[13] It is clear that purveyors of pornography, for example, will be galvanized by the discovery of a hoard of titillating photos[14] in TUPA. Best of all, the Universal Library will be there as well, because TUPA will have a picture of every page in every book ever published.

This entire project could be started immediately with a front-end capitalization of $20 million—give or take a few million—for computers, laser equipment, computer-program development, and film-chemistry engineering. We would also need a sustaining budget of $3 million for computer time, film, and salaries for the photo interpreters. This level of funding is trivial compared to the Defense Department's and is well worth the investment if you consider the potential payoff militarily. TUPA would include pictures of every top-secret Soviet military document, present and future.

Project TUPA could be our next great endeavor, rivaling the Apollo and Manhattan projects. But we must act now because, as the Russians will soon realize, if they haven't already, TUPA contains photos of every top-secret American military document, too.

I was (and still am) enormously intrigued by the concept of TUPA. TUPA is a generalization of the Universal Library that goes far beyond the Universal Library. As I mentioned in my "Last Word" essay, TUPA contains the Universal Library as a subset because in TUPA there is an image of every page of every book in the Universal Library. And not just in English, but in every language that has ever existed, exists now, or will exist, on Earth as well as on *every* other planet in the entire universe that has, does now, or will support intelligent life with a written language. The Great Library of Alexandria, Egypt, the largest collection of books in the ancient world that was destroyed numerous times, from when it was accidently burned in 48 B.C. by Caesar's troops until its final destruction in A.D. 642 by Arab invaders, would be lost no more. TUPA contains all possible illustrated books, too, something not in the original Universal Library. And not only that, TUPA contains itself as a subset, because in TUPA there are images of ever more slightly reduced versions of every image in TUPA.

Some readers of *Omni* were just as excited by TUPA as I was, in fact, and wrote to the magazine to ask just why in heck the U.S. government wasn't going forward with Project TUPA. As much as I love the concept of TUPA, however, those letters caught me by surprise. After all, while it is true that TUPA violates no laws of physics, there is nonetheless a very good *mathematical* reason why it will forever remain out of reach—see Challenge Problem I.4. On the other hand, maybe those readers were simply returning the joke!

Okay, while I take the position that even though a good nighttime campfire story is always fun to hear while roasting marshmallows on a stick, or that conceits like TUPA can stimulate some fun talk, in the end, such amusements, when matters finally turn to serious business, are simply no substitute for disciplined, logical, analytical thinking. Since you're reading this, I take it that you agree with me. So, with no further delay, let's get started. Here are your first four challenge problems, to give you a feeling for the level of the mathematics in this book. The first one, in particular (it's not computationally difficult), reflects, I believe, Feynman's true feelings about the non-trivial nature of mathematics.

Challenge Problems

CP I.1. In a letter dated July 7, 1948, to the physicist Hans Bethe, winner of the Nobel Prize in 1967, Feynman outlined his latest research efforts in quantum electrodynamics. At one point he said it all hinged on the use of what he called a "great identity": $\frac{1}{ab} = \int_0^1 \frac{dx}{[ax + b(1 - x)]^2}$, where a and b are constants. Assuming a and b are real but otherwise arbitrary constants, explore the truth of Feynman's identity. In particular, explain the following "puzzle": if a and b have opposite signs, then obviously $\frac{1}{ab} < 0$, but just as obviously, the integral is *never* less than zero for *any* real a and b since the integrand is a real quantity squared, and so is nowhere negative. Clearly, *something* more needs to be said here. What's going on?

CP I.2. Here's another purely mathematical problem that, even though it is "just" high school algebra and trigonometry, almost always astonishes even professional mathematicians. Just about all (that includes me) who first see this problem initially think it is impossible to do—but that isn't so! In addition, it has a technical application to a famous physics problem first solved by Einstein in 1905 (the reflection of light from a moving mirror). Starting with the equation $\sin(\alpha) - \sin(\beta) = -k \sin(\alpha + \beta)$, where k is a constant and both α and β are in the interval 0 to $\pi/2$, solve for β as a function of k and α. Despite its convoluted appearance, this is not a transcendental equation. I think Feynman would have had fun working through this calculation, and I hope you do, too, but fair warning: prepare yourself for at least some mental exertion (there is significantly more computation involved than in the first problem). As a partial check on your general result, notice that in the case of $k = 0$, your answer should obviously reduce to $\beta = \alpha$.

CP I.3. When I was a freshman at Stanford I did well enough during the first two terms of calculus to be allowed to transfer into the honors section of the course. (That's when I found out *lots* of my fellow freshmen were at least as good at math as I was!) One of the homework problems in that course (Math 53, Spring Quarter 1959) was the following: prove that the shortest path length between any two given points in a plane is that of the straight line segment connecting the two points. This is, of course, "obvious" to anybody with a body temperature above that of an ice cube, but the point of the problem was to construct a proof. I remember sitting at my desk in the dorm one night wondering just how I might attack this problem—and then I hit upon what I thought an incredibly clever idea. Obviously the professor intended us to use what we had been discussing in lecture, and that just happened to be the general formula for the arc length of a smooth curve (*smooth* means the curve has a tangent at every point, which means the curve has a derivative at every point). In fact, if we have the curve $y = y(x)$, then the length L of the curve segment connecting points A and B, with coordinates (x_A, y_A) and (x_B, y_B), respectively, in the usual Cartesian coordinate system, with $x_A \leq x_B$, is given by $L = \int_{x_A}^{x_B} \sqrt{1 + \left(\dfrac{dy}{dx}\right)^2} \, dx$. Well, I reasoned, the length L is *independent* of the particular choice of coordinate axes, that is, we can choose the orthogonal x- and y-axes to be at any orientation we wish (and place the origin anywhere we want, too) and we'll always get the same L. (Physicists call L a *physical invariant*, while the choice of coordinate axes is a *mathematically arbitrary* decision.) In particular, then, let's pick the x-axis to be such that $y_A = y_B$, which we clearly can always do. Then the straight line segment joining A and B will have zero slope, and so $\dfrac{dy}{dx} = 0$ *always*. But clearly, "$\dfrac{dy}{dx} = 0$ *always*" is just what is needed to minimize the integrand *at every point in the interval of integration*, and so L itself will be minimized. I recall I wrote that all up and finished it off in a flourish with a big "QED" (just to impress the professor with my cultural sophistication and mastery of Latin!). Alas, my homework paper came back the next week with a big red cross through my brilliant proof (the word **NO** was also quite prominently displayed), along with a score of 1 (for trying) out of 10. I was too embarrassed by all that red ink to ask where I had gone wrong (which of course was a silly reaction). So, there's *your* challenge: explain where I went wrong fifty years ago. *Note:* In my book *When Least Is Best* (Princeton, N.J.: Princeton University Press, 2004, corrected edition 2007), written when I had become a somewhat more sophisticated fellow than I was in 1959, I solve this problem (pp. 238–239) using the Euler-Lagrange formulation of the calculus of variations. And in my book *Dr. Euler's Fabulous Formula* (Princeton, N.J.: Princeton University Press, 2006) I solve it again, that time using Fourier analysis (pp. 181–187). But surely my Math 53 professor didn't expect his *freshman* students to use either of those approaches. At least I don't think he did.

CP I.4. (a) A black-and-white $m \times n$ digital TUPA image consists of mn pixels, each with k bits of gray-level coding ($2^k - 1 \Longrightarrow$ whitest white and $0 \Longrightarrow$ blackest black). How many such pictures are in TUPA? (b) If $mn =$ ten million (that is, if a black-and-white TUPA picture is ten megapixels, comparable to what, as I write, a good digital camera produces), and if $k = 11$ bits (there are $2, 048$ gray levels), a black-and-white TUPA image would be a very crisp, ultra-sharp, high-resolution photo. How many such images are in TUPA? (c) If we had a computer that could automatically generate one hundred trillion of these TUPA images per nanosecond, then what fraction of TUPA would have been generated by now since the start of time at the instant of the Big Bang fifteen billion years ago?

(The answer to the buckled railroad track problem is several yards. Are you surprised? For an analysis, see the Solutions section at the end of the book.)

Notes and References

1. I never personally talked with Feynman, but when I was in high school my father took me to a public lecture Feynman gave one weekend at Caltech. It must have been sometime in 1957 or so. Brea, my hometown in 1950s Southern California, is only an hour's drive from Pasadena, and my father (holder of a PhD in chemistry, but with scientific interests that ranged far beyond chemistry) had heard that Feynman was "the next Einstein." So, off we went. That experience gave me a firsthand encounter with Feynman's personality. I had never before heard anyone like Feynman. He certainly wasn't boring, dull, or pompous! He talked like a New York City wiseguy, cracked jokes, and clearly had a good time. I recall being initially shocked (and then enormously entertained) by how irreverent he was during the talk, which long after I recognized as presenting material that later appeared in his famous, equally irreverent *Lectures on Physics* (Addison-Wesley, 1963). It is in volume 2 of *Lectures* that you'll find (p. 7–2) this comment by Feynman, which I think correctly illustrates his real feelings about math: "Now we come to a

miraculous mathematical theorem which is so delightful we shall leave a proof of it for one of your courses in mathematics." The reference is to the Cauchy-Riemann equations from complex variable theory (although Feynman didn't tell his young students that), about which you can find more in my book *An Imaginary Tale: The Story of $\sqrt{-1}$* (N.J.: Princeton University Press, Princeton, 1998 [corrected printing 2007], pp. 191–194).

2. For a Feynman comment on EEs, see pp. 23-6 and 23-7 in volume 1 of *Lectures*. He does make an apology of sorts to electrical, mechanical, and civil engineers on pp. 16-8 to 16-10 of volume 2 of *Lectures* when he describes his impressions of Boulder Dam (which Feynman rightfully thought to be an engineering marvel). On at least one occasion Feynman got back at least as good as he gave. The great probabilist Mark Kac (1914–1984) once gave a lecture at Caltech, with Feynman in the audience. When Kac finished, Feynman stood up and loudly proclaimed, "If all mathematics disappeared, it would set physics back precisely one week." To that outrageous comment, Kac shot back with that yes, he knew of that week; it was "Precisely the week in which God created the world."

3. I have taken the two opening quotes from Lax's talk; see Peter D. Lax, "Mathematics and Physics" (*Bulletin of the American Mathematical Society*, January 2008, pp. 135–152). The quotation from Lax himself repeats what the great English mathematician J. J. Sylvester (1814–1897) wrote decades earlier in an 1879 paper, "The object of pure Physic is the unfolding of the laws of the intelligible world; the object of pure Mathematic that of unfolding the laws of human intelligence."

4. Edward Witten, "Magic, Mystery, and Matrix" (*Notices of the American Mathematical Society*, October 1998, pp. 1124–1129).

5. See, for example, my book *Time Machines: Time Travel in Physics, Metaphysics, and Science Fiction*, 2nd ed. (New York: Springer, 1999, pp. 19, 80–84, and 489–495), and Wolfgang Rindler, "Gödel, Einstein, Mach, Gamow, and Lanczos: Gödel's Remarkable Excursion Into Cosmology" (*American Journal of Physics*, June 2009, pp. 498–510).

6. Nicholas Metropolis, "The Age of Computing: A Personal Memoir" (*Daedalus*, Winter 1992, pp. 119–130). Metropolis will appear again, in a central role, in Chapter 3.

7. In volume 2 of Feynman's *Lectures*, p. 41–12.

8. See my book *Oliver Heaviside: The Life, Work, and Times of an Electrical Genius of the Victorian Age* (New York: IEEE Press, 1988; repr., Baltimore: Johns Hopkins University Press, 2001, pp. 25–26). Since Heaviside's time, better

methods for locating breaks have been developed. One of the grandmasters of science fiction, Robert Heinlein (1907–1988), used one of the new ideas in his first published story, "Life-Line" (*Astounding Science Fiction*, August 1939). Heinlein was no mere "penny-a-word" pulp magazine hack; he was in the top 10% of the U.S. Navy Academy's Class of 1929, and took graduate courses in physics and mathematics at UCLA after his early forced retirement (because of tuberculosis) from the Navy in 1934. Heinlein used his technical background to draw an analogy in his story between a space-time worldline and an electrical cable. The beginning and ending points in space-time of the worldline of a person (birth and death) are associated with breaks (faults) in the cable. By sending a brief pulse through the cable and measuring the time delay until the arrival of the echoes produced by any fault discontinuities in the cable, a technician can both detect and accurately locate faults. In a similar manner, Heinlein's story gadget locates the birth and death "faults" along a person's worldline. Knowledge of the death fault, in particular, causes financial chaos in the life insurance business, and an examination of that tension (not weird physics) is the point of the story.

9. Paul Nahin, "Last Word" (*Omni*, April 1982). (Please note the publication month.)

10. The monkey-and-typewriters parable can be traced back to a paper on statistical mechanics that the French scientist Émile Borel published in a 1913 physics journal (the concept of the Universal Library is even older, however— see the next note). The idea has reappeared many times since; for example, the English mathematical physicist A. S. Eddington used it in his 1927 book *The Nature of the Physical World*, as did Sir James Jeans in his 1930 book *The Mysterious Universe*. The American writer Russell Maloney updated the concept in his elegant 1940 short story "Inflexible Logic," published in The *New Yorker*. (Maloney's tale is reprinted in both the fourth volume of James R. Newman's *The World of Mathematics* and Clifton Fadiman's *Fantasia Mathematica*.) Three years later, in October 1943, the fictionalized idea moved from the glossy, high-class *New Yorker* to the cheap wood-pulp pages of *Astounding Science Fiction*, in Raymond F. Jones's "Fifty Million Monkeys." In this poorly written super-science tale the entire universe is doomed. To find the solution to the problem—which has eluded all human scientists—the hero builds a random machine to "try everything" in the spirit of Lully's machine (see the next note).

11. Lasswitz's short story "The Universal Library" was originally published in a 1901 German book, and its translation by Willy Ley is in Clifton Fadiman, *Fantasia Mathematica* (New York: Simon and Schuster, 1958). A few years later the Argentine writer Jorge Luis Borges gave a very detailed description of the Universal Library in his short story "The Library of Babel," collected in

Ficciones (New York: Grove Press, 1962). As Ley explains in a postscript to Lasswitz's story, Lully's thirteenth-century version of the Universal Library was actually in the form of a *constructable machine* that could automatically generate all possible hypotheses. This idea was elaborated in a famous 1947 book by physicist George Gamow, *One Two Three ... Infinity* (Ley's postscript gives a nice summary of Gamow's machine that, if you read it carefully, makes it plain that it isn't *quite* the Universal Library itself that is generated). More recently, Lully's idea was given a humorous treatment by Philip Cole in his paper "The Hypothesis Generating Machine" (*Epidemiology*, May 1993, pp. 271–273).

12. "Small flecks of brightness" is, of course, a terribly awkward phrase. My original draft of this essay simply defined a digital black-and-white TUPA image as an $m \times n$ array of k-bit numbers. That is, as mn *pixels* (picture elements) with k-bits of gray level (2^k levels, from 0 to $2^k - 1$). With some additional bits per pixel we could easily add color to TUPA, too, and so TUPA would contain not only all possible books but also the images of all possible paintings. The editors at *Omni*, however, decided that bits and pixels were "too technical" and substituted "small flecks of brightness." Still, since 1982 was back in the personal computer dark ages, and not today's world, in which six-year old, computer-savy kids routinely blast through Xbox360 or Playstation 3 video games on HD displays, maybe those long-ago editors were right.

13. There is just one image in TUPA in which every last pixel is the blackest black. But there are *many more* images in which almost all the pixels are the blackest black or nearly so, and those images too could be legitimately called "lens cap on" pictures.

14. Since *Omni* was published under the corporate umbrella of *Penthouse*, I had hopes for a while that *Penthouse* would hire me as a consultant, specifically to construct this exciting subset of TUPA photos for the magazine's centerfold feature. And, since TUPA is gender-neutral, TUPA also contains super-sizzling images that would make the famous centerfold shot of Burt Reynolds in the April 1972 *Cosmopolitan* look tame. Alas, neither magazine ever called.

NUMBER-CRUNCHING

1
FEYNMAN MEETS FERMAT

For my money Fermat's theorem is true.
    —Richard Feynman, after calculating that the probability the
theorem is false is "certainly less than 10^{-200}..."

1.1 The Physicist as Mathematician

In the Introduction I mentioned that we will often use a computer in this book. Indeed, modern computer software is a powerful tool for mathematical physicists, and in this first chapter I want to give you a simple illustration of this. (I have strong personal views on the usefulness of computers in society in general, and not just as a tool for mathematical physicists. In Chapter 8 of this book are illustrations of some of those views using the device of short fictional stories published more than thirty years ago.) As the epigraph to this chapter suggests, my example involves both Feynman and the famous Last Theorem of Fermat. Now, I also told you in the Introduction that physicist Feynman was a highly skilled mathematician, and to illustrate that claim, let us consider the following amusing tale.

Every year the best college undergraduate math students in America compete (as multiperson teams representing each participating school) in a national test called the Putnam Exam. In addition to naming the top school teams, the five top scoring individuals are awarded the title of Putnam Fellow. The Putnam Exam is very difficult, with a median individual score in some years being 1 (yes, *one*) out of a possible 120. That means an awful lot of people get zeros!

The Putnam was in its second year of existence when Feynman was a senior physics major at MIT in 1938–39. When the MIT math department discovered, to its horror, that it didn't have enough good math majors to complete its team, the department turned in desperation to Feynman (who, you won't be surprised to learn, already had a reputation). Happy to help, Feynman joined the team and took the exam (indeed, he turned his paper in before the allowed time was up)—and was soon after named a Putnam Fellow. (The MIT team placed second, behind Brooklyn College, and it's clear that Feynman did his part.) The math people at Harvard were sufficiently impressed that they offered a graduate scholarship in mathematics to Feynman, but he had already committed to starting a Princeton PhD in physics. Still, while first a physicist, Feynman always remained an imaginative fellow with his mathematics, too, as you'll see by the end of this chapter.

1.2 Fermat's Last Theorem

Now, before continuing with Feynman, who was Fermat? The French lawyer Pierre de Fermat (1601–1665) was also a mathematical genius who loved to make numerous conjectures. His most famous one is that there are no integer solutions x, y, z to $x^n + y^n = z^n$ for any given integer $n \geq 3$. (There is, of course, an *infinity* of solutions for $n = 2$, as all high school geometry students learn when they first encounter the Pythagorean theorem.) Fermat never published a proof of this conjecture, although he did publish a proof for the single, special case of $n = 4$. His conjecture became famous for two reasons: (1) at some time around 1637 Fermat thought he had a general proof and wrote that claim (but not the proof—it was too long, he said) in the margin of a page in one of his personal books, and (2) for centuries after Fermat not even the greatest mathematicians could figure out how he might have done it.[1] All his other conjectures were eventually settled (he was almost always right), and so the final remaining, unproven $x^n + y^n = z^n$ conjecture became known as his "last" one.

All through the eighteenth and nineteenth centuries, proofs for additional special values of n were found: ($n = 3$ in 1753, $n = 5$ in 1825, $n = 14$ in 1832, and $n = 7$ in 1839), but, as with Fermat's own $n = 4$ proof, which used his newly developed method of *infinite descent*, each new value of n required new, special, highly ingenious tricks. There seemed to be no common, *general* approach that would work for all $n \geq 3$. So difficult did this problem appear that the writers of the television show *Star Trek: The Next Generation* worked the conjecture into one of the episodes ("The Royale"); in that 1989 episode we learn that Captain Picard retreats, when he can, from the pressures of starship command by attempting to prove Fermat's Last Theorem. Since the show was set in the twenty-fourth century, it seems that fans must have thought another four hundred years at least would pass before success. And then, just six years after "The Royale" was broadcast, the conjecture was suddenly proved by Andrew Wiles (born 1953) at Princeton. Feynman died in 1988 and so missed this dramatic event,[2] but he too had tried his hand at the conjecture, although he used an approach that no professional mathematician would think worthy of even a single look.

1.3 "Proof" by Probability

In his brilliant historical treatment of the development of quantum electrodynamics, a branch of physics thought to be a masterpiece of human intellect, the Brandeis University physicist-historian Silvan Schweber presents a sketch of Feynman's unpublished *probabilistic* analysis of Fermat's Last Theorem.[3] That is, Feynman didn't try to formally prove the conjecture but rather tried to show it is statistically improbable that it is false. (It is from Feynman's analysis that the opening quotation comes.) Mathematicians are rightfully unimpressed with such an approach—Feynman may well have been simply having fun with it—and they usually point to a famous historical counter example where the "obviously true" turned out to be utterly false. That is, something that everybody thought had probability 1 turned out to actually have probability 0. The "experimental" (that is, computed numerical) evidence for the "obviously true" conclusion is so compelling, however, that Feynman would no doubt have accepted it even faster than did the mathematicians. So, before turning to Feynman's "proof" of Fermat's conjecture, let me describe this earlier counterexample.

Let $\pi(x)$ be defined as the number of primes that are less than or equal to x; for example, $\pi(10) = 4$ because 2, 3, 5, and 7 are the four primes less than or equal to 10. By direct calculation we can, in principle, determine $\pi(x)$ for any given x; for example, $\pi(1,000) = 168$ and $\pi(100,000,000) = 5,761,455$. By looking at the known values of $\pi(x)$, the German mathematician Carl F. Gauss (1777–1855) noted, in an 1849 letter, that the behavior of $\pi(x)$ is much like that of the logarithmic integral $\int \frac{du}{\ln(u)}$. Today we approximate $\pi(x)$ with $Li(x) = \int_2^x \frac{du}{\ln(u)} + 1.045 \cdots$. For example, $Li(1,000) = 177.6$ and $Li(100,000,000) = 5,762,209.4$.[4] In these two cases, $x = 1,000$ and $x = 100,000,000$, $d(x) = Li(x) - \pi(x) > 0$, and in fact, for every value of x for which $d(x)$ has ever been calculated, it always turns out that $d(x) > 0$. There is no known specific value of x for which $d(x) < 0$. That is, $Li(x)$ *apparently* always gives an *over*estimate of $\pi(x)$. Very dramatic, in addition, is the *growth* of $d(x)$ with increasing x; for example, $d(10^{18})$ is nearly $22,000,000$. Based on this impressive numerical "evidence" it was commonly believed that $d(x) > 0$ for all x. As a physicist who always looked to physical experiment to guide his theoretical work, Feynman would almost surely have agreed with this conjecture.

Then in 1914 the English mathematician John Littlewood (1885–1977) astonished the mathematical world by proving that $d(x)$ changes sign *infinitely many times* as x increases without bound! Littlewood's proof was an *existence* proof, in the sense that while he showed $d(x)$ switches back and forth between positive and negative values infinitely many times, he could not calculate for what values of x the sign switches occur, not even just the x at which $d(x)$ would first become negative.

To this day it is not known where $d(x)$ first switches from positive to negative, but in 1933 the South African mathematician Stanley Skewes (1899–1988) published an upper bound on the value of x at which the first sign switch occurs: it is the absolutely unimaginably large number (it is even larger than the number of pictures in TUPA, a claim you should confirm)

$$e^{e^{e^{79}}} \approx 10^{10^{10^{34}}}.$$

This number is called the *first Skewes number*, and its calculation (under the direction of Littlewood) earned Skewes a 1938 PhD from Cambridge. Calling this number X_1, as did Skewes, what makes X_1 particularly interesting for us in this discussion is that Skewes had to assume the truth of the *Riemann hypothesis* (RH). The RH says that if one calculates all the complex zeros of the zeta function $\zeta(s)$, defined as

$$\zeta(s) = \sum_{n=1}^{\infty} \frac{1}{n^s},$$

then every single one of them has a real part of $1/2$. That is, for every complex s such that $\zeta(s) = 0$ it will be that $s = \frac{1}{2} + t\sqrt{-1}$, where t is some real number. (I'll tell you why the RH is interesting in just a moment.) The RH is named after the German mathematician Bernhard Riemann (1826–1866), who conjectured it in 1859.

Littlewood's existence proof was also based on assuming the truth of the RH, but that assumption isn't really required for his end result. That is, $d(x)$ will switch signs infinitely many times whether the RH is true or not. This assumption *does* make a difference in Skewes's original proof, however, and in 1955 he developed a significantly improved proof that avoided the "RH is true" assumption and calculated a new upper bound on where the first switch in the sign of $d(x)$ occurs.[5] This is called the *second Skewes number*, X_2, which is immensely larger than the already huge X_1:

$$X_1 \ll X_2 = 10^{10^{10^{1000}}}.$$

The role of the RH in all this is ironic because the available numerical "evidence" for the truth of the RH is, like the pre-1914 "evidence" for the behavior of $d(x)$, overwhelming, and yet the RH remains perhaps the most important unsolved problem in mathematics. It is important because there are numerous theorems in mathematics, all of which mathematicians believe to be correct, that are based on the assumed truth of the RH. If the RH is someday shown to be false, then the existing proofs of all those other theorems collapse, and they will have to be revisited and new proofs (hopefully) found.

The numerical "evidence" for the RH I'm referring to is that hundreds of *billions* (*trillions*, in fact) of complex zeros of $\zeta(s)$ have been computed and every last one of them does indeed have a real part of precisely $1/2$. But all it would take to disprove the RH is the discovery of just one zero with a real part $\neq 1/2$. (See Chapter 9 for a bit more on this point.) All professional mathematicians know this and would absolutely reject any sort of Feynman probability-of-truth analysis as meaningless—and yet I believe just about every mathematician on Earth *knows*, in his or her heart, that the RH just *has to be* true. After all, with apologies to the memory of the late Carl Sagan, look at all those *billions and billions* of zeros with exactly the correct real part. To their credit (and occasional embarrassment), the inherent, logical contradiction between these two positions is not lost on mathematicians.

Now, what did Feynman do with this idea of using probability to study Fermat's Last Theorem?

1.4 Feynman's Double Integral

In his history of QED, Schweber writes, "Among Feynman's papers is an undated two-page manuscript on Fermat's theorem ... which illustrates the meaning of 'more sure, less sure' in Feynman's world. He first asks "What is the prob[ability] N is a perfect nth power?" Feynman's approach to answering this was to start by mapping the unit interval defined by two large consecutive integers (N and $N + 1$) into the shorter interval of the nth roots of the unit interval (I'll tell you why he did this in just a moment). Since such a mapping is monotonic, the end points map into the end points, and all the interior points in the unit interval map into all the interior points of the shorter interval. To determine the length of the shorter interval, Feynman wrote it as (where I'll use l for the length)

$$l = (N + 1)^{1/n} - N^{1/n} = \left[N \left(1 + \frac{1}{N} \right) \right]^{1/n} - N^{1/n}$$

$$= N^{1/n} \left(1 + \frac{1}{N} \right)^{1/n} - N^{1/n},$$

or

$$l = N^{1/n} \left[\left(1 + \frac{1}{N} \right)^{1/n} - 1 \right].$$

If we recall the power series expansion

$$(1+x)^a = 1 + ax + \frac{a(a-1)}{2}x^2 + \cdots,$$

and then set $x = \frac{1}{N}$ and $a = \frac{1}{n}$, we have

$$l = N^{1/n} \left[\left(1 + \frac{1}{n} \cdot \frac{1}{N} + \frac{\frac{1}{n}\left(\frac{1}{n} - 1\right)}{2} \cdot \frac{1}{N^2} + \cdots \right) - 1 \right].$$

For large N this becomes

$$l \approx \frac{N^{1/n}}{nN},$$

where we keep just the first term of the series. For example, if $N = 1,000$ and $n = 4$, then $l = 0.0014059$, while $(1,001)^{1/4} - (1,000)^{1/4} = 0.0014053$. Pretty close.

Feynman then wrote in his manuscript, "Probability that N is a perfect nth power is $\dfrac{N^{1/n}}{nN}$ [and] therefore probability $x^n + y^n$ is perfect nth power is $\dfrac{(x^n + y^n)^{1/n}}{n(x^n + y^n)}$." The second half of this claim certainly follows from the first half (just set $N = x^n + y^n$), but where does the first half come from? Feynman himself left no record, but what I think he had in mind is simply this: the shorter, mapped interval of length l includes an integer with probability $l/1$ because the integers are spaced—dare I actually write this?—unit distance apart. You may wonder about the mathematical logic of all this, but please remember, I am only offering here a personal reconstruction of Feynman's thinking. In any case, Schweber goes on to report that Feynman then writes, "Therefore the total probability any $x^n + y^n$ is perfect nth power [that is, is z^n] for $x > x_0$ and $y > y_0$ is equal to

$$\int_{x_0}^{\infty} \int_{x_0}^{\infty} \frac{1}{n}(x^n + y^n)^{-1+\frac{1}{n}} \, dx \, dy = \frac{2}{nx_0^{n-3}} c_n$$

where

$$c_n = \frac{1}{2} \int_0^\infty \int_0^\infty (u^n + v^n)^{-1+\frac{1}{n}} \, du \, dv."$$

Feynman is (I believe) *integrating* over all u and v rather than *summing* over all *integers* u and v simply because integrals are (generally) easier to do than are sums.

Now, some problems. Why the outer, y-integral doesn't have a lower limit of y_0 instead of x_0 escapes me; I suspect it is either a typo in Schweber's book or an accurate reporting of a Feynman slip in writing. In any case, we never hear from y_0 again. Also, why Feynman writes c_n with a $1/2$ factor escapes me, too, as opposed to just writing

$$\int_{x_0}^\infty \int_{x_0}^\infty \frac{1}{n} \left(x^n + y^n \right)^{-1+1/n} \, dx \, dy = \frac{1}{nx_0^{n-3}} c_n,$$

where

$$c_n = \int_0^\infty \int_0^\infty (u^n + v^n)^{-1+1/n} \, du \, dv.$$

But there is a more serious problem here than just factors of $1/2$ and/or 2: the lower limits on the c_n double integral are wrong! They should be 1s, not 0s. I'll let you go through the (easy) details of changing variables to show this. Simply write

$$s = \frac{x - x_0}{x_0} \quad \text{and} \quad t = \frac{y - x_0}{x_0},$$

which of course maps $x_0 \to 0$ and $\infty \to \infty$. The result is

$$\int_{x_0}^\infty \int_{x_0}^\infty \frac{1}{n} \left(x^n + y^n \right)^{-1+1/n} \, dx \, dy$$

$$= \frac{1}{nx_0^{n-3}} \int_0^\infty \int_0^\infty [(s+1)^n + (t+1)^n]^{-1+1/n} \, ds \, dt.$$

Then, with the obvious second change of variables to $u = s + 1$ and $v = t + 1$, and canceling the n on both sides, the conclusion is

$$\int_{x_0}^{\infty} \int_{x_0}^{\infty} (x^n + y^n)^{-1+1/n} \, dx \, dy$$

$$= \frac{1}{x_0^{n-3}} \int_{1}^{\infty} \int_{1}^{\infty} (x^n + y^n)^{-1+1/n} \, dx \, dy, \tag{1.4.1}$$

where I've used x and y on both sides of (1.4.1) instead of u and v on the right-hand side because, after all, these are all just dummy variables of integration.

And so at last we come to the real purpose of this entire discussion. Feynman's probabilistic analysis of Fermat's Last Theorem would have no mathematical interest at all but for the fact it was Feynman who cooked it up, but it does offer me the opportunity to show you how modern computer software can quickly and easily check a theoretical result via numerical calculations that would be just plain awful to do by other means. If you look again at note 4, where I showed you how to write some easy MATLAB code to evaluate $Li(x)$, then you should be able to see how the following code (named **fdi.m** for *Feynman's double integral*) evaluates (for specified values of x_0 and n) the right-hand side (*rhs* in the last line of code) of (1.4.1). In that line, *inf* is MATLAB's way of expressing ∞.

---

fdi.m
```
syms   x   y   n
x0=input('x0=?');
n=input('n=?');
f=(x^n+y^n)^(-1+(1/n));
rhs=vpa(int(int(f,x,1,inf),y,1,inf)/(x0^(n-3)))
```

Feynman ended his manuscript with some comments on the behavior of the c_n integral (with the $1/2$ factor and the zero lower limits), noting it diverges for both $n = 2$ and $n = 3$ (in this last case the rhs of

(1.4.1) is independent of x_0), and so the $n = 4$ case is the first case of interest. What follows is a sample of results from running **fdi.m** for some various values of n and x_0. The probability of there being integer solutions to $x^n + y^n = z^n$ does decrease with increasing n, but I really don't believe a mathematician would be convinced by Feynman's argument. This is, I think, enough about Feynman, but whatever you might think about how he attacked Fermat's Last Theorem, you must admit it is a colorful illustration of what an original and agile mind he had.[6]

n	x_0	rhs of (1.4.3)
2	1	∞
4	1	0.8308962
5	2	0.0667596
6	3	0.0048266
10	1	0.0266076
15	1	0.0093240

1.5 Things to Come

This chapter on Feynman and Fermat has given an elementary illustration of symbolic manipulation by computer, but of even greater use to mathematical physicists is a computer's ability to perform massive amounts of pure number crunching. As a quick example of this, and as a teaser for what's to come in later chapters, consider Figure 1.5.1, the so-called *phase-plane portrait*, over the interval $0 \leq t \leq 15.9$, of the second-order differential equation $\dfrac{d^2x}{dt^2} + x - \dfrac{1}{6}x^3 = 3\sin(3\omega t)$ when $x(0) = \dfrac{dx}{dt}\Big|_{t=0} = 0$ and $\omega = 0.9284513988$. This is a special case of the *Duffing equation*, named for the German engineer Georg Duffing (1861–1944), who published his nonlinear differential equation in 1918 as part of studies on vibrating machinery. The Duffing equation occurs in many modern applications of physics, engineering, and the mathematics of chaos theory. This type of plot, created by MATLAB, is discussed in Chapter 4. This particular plot, known to computer scientists and applied mathematicians as *Murphy's Eyeballs*, gets its name from the infamous "Murphy" of Murphy's Laws (which number in the

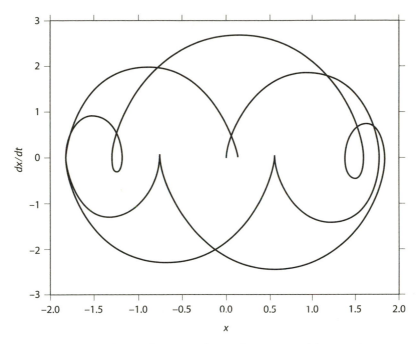

FIGURE 1.5.1. The Eyes of Murphy Are Watching You!

dozens, two of which are "*If anything can go wrong, it will,*" and "*Even if nothing can possibly go wrong, something will still go wrong*"). They are immutable laws of nature, experimentally confirmed every day by all who write complex computer programs. If you look at the plot carefully (it took me three attempts to get the MATLAB code running properly!), you can see Murphy's grin of anticipation at pouncing on your slightest misstep.

1.6 Challenge Problems

CP 1.1. Calculate the probability of randomly plucking a single needle (a cylinder 2 inches long with a diameter of 0.01 inches) out of a spherical haystack the size of the Earth. Take the radius of the Earth to be 3,960 miles. Compare this probability to that of randomly picking one of te Riele's integers[7] from somewhere in a sequence of integers of length 10^{180} that is itself somewhere in the interval $6.62 \cdot 10^{370}$ to $6.69 \cdot 10^{370}$.

CP 1.2. In James Gleick's outstanding biography of Feynman, *Genius* (New York: Pantheon, 1992, p. 47), we read the line, "If a boy named Morrie Jacobs [a childhood friend] told him that the cosine of 20 degrees multiplied by the cosine of 40 degrees multiplied by the cosine of 80 degrees equaled exactly one-eighth, he would remember that curiosity the rest of his life." I was reminded of this when I received a letter in early 2008 from Mr. Don G. Olmstead of Pearl, Mississippi, who had just read my book, *An Imaginary Tale: The Story of $\sqrt{-1}$* (Princeton, N.J.: Princeton University Press, 1998 [corrected paperback edition 2007]). Don wrote to tell me that he had succeeded in deriving this result (using Euler's identity), which he called *Feynman's equality* (although maybe it should really be called the *Jacobs-Feynman equality*), but it had been just a bit more difficult to do than he had initially thought it would be. As part of his proof he had found it helpful to first establish a new, preliminary result (which he was far too modest to name but which I'm going to call *Olmstead's equality*), that is, $\cos(20°) = \cos(40°) + \cos(80°)$. So, that's your challenge here: first derive Olmstead's equality, and then use it to establish Feynman's equality of $\cos(20°)\cos(40°)\cos(80°) = 1/8$. *Hints:* For Olmstead's equality start by writing $\cos(\mu) = \cos(2\mu) + \cos(4\mu)$, manipulate this into a *cubic* in $\cos(\mu)$, and then solve the cubic to conclude $\mu = 20°$. For Feynman's equality, recall Euler's identity $e^{i\mu} = \cos(\mu) + i\sin(\mu)$, $i = \sqrt{-1}$, and so $\cos(\mu) = \dfrac{e^{i\mu} + e^{-i\mu}}{2}$.

CP 1.3. Finally, here's one last calculation for you that I think Feynman would have loved. Consider the following claim: for a and b non-negative but otherwise arbitrary constants we can write $\int_0^{2\pi} \sqrt{a^2 \sin^2(t) + b^2 \cos^2(t)}\, dt \geq \sqrt{4\pi \left\{ \pi ab + (a-b)^2 \right\}}$ (with obvious equality when $a = b$). In my book, *When Least Is Best* (Princeton, N.J.: Princeton University Press, 2004 [corrected paperback edition 2007]), I give a *geometric* proof of this inequality, based on dissecting and then rearranging an ellipse and then applying the isoperimetric theorem (a closed, non-self-intersecting curve of given length that encloses the maximum possible area forms a circle), which is proven in that book (pp. 251–259). In that book the challenge I made to readers was to find a purely *analytical* proof. The challenge stood unanswered until August 2008, when M. W. Perera (who lives near Colombo, Sri Lanka) sent me a beautifully simple and elegant derivation. Indeed, Perera's proof improves on the above inequality by showing how to make the integral's lower bound on the right-hand side even larger. Can you repeat Perera's achievement? *Hint:* You may find the following power series expansion of the complete elliptic integral of the second kind useful: if $0 \leq k \leq 1$ with $n = \dfrac{1 - \sqrt{1 - k^2}}{1 + \sqrt{1 - k^2}}$, then

$$\int_0^{\pi/2} \sqrt{1 - k^2 \sin^2(t)}\, dt$$

$$= \frac{\pi}{2(1 + n)} \left[1 + \left(\frac{1}{2}\right)^2 n^2 + \left(\frac{1}{2 \cdot 4}\right)^2 n^4 \right.$$

$$\left. + \left(\frac{1 \cdot 3}{2 \cdot 4 \cdot 6}\right)^2 n^6 + \left(\frac{1 \cdot 3 \cdot 5}{2 \cdot 4 \cdot 6 \cdot 8}\right)^2 n^8 + \cdots \right].$$

1.7 Notes and References

1. It is now believed by historians of science that at one time Fermat may have thought he had a general proof for his conjecture but later found it to be in error. It was only after his death that the famous margin claim became known, a claim most likely left unerased by Fermat simply because *why would he bother to correct a personal note to himself that he never published?* Very likely he had quickly forgotten he had even made the now famous margin entry. Do you remember every note you've ever scribbled to yourself? I'd bet not! Adding credence to this line of thought is that in 1638 he started challenging other mathematicians to prove the special case of $n = 3$. In Fermat's day the issuing of a challenge implied that the issuer had himself already solved the challenge—but why would Fermat have been concerned about $n = 3$ if he had a proof for *all* $n \geq 3$? In fact, he never published an $n = 3$ proof either, and so there is some question whether he had really solved even that special case!

2. Wiles's proof is well over a hundred pages long and uses very deep mathematics invented long after Fermat's death. The entire proof has been looked at by only a small number of number theory specialists, and their

pronouncement of success is accepted by everybody else. The status of Wiles's proof is very different from the traditional situation, that is, as a constantly read, reread, and perpetually vetted analysis by generation after generation of students. The romantic notion that just maybe Fermat actually did have what he scribbled in his book margin, "a truly marvelous demonstration," is therefore hard to give up. The late science fiction writer Arthur C. Clarke's last novel, *The Last Theorem* (New York: Del Rey, 2008), plays with this idea, with the imagined discovery of a three-page proof by a young supergenius who thereby attracts the attention of the National Security Agency. If that isn't exciting enough for diehard science fiction fans, this novel also features an alien invasion fleet streaking toward Earth at near light speed.

3. Silvan S. Schweber, *QED and the Men Who Made It: Dyson, Feynman, Schwinger, and Tomonaga* (Princeton, N.J.: Princeton University Press, 1994, p. 464).

4. Notice, too, that $Li(2) = 1.045$ while $\pi(2) = 1$ (the only prime less than or equal to 2 is 2). You can find tables of $Li(x)$ in the literature, but with modern computer software it is easy to generate values yourself. For example, MATLAB's Symbolic Toolbox has the wonderful command *int* (for *integrate*), with the following syntax: the three lines of code

```
syms    u    a    b
f=1/ln(u);
int(f,u,a,b)
```

generate the *symbolic* integration of $\int_a^b f(u)\,du = \int_a^b \frac{du}{\ln(u)}$. The first line simply tells MATLAB that u, a, and b are to be treated as *symbolic variables*. If we replace the third line with $vpa(int(f,u,2,1000)+1.045)$, then MATLAB calculates (using however many digits that have been previously specified—I used 32) the *variable-precision arithmetic* value of

$$Li(1000) = \int\limits_{2}^{1000} \frac{du}{\ln(u)} + 1.045 = 177.6.$$

5. In his 1955 paper, published in the *Proceedings of the London Mathematical Society*, Skewes has a provocative line of acknowledgement for Littlewood: "I wish in conclusion to express my humble thanks to Professor Littlewood, but for whose patient profanity this paper would never have become fit for publication." One can only wonder at just how Littlewood expressed himself to Skewes!

6. In a 1985 note to Schweber, Feynman wrote that a Caltech colleague, number theorist Morgan Ward (1901–1963), had told him "the same argument would show that an equation like $x^7 + y^{13} = z^{11}$ (powers prime to each other) would be unlikely to have integer solutions—but they do, an infinite number of them!" So, from the date of Ward's death it is clear that Feynman's unpublished work predates a paper by the mathematicians Stanislaw Ulam and Paul Erdös, "Some Probabilistic Remarks on Fermat's Last Theorem" (*Rocky Mountain Journal of Mathematics*, Fall 1971, pp. 613–616).

7. Since the work of Skewes, the upper bound on where $d(x)$ first switches sign has been greatly reduced. Skewes was still alive, for example, when the Dutch mathematician Herman te Riele (born 1947) showed in 1987 that, somewhere between $6.62 \cdot 10^{370}$ and $6.69 \cdot 10^{370}$ (bounds vastly less than X_1), there is a stretch of more than 10^{180} consecutive integer values of x where $d(x) < 0$. This is a long stretch, to be sure, but finding it by random searching would be far more difficult than would be finding (by random searching) a single needle in a haystack the size of the Earth (see Challenge Problem 1.1)!

2
JUST FOR FUN: TWO QUICK NUMBER-CRUNCHING PROBLEMS

The human mind is seldom satisfied, and is certainly never exercising its highest functions, when it is doing the work of a calculating machine.
—James Clerk Maxwell, in an 1870 address to the British Association for the Advancement of Science

2.1 Number-Crunching in the Past

Heroic calculation of certain particular numbers (π, e, and $\sqrt{2}$, for example) has long had a special attraction for persons blessed with genius-level intellects—that is, for people who, you would think, should have had far better things to do! In my *An Imaginary Tale: The Story of $\sqrt{-1}$* (Princeton, N.J.: Princeton University Press, 1998, 2007, 2010), for example, I mention Newton's calculation of π to sixteen decimal places, an achievement the great man felt obliged to explain with the words, "I had no other business at the time."

I used the Newton story as a lead-in to the incredible tale of how the Yale physicist Horace Scudder Uhler (1872–1956) came to hand-calculate i^i (where $i = \sqrt{-1}$) $= e^{-\pi/2}$ to 135 decimal places in 1947, just as people were starting to think of electronic computers like MANIAC-I (about which I'll tell you more in the next chapter). I then showed readers how MATLAB could do the same calculation today with a single line of code in the time it takes to push a button. That discussion was all "just for fun." For your additional amusement, then, in the same spirit of fun, here's yet another curious episode from the precomputer era of number-crunching.

In his posthumously published 1831 book, *Analyse des equations déterminées*, the French mathematical physicist Jean Baptiste Joseph Fourier (1768–1830), of Fourier series and integral fame, calculated the one real root of $x^3 - 2x - 5 = 0$ to more than thirty digits.[1] That amazing calculation cost Fourier an enormous amount of labor.[2] After reading Fourier's book, the English mathematician Augustus De Morgan (1806–1871) thought Fourier's calculation could be the basis for a really neat exercise for his students at University College, London. (De Morgan tells this story in his posthumously published book, *A Budget of Paradoxes* [1872].)

Using a recently published algorithm of Horner,[3] De Morgan challenged his students of 1841 to beat Fourier's digit count, all to be done as a Christmas vacation exercise! (Who says mathematics professors aren't old softies when it comes to the well-being of their students?) The students didn't fail their teacher, with several producing calculations (that agreed) out to an astounding fifty decimal places. So impressed by that work was De Morgan that in 1848 he repeated the homework assignment, with the challenge now being to beat the class of 1841. Several students did just that, too, with one of them calculating the root to Fourier's cubic out to a prodigious 101 decimal places![4]

But, of course, De Morgan then asked himself the obvious question: Were all those digits correct? To answer that, De Morgan came up with an ingenious test, one that is actually doubly ingenious: it gave high credibility to the student's digits being correct *and* it created yet another homework problem for the lucky student, who (not De Morgan) then did all the incredibly grubby checking of arithmetic! What De Morgan did was ask the student to now calculate the real-valued solution to $y^3 - 90y^2 + 2500y - 16,000 = 0$. As it turns out, the real-valued solution to Fourier's cubic is 2.0945514815423265..., while the real-valued solution to De Morgan's cubic is 9.0544851845767340..., a solution he could write down by inspection from the (if correct) solution to Fourier's cubic. De Morgan could do that because of an intimate connection between the two cubics. If the student's two solutions satisfied De Morgan's connection, then De Morgan could have very high confidence that both solutions were in fact correct.

Can you look at the two above solutions and see that connection? Probably not—De Morgan had, in fact, gone to some length to hide the connection—but give it a try before I reveal it to you in just a bit. Before I show you De Morgan's clever idea, however, let me show you how MATLAB easily solves both cubics. Indeed, MATLAB reduces both problems to routine elementary computer programming exercises that require less time and work than it takes to walk across the street. You'll recall from Chapter 1 that MATLAB can manipulate symbolic quantities as well as numeric ones, using its Symbolic Toolbox. The following brief, simple code does the job for Fourier's cubic:

```
digits(120)
syms x
S=solve('x^3-2*x=5');
vpa(S(1,1))
```

Here's how it works.

The first line tells MATLAB to do all its calculations using 120 digits; I picked the value 120 somewhat arbitrarily; the only requirement was to be larger than the 101 of De Morgan's student in 1848. The second line defines x as a symbolic variable. The third line invokes the *solve* command to find the symbolic expression for x (not the numerical value) that is the symbolic solution to the equation between the two quotation marks. For a cubic equation there are, of course, three such expressions, one for each of the three roots (two of which are complex). That means S holds three expressions, and so S is actually a column vector with three elements. The real-valued expression is the first element, $S(1, 1)$. The fourth line of code uses the variable precision arithmetic command (*vpa*) to numerically evaluate the real-valued expression with 120 digits of precision.

The result, for Fourier's cubic, produced in less than 0.025 seconds on my quite ordinary home computer, with 118 correct digits to the right of the decimal point, is:

2.09455 14815 42326 59148 23865 40579 30296 38573 06105
62823

91803 04128 52904 53121 89983 48366 71462 67281 77715
  77578

60839 52118 90629 634.

Replacing the second line of code with *syms y* and the third line of code
with

S=solve('y^3-90*y^2+2500*y-16000=0')

produces a real solution to De Morgan's cubic, with 117 correct digits
to the right of the decimal point, of:

9.05448 51845 76734 08517 61345 94206 97036 14269 38943
  71760

81969 58714 70954 68781 00165 16332 85373 27182 22842
  24213

91604 78810 93703 65.

So, have you spotted the connection between the two solutions? *The
entire digit sequence for De Morgan's solution is the 9s complement of the digit
sequence for Fourier's solution (starting with the first digit to the right of the
decimal point), shifted one place to the left.*

Here's why. Knowing that the solution to Fourier's cubic is just a bit
larger than 2 (you can see this literally by inspection), De Morgan then
defined $y = 10(3 - x)$. We can replace the 3 with 2.9999999999..., and
so the $(3 - x)$ factor will be a value between 0 and 1, with every digit
in the result obviously being a complement of 9. And of course the 10
factor will shift that value one decimal place to the left. If you substitute
y into Fourier's cubic (that is, replace x with $3 - \frac{1}{10}y$), you get—big
drum roll—De Morgan's cubic! Try it and see.

De Morgan knew, of course, that this was all an enormous expen-
diture of energy for little (if any) real purpose. He himself wrote of it,
in fact, as simply an example "of computation carried to paradoxical
length, in order to illustrate a method" (Horner's method). I'll admit
that there is a certain bravado to it all that the MATLAB code fails to
capture. As for me, though, if given a choice, I would much prefer the
MATLAB approach, and De Morgan can keep the bravado!

2.2 A Modern Number-Cruncher

In an earlier book (*Mrs. Perkins's Electric Quilt*) I challenged readers
to calculate the input resistance R of the infinite "ladder" of resistors shown in Figure 2.2.1, where the resistor values are in ohms. This
infinite circuit was intentionally created because it is not solvable with
the trick I used in *Mrs. Perkins's Electric Quilt* to solve the special case
where all the resistors are equal. That special case is famous today precisely because it *can* be solved in closed form; if all the resistors are one
ohm, for example, the input resistance is $\dfrac{1+\sqrt{5}}{2}$ ohms.[5] But the trick
doesn't work on the circuit of Figure 2.2.1, and so, based on some brief,
far too hasty calculations, I was reduced to speculating that the answer
is the rational value of 21/11 ohms. **This is not correct!**[6] A reader of
that book, Ruediger Klostermeyer, a professor of electrical engineering
and computer science at the Stralsund University of Applied Sciences,
Germany, promptly set me straight with an e-mail, and what follows in
this section is how to correctly solve the problem. What makes this an
interesting problem for this book is that while we will arrive at a beautiful analytical formula for the answer, it is a real number-crunching
challenge to *numerically* evaluate because that formula is infinitely long!

In Figure 2.2.2 I've redrawn Figure 2.2.1, with α denoting the factor
by which the horizontal resistors decrease in value from left to right
and $1/\alpha$ denoting the factor by which the vertical resistors increase in
value from left to right. For the original challenge problem, $\alpha = 1/10$
in this generalized version. Let's write $R_k(\alpha)$ as the input resistance
to the finite version of Figure 2.2.2 when there are k stages (a stage is
one horizontal resistor plus one vertical resistor). Obviously, $R_1(\alpha) = 2$,
and $\lim_{k\to\infty} R_k(\alpha) = R < 2$. What I'll do do next is work out explicit
formulas for $R_k(\alpha)$ for the first few values of k, and from them we'll be

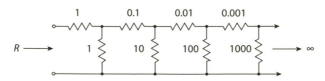

FIGURE 2.2.1. The original ladder

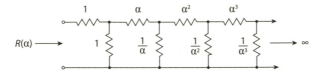

FIGURE 2.2.2. The original ladder, generalized

able to see the solution we are after as $k \to \infty$. To keep things as simple as possible, let's agree to represent a horizontal resistor as a resistor in units of ohms but to represent a vertical resistor by its equivalent admittance (that is, we'll replace a vertical resistor with its reciprocal in units of what electrical engineers used to amusingly call the *mho*; that's *ohm* spelled backward!). Figure 2.2.3 shows the two-stage ($k = 2$) case.

Since the *vertical* α in Figure 2.2.3 is an admittance, it has resistance $1/\alpha$, which adds to the horizontal α (a resistance) to give a total resistance of $\alpha + 1/\alpha$, which is the admittance $\dfrac{1}{\alpha + \frac{1}{\alpha}}$ in parallel with unity admittance, as shown in Figure 2.2.4. Since admittances in parallel add, then Figure 2.2.4 becomes Figure 2.2.5, which allows

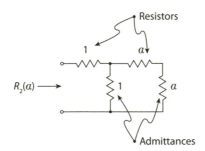

FIGURE 2.2.3. Two-stage approximation

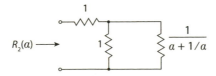

FIGURE 2.2.4. Vertical elements are admittances

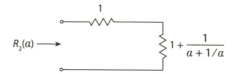

FIGURE 2.2.5. The reduced Figure 2.2.4

us to immediately write the input resistance as

$$R_2(\alpha) = 1 + \cfrac{1}{1 + \cfrac{1}{\alpha + \frac{1}{\alpha}}}. \tag{2.2.1}$$

If you repeat this entire line of argument, of alternating back and forth between horizontal resistances and vertical admittances, then it is pretty short work to show that

$$R_3(\alpha) = 1 + \cfrac{1}{1 + \cfrac{1}{\alpha + \cfrac{1}{\alpha^2 + \frac{1}{\alpha^2}}}} \tag{2.2.2}$$

and

$$R_4(\alpha) = 1 + \cfrac{1}{1 + \cfrac{1}{\alpha + \cfrac{1}{\alpha^2 + \cfrac{1}{\alpha^3 + \frac{1}{\alpha^3}}}}}. \tag{2.2.3}$$

And now we see the general pattern. The input resistance R to the *infinite* ladder circuit in Figure 2.2.1 is the *infinite* continued fraction

$$R(\alpha) = 1 + \cfrac{1}{1 + \cfrac{1}{\alpha + \cfrac{1}{\alpha^2 + \cfrac{1}{\alpha^3 + \cfrac{1}{\alpha^4 + \cdots}}}}}, \tag{2.2.4}$$

and R for Figure 2.2.1 is the value of $R(0.1)$. There is a wonderful theorem in analysis that tells us that this particular value for R is irrational, and so my speculation in *Mrs. Perkins's Electric Quilt* that $R(0.1)$ is a rational value was embarrassingly incorrect.[7]

We can hand-calculate the numerical value of the $R_k(\alpha)$ as accurately as we wish for a given value of α, for the first (very) few values of k, but as k increases, that very quickly becomes very tedious. Using MAT-LAB's ability to perform high-precision calculations (using the *digits* command), however, we can avoid round-off errors even as we calculate $R_k(\alpha)$ out to, say, thirty decimal places or more for large k. Here's a simple MATLAB code, **cf.m**, that does the job, where all calculations are carried out with the massive overkill of 100 digits of precision (thirty-two digits is the MATLAB default value).

```
cf.m
clear
digits(100)
a=0.1;
k=input('How many stages (at least 2)?');
for j=1:k-1
    alpha(j)=a^j;
end
d=alpha(k-1);
d=d+1/d;
d=1/d;
for loop=k-2:-1:1
    d=alpha(loop)+d;
    d=1/d;
    d=alpha(loop)+d;
    d=1/d;
end
1+1/(1+d)
```

When run for several selected values of k, the results were, for $\alpha = 0.1$ (the value set in the third line of the code):

$R_2 = 1.909909909909910$

$R_3 = 1.901874123079695$

$R_4 = 1.901079296951190$

$R_5 = 1.900999900982170$

$R_{10} = 1.900991080267494$

$R_{20} = 1.900991080179288$

$R_{50} = 1.900991080179288$

As you can see from these computed values, $R_k(0.1)$ converges very quickly. One nice feature of **cf.m** is that it allows us to easily experiment with different values of α. For example, suppose that we have the "mirror-image" infinite ladder version of Figure 2.2.1 shown in Figure 2.2.6. Now $\alpha = 10$ ($a = 10$ in the code), and after recomputing we see that $R_k(10)$ converges even more quickly and that, while it is still true that $R(10) < 2$, we also see that $R(10) \neq R(0.1)$.

$R_2 = 1.909909909909910$

$R_3 = 1.909909099171985$

$R_4 = 1.909909099171984$

$R_5 = 1.909909099171984$

$R_{10} = 1.909909099171984$

$R_{20} = 1.909909099171984$

$R_{50} = 1.909909099171984$

Let me end here on a philosophical note. Some readers may feel that saying (2.2.4) is a solution is a bit of a cheat because it involves an infinite process. But consider this: for Feynman's infinite ladder (see note 5 again) made entirely from one-ohm resistors, where we obtained a closed-form result for the input resistance of $\dfrac{1 + \sqrt{5}}{2}$, what does writing $\sqrt{5}$ really signify? We are so used to writing square-root signs that we've forgotten that it is simply shorthand for an *infinite process*!

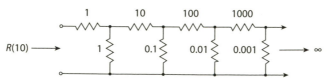

FIGURE 2.2.6. 'Mirror-image' version of Figure 2.2.1

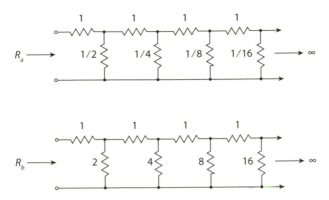

FigURE 2.3.1. Two curious infinite ladders

The square root of any non-square integer is irrational, in fact, and the numerical calculation of an irrational number is a never-ending process. We always have to stop, as a practical matter, after calculating a finite number of digits. So, an explicit infinite continued fraction solution for the infinite ladder of Figure 2.2.1 is, in my view, just as valid as is writing a square-root sign.

2.3 Challenge Problem

CP 2.1. Calculate the input resistance for each of the two infinite resistor ladders in Figure 2.3.1, giving each value correct to fourteen decimal places. *Hint:* Let the vertical resistances have the resistance values of $\alpha, \alpha^2, \alpha^3, \alpha^4$, etc., find an infinite continued fraction expression for the input resistance $R(\alpha)$ using the resistance/admittance trick, and then write a computer code to numerically evaluate $R_a = R(0.5)$ and $R_b = R(2)$.

2.4 Notes and References

1. There are three roots to any cubic with real coefficients; either all three are real, or exactly one is real and the other two are a complex conjugate pair. As shown in my *An Imaginary Tale* (p. 20), the three solutions to $x^3 = px + q$, where p and q are both non-negative, are all real if $q^2/4 - p^3/27 < 0$ (in

Fourier's cubic, $p = 2$ and $q = 5$). But, for Fourier's cubic, the left-hand side of this inequality is positive, and so only one root is real.

2. You can read for yourself exactly what Fourier did because his book is available (in French) on the Web as a Google book scan (see, in particular, its pp. 209–217).

3. William George Horner (1786–1837) was an English schoolteacher who discovered his method for calculating the roots of high-order polynomials by some unknown process. It was published in the *Philosophical Proceedings of the Royal Society of London* in 1819, whereupon it apparently became all the rage, at least for a while. Although there is no evidence of his work being anything but an independent discovery, his method is essentially that used by a Chinese mathematician in the mid-thirteenth century, which in turn was an extension of a procedure dating back to the second century B.C. I personally think the method to be awkward if not downright ponderous and, if you can find it, I think you'd agree. I could only find it in an old science text, buried in the storage stacks of the University of New Hampshire's Dimond Library and last checked out more than fifty years ago: J. W. Mellor, *Higher Mathematics for Students of Chemistry and Physics* (New York: Longmans, Green and Co., 1931, pp. 363–367).

4. In 1851 yet another of De Morgan's students calculated the root out to more than 150 decimal places. There is, as far as I have been able to determine, no record that De Morgan tortured any more of his students with this problem after 1851.

5. See *Mrs. Perkins's Electric Quilt* (Princeton, N.J.: Princeton University Press, 2009, pp. 24–43) for an extended discussion of the trick and its pre-Feynman history. You'll also find it, more briefly, in Chapter 8 of this book (Section 8.3). The all-are-equal infinite resistor ladder is famous mostly because Feynman included it in his even more famous *Lectures on Physics* that I mentioned in note 1 of the Introduction.

6. See *Mrs. Perkins's Electric Quilt*, p. 334.

7. See, for example, E. Hairer and G. Wanner, *Analysis by Its History* (New York: Springer, 1996, pp. 76–77). Not all infinite continued fractions are irrational, however. As a counterexample, consider the rational 2 and its infinite continued fraction form:

$$2 = 1 + 1 = 1 + \frac{2}{2} = 1 + \cfrac{2}{1 + \cfrac{2}{1 + \cfrac{2}{1 + \cdots}}}.$$

3 COMPUTERS AND MATHEMATICAL PHYSICS

> It seemed to me that Fermi was always calculating something. It was Fermi's view that nature revealed itself through experiments devised to test it. You may construct a theory to explain what was going on, but unless the numbers come out right you can't be sure the theory is right. So you have to do a lot of calculating.
> —H. L. Anderson, remembering the great Italian-born physicist Enrico Fermi

3.1 When Theory Isn't Available

The opening quotation gets right to the heart of mathematical physics, with its passion for calculation. If you can't calculate the value of whatever it is you are talking about *and get an answer in agreement with experiment*, then the conclusion is that there is at least *something* (maybe even a lot) you don't understand. Being able to calculate something that nobody else has been able to is the fantasy dream of all young mathematical physicists. (Being able to simply understand what some young person has just calculated for the first time is the dream of all *old* mathematical physicists!) The lofty position of calculation in physics is illustrated by the great American theoretical physicist Julian Schwinger (1918–1993), who shared—along with Feynman and the Japanese physicist Sin-Itiro Tomonaga (1906–1979)—the 1965 Nobel Prize in physics for their work in quantum electrodynamics. Schwinger's ability to calculate was greatly admired by his contemporaries; as one of his students wrote, "Isidor Rabi, who discovered Schwinger and brought him to Columbia University [Rabi (1898–1988) was a physics professor at Columbia who, in 1944, received the Nobel Prize], generally had a poor opinion of theoretical physicists. But Rabi was always very impressed with Schwinger because in nearly every paper, he 'got the numbers out' to compare with experiment. Even in his most elaborate field-theoretic papers he was always concerned with making contact with the real world."[1]

An exact, perfect, beautiful solution to elegant equations is the gold standard for mathematical physics, but often that isn't possible. Sometimes there are simply no equations to solve or, if there are, nobody (yet) knows how to solve them. In such cases, fortunately, the analyst still has one last tool available, the modern electronic computer and its enormously powerful software. If you read many of the physics papers published before the invention of the electronic computer you'll be impressed by how inventive the authors of those papers could be in obtaining numerical answers to their questions, with either ingenious approximations or just downright laborious, tedious calculations that traded endurance for elegance. They did that because they had to. We don't.

In this chapter I'll describe four problems in which the use of a computer completely avoids the necessity of obtaining a theoretical result. In the first three examples, it *will* be possible to get theoretically exact answers, and that will allow us to evaluate just how close (or not) the computer solutions, using the intuitive Monte Carlo approach, come to being correct. The fourth example is based on one of the very first serious computer simulation "experiments" done by Fermi and his colleagues at the Los Alamos Scientific (now National) Laboratory, New Mexico, in the early 1950s, which produced a result so surprising that it was called a "paradox" for the next ten years.

3.2 The Monte Carlo Technique

One of the first sophisticated uses of the electronic computer was to perform high-speed simulations of a complex physical process by using random numbers. This approach is called *Monte Carlo*, and I have written about it in two earlier books.[2] Long before Monte Carlo got its romantic name, however, Enrico Fermi (1901–1954) was using a hand-calculation variety of it as early as 1934, in the neutron diffusion work that resulted in his 1938 Nobel Prize in physics.[3] So, how does Monte Carlo work? To answer that, here are a couple of computer examples for you, from pure mathematics, as warmups for the Monte Carlo *physics* problem that will be a later example. Both of these Monte Carlo examples take advantage of a idea expressed by the noted Russian mathematician Vladimir Arnol'd (1937–2010): "Mathematics is that part of physics where experiments are cheap."

To start, imagine an equilateral triangle with side lengths 2, as shown in Figure 3.2.1. If we pick a point "at random" from the interior of the triangle, what is the probability that the point is no more distant than $d = \sqrt{2}$ from each of the triangle's three vertices? The shaded

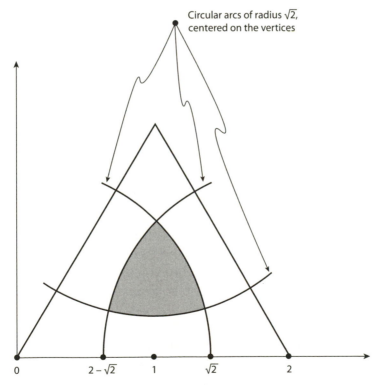

FIGURE 3.2.1. The points in the shaded region are all the points within $d = \sqrt{2}$ of all three vertices of the equilateral triangle

region in the figure is where all such points are located. There is nothing special about the $\sqrt{2}$ other than it will make some of the theoretical calculations we'll do, to check the Monte Carlo computer code, particularly simple to perform. We could, however, still solve the problem for different values of d. The exact theoretical answer (for $d = \sqrt{2}$) is

$$\frac{\pi}{2\sqrt{3}} + 1 - \sqrt{3} = 0.1748488\ldots \qquad (3.2.1)$$

The theoretical calculation of (3.2.1) requires mostly only high school geometry, plus one step that I think requires a simple freshman calculus computation. So, this isn't close to being a terribly hard problem; you can decide that for yourself, however, as it is Challenge Problem 3.1! But what if you can't see how to do those calculations, or what if you just want a numerical answer right now, and don't want to be bothered with doing a formal analysis? In those cases, Monte Carlo can be the ticket to success.

The fundamental question we have to answer before we can write even the first line of computer code is, what does it mean to say we have selected a point "at random" from the interior of the equilateral triangle? The words "at random" are generally interpreted to mean that if we look at any tiny area ΔA in the triangle's interior, and if the triangle has total area A, then the probability of the point being somewhere in that tiny area is $\Delta A / A$. (This is the fundamental assumption that underlies what is called *geometric probability*.) Most high-level computer languages (such as MATLAB) implement a uniform random number generator that, on each call to it, returns a number from a uniform distribution over the interval 0 to 1. For example, if we wanted to generate the coordinates (x, y) of a point picked "at random" from the interior of a square with side length s we could just write (in MATLAB) the two commands **x=s*rand** and **y=s*rand**. But what do we write to generate the coordinates (x, y) for a point picked "at random" from the interior of our triangle?

Here's one simple way to do that. If you look at Figure 3.2.2 you'll see a rectangle of width 1 and height $\sqrt{3}$. We can generate points picked "at random" from the interior of that rectangle by writing (in MATLAB) **x=rand** and **y=sqrt(3)*rand**. But physically this is the same as picking a point "at random" from our triangle. To understand this,

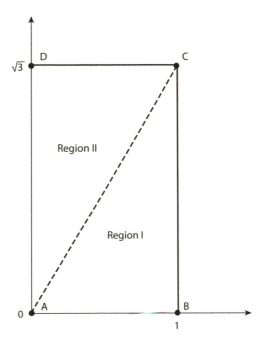

FIGURE 3.2.2. Our triangle reassembled as a rectangle

notice the following about the rectangle: it has a diagonal AC of length 2, which divides the rectangle into two regions which I've labeled as I (ABC, below the diagonal) and II (ACD, above the diagonal). If (in your mind) you pick up region II, flip it over in space, and then place it next to region I so that points A and D of region II are now on top of points C and B of region I, respectively (point C of region II now has coordinates $(2,0)$), then you should see that we have our equilateral triangle with side lengths 2. Thus, if we have selected a point "at random" from the interior of the rectangle, that point can also be thought of as being selected "at random" from the interior of the triangle.

That is, we can think that if we make one last adjustment of the coordinates of our randomly selected point. If x and y are such that the point is in region I, we are done. If, however, x and y are such that the point is in region II, then, because of the flipping and repositioning of region II, we must recalculate x and y with the rules $x \to 1 + x$ and $y \to \sqrt{3} - y$. To check which region of the rectangle the point is in, all we have to do is calculate the value of y/x; if $y/x < \sqrt{3}$, the point is in region I (and otherwise the point is in region II). With all this in mind, you should now be able to follow the MATLAB code **all3.m** shown below (the line numbers are not part of MATLAB but serve simply as reference tags), which randomly generates ten million points uniformly distributed over the interior of an equilateral triangle with side lengths 2, and which ends with the value of the variable *total* equal to the number of those points within distance $d = \sqrt{2}$ of all three vertices. (Lines 07, 08, and 09 calculate the distances squared of each point from the vertices, and line 10 compares those distances with 2, an approach that avoids computing thirty million square roots!) The code ran on my quite ordinary home computer in less than six seconds and produced an estimate for the probability of the event of interest of 0.1748068. This compares well, I think, with the theoretical value I gave you earlier in (3.2.1).

```
all3.m
01      h=sqrt(3);total=0;
02      for loop=1:10000000
03          x=rand;y=h*rand;
04          if (y/x)>h
05              x=1+x;y=h-y;
06          end
07          d1=x^2+y^2;
08          d2=(x-2)^2+y^2;
09          d3=(x-1)^2+(y-h)^2;
10          if d1<2&&d2<2&&d3<2
11              total=total+1;
12          end
13      end
14      total/10000000
```

As a second example of the Monte Carlo method, and of geometric probability, suppose we pick two independent points at random along the unit interval. Specifically, let X_1 be picked (uniformly) from somewhere between 0 and 0.5, and X_2 be picked (uniformly) from somewhere between 0.5 and 1. What is the probability that $X_2 - X_1 < 1/3$? The theoretical answer is $2/9 = 0.2222..$, and I'll show you how to calculate that in just a moment, but for now suppose we just want a fast answer from the computer? The MATLAB code **space.m** does the job, and I think its logic is pretty transparent. It simulates the above process ten million times, and keeps track of how many of those times result in $X_2 - X_1 < 1/3$. The computer's estimate of the answer is $0.22211...$, which is fairly close to the exact answer.

```
space.m
total=0;f=1/3;
for loop=1:10000000
    X1=rand/2;X2=0.5+rand/2;
    if X2-X1<f
        total=total+1;
    end
end
total/10000000
```

Figure 3.2.3 shows where the 2/9 comes from. The region of all possible values of X_1 and X_2 is the outlined square (what mathematicians call the *sample space* of the problem) with total area 1/4, on which is plotted the (straight) line $X_2 - X_1 = 1/3$. The subset of all possible values of X_1 and X_2 that satisfy the condition $X_2 - X_1 < 1/3$ is the shaded (triangular) region below the line, and its area is easily calculated to be 1/18. So, since X_1 and X_2 are each uniformly distributed and are independent, then each differential area in the sample space has the same probability as any other differential area of the same size. This means that the probability of the shaded area is simply the ratio of its area to the area of the entire sample space (notice that this calculation correctly assigns a probability of 1 to the entire sample space). That is, the probability of $X_2 - X_1 < 1/3$ is $\dfrac{1/18}{1/4} = \dfrac{4}{18} = \dfrac{2}{9}$.

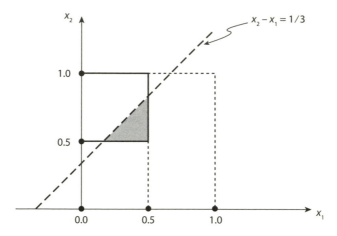

FIGURE 3.2.3. A geometric probability problem

3.3 The Hot Plate Problem

Now, with those purely mathematical problems as Monte Carlo warmups, I'll show you how the approach can be applied to a real physics problem. Suppose we have a square plate of some material of some constant thickness. Suppose further that we define our unit of distance so that the square plate is a *unit* square, that is, we imagine that the two faces of the plate are the square $0 \leq x \leq 1$, $0 \leq y \leq 1$. Imagine too that the two faces are insulated (covered with a material that has zero conductivity of heat), while the four edges of the plate can each individually be set at a given independent temperature. For example, the bottom edge, defined by $0 \leq x \leq 1$, $y = 0$, might be set at 100°C, and the other three edges might all be set at 0°C. We could then ask all sorts of questions about the temperature of the plate. Most generally, given $u(x, y, z, 0)$ for all x, y, and z, we could ask what is $u(x, y, z, t)$, the temperature at the point in the plate with coordinates (x, y, z) at any time $t > 0$? Right now I'll ask you to think about a slightly less general question: What is the *steady-state* temperature at the *middle* of the plate, that is, what is the value of $u(\frac{1}{2}, \frac{1}{2}, z, \infty)$?

The fundamental physics that is behind the answer to our questions is the so-called *heat* (or *diffusion*) *equation*. If k denotes what is called the *thermal conductivity* (or *diffusivity*) of the plate material (which might,

most generally, be $k = k(x, y, z)$), then it can be shown[4] that we have
the partial differential equation

$$\frac{\partial u}{\partial t} = k \left(\frac{\partial^2 u}{\partial x^2} + \frac{\partial^2 u}{\partial y^2} + \frac{\partial^2 u}{\partial z^2} \right). \tag{3.3.1}$$

In the steady state we have, by definition, no temperature variation
with time, and so $\frac{\partial u}{\partial t} = 0$. Also, because the two faces of the plate are
insulated, we know there will not be any steady-state heat flow in the
z-direction (the direction normal to the faces), and so there will be no
variation of the plate temperature in that direction (there is a *flow* of
heat only if there is a temperature gradient). That is, $\frac{\partial u}{\partial z} = 0$ for all x
and y, and so too $\frac{\partial^2 u}{\partial z^2} = 0$, and thus for any given (x, y) the steady-state
temperature of the plate is independent of z. Thus, our mathematical
problem is that of finding the temperature function $u(x, y, z) = u(x, y)$
that satisfies the simplified heat equation (simplified, yes, but still so
important that it has its own name: *Laplace's equation*, after the French
mathematical physicist Pierre Simon Laplace [1749–1827])

$$\frac{\partial^2 u}{\partial x^2} + \frac{\partial^2 u}{\partial y^2} = 0 \tag{3.3.2}$$

subject to the *boundary* (edge) *conditions* of

$$u(0, y) = 0 \text{ for the left vertical edge,} \tag{3.3.2a}$$

$$u(1, y) = 0 \text{ for the right vertical edge,} \tag{3.3.2b}$$

$$u(x, 1) = 0 \text{ for the top horizontal edge,} \tag{3.3.2c}$$

$$u(x, 0) = T(100°) \text{ for the bottom horizontal edge.} \tag{3.3.2d}$$

Mathematicians call this sort of problem, that of solving Laplace's
equation in a region of space with prescribed conditions for $u(x, y)$
on the region's boundary, a *Dirichlet problem*, after the German mathematician Peter Dirichlet (1805–1859). Related problems, which I'll not
discuss here, have prescribed conditions for $\frac{\partial u}{\partial x}$ and/or $\frac{\partial u}{\partial y}$ on the
region's boundary and are called *Neumann problems*, after the German

mathematician Carl Neumann (1832–1925). By the way, it is not necessary that the region be finite in extent, as it is in our hot plate problem. In Challenge Problem 3.2 I ask you to try your hand at solving Laplace's equation in a region that, along one dimension, extends to infinity.

I'll soon show you how to solve (3.3.2) for our $u(x, y)$, in general, but first let me show you a clever way to determine the specific value of $u\left(\frac{1}{2}, \frac{1}{2}\right)$ without having to calculate $u(x, y)$ for all x and y. Suppose we have obtained, by some means, the solution $u(x, y) = u_1(x, y)$ for Laplace's equation

$$\frac{\partial^2 u}{\partial x^2} + \frac{\partial^2 u}{\partial y^2} = 0$$

with the boundary conditions $u_1(0, y) = 0, u_1(1, y) = 0, u_1(x, 1) = 0$, and $u_1(x, 0) = T$, and another solution $u(x, y) = u_2(x, y)$ with the different boundary conditions $u_2(0, y) = T, u_2(1, y) = 0, u_2(x, 1) = 0$, and $u_2(x, 0) = 0$. Then the sum $u(x, y) = u_1(x, y) + u_2(x, y)$ also satisfies Laplace's equation (it's *linear!*) with boundary conditions that sum too; $u(0, y) = u_1(0, y) + u_2(0, y) = 0 + T = T, u(1, y) = u_1(1, y) + u_2(1, y) = 0 + 0 = 0, \quad u(x, 1) = u_1(x, 1) + u_2(x, 1) = 0 + 0 = 0$, and $u(x, 0) = u_1(x, 0) + u_2(x, 0) = T + 0 = T$. Now, physically, the temperature in the middle of the plate is the same for u_1 and u_2 (by symmetry, the middle of the plate is uniquely unaffected by which edge happens to be the single edge at temperature T), and so

$$u\left(\frac{1}{2}, \frac{1}{2}\right) = u_1\left(\frac{1}{2}, \frac{1}{2}\right) + u_2\left(\frac{1}{2}, \frac{1}{2}\right) = 2u_1\left(\frac{1}{2}, \frac{1}{2}\right).$$

If you repeat this argument in the obvious way for the case of all four edges at temperature T, then you should see that the steady-state temperature in the middle of the plate is

$$u\left(\frac{1}{2}, \frac{1}{2}\right) = 4u_1\left(\frac{1}{2}, \frac{1}{2}\right).$$

But, again arguing physically, if all four edges are fixed at temperature T, then the *entire plate*, and in particular the middle point, would in the steady state be at temperature T. Thus, our result is that, with one edge at temperature T and the other three at zero, we have

$$u_1\left(\frac{1}{2},\frac{1}{2}\right) = \frac{1}{4}u\left(\frac{1}{2},\frac{1}{2}\right) = \frac{1}{4}T.$$

For the case of $T = 100°$, then, we conclude that the middle point will be at temperature $25°$ if we fix one edge (any edge) at $100°$ with the remaining three edges at $0°$.

Is this correct? To convince you it is, in the next section I'll calculate the exact answer by analytical means for *any* point in the plate, and then use that general answer to specifically calculate $u\left(\frac{1}{2},\frac{1}{2}\right)$ and verify that it is indeed $25°$. Later I'll show you yet two more ways to get $u\left(\frac{1}{2},\frac{1}{2}\right)$, in fact $u(x,y)$ in general, with iteration and Monte Carlo computer codes. We can use the theoretical $u(x,y)$ we are about to calculate to check on how well (or not) the codes work.

One last, quick item before I start the theoretical derivation of $u(x,y)$: I want to show you one more example of the use of physical reasoning to get a mathematical result. When analyzing a solution to a differential equation, for a given set of boundary conditions, mathematicians routinely worry about its *uniqueness* (or not). For our hot plate problem, however, we can be assured that if we find *a* solution, then we have found *the* solution. We can "physically prove" this as follows. What we have is Laplace's equation

$$\frac{\partial^2 u}{\partial x^2} + \frac{\partial^2 u}{\partial y^2} = 0,$$

with the boundary conditions $u(0,y) = 0, u(1,y) = 0, u(x,1) = 0$, and $u(x,0) = T$. Let's now suppose that someone claims to have found the solution $u(x,y) = u_1(x,y)$, and (with the *same boundary conditions*) somebody else claims to have found the solution $u(x,y) = u_2(x,y)$. Our claim here is that $u_1(x,y) = u_2(x,y)$ for *all* x and y, and not just on the boundary. Here's the proof. Since Laplace's equation is linear, then it must be true that $u_3(x,y) = u_2(x,y) - u_1(x,y)$ is also a solution with the boundary conditions

$$u_3(0,y) = u_2(0,y) - u_1(0,y) = 0 - 0 = 0,$$
$$u_3(1,y) = u_2(1,y) - u_1(1,y) = 0 - 0 = 0,$$

$$u_3(x, 1) = u_2(x, 1) - u_1(x, 1) = 0 - 0 = 0,$$
$$u_3(x, 0) = u_2(x, 0) - u_1(x, 0) = T - T = 0.$$

That is, $u_3(x, y)$ is the steady-state temperature of the plate with all four edges fixed at $0°$. We know, physically, this means $u_3(x, y) = 0°$, that is, the entire plate, in the steady state, is at $0°$. Thus, $u_3(x, y) = u_2(x, y) - u_1(x, y) = 0°$, and so $u_1(x, y) = u_2(x, y)$, and our uniqueness proof is done. This argument may not please an absolutely pure mathematician, but physicists and engineers (and most mathematicians, too) will be convinced.

3.4 Solving the Hot Plate Problem with Analysis

To start our theoretical analysis for $u(x, y)$, we'll take $u(x, y)$ as the product of a function of x alone, with a function of y alone, that is, we'll write

$$u(x, y) = X(x)Y(y). \tag{3.4.1}$$

This argument, that we can so separate the two spatial variables, is an assumption that we are free to make (we can assume anything we wish that doesn't violate the rules of mathematics, with the trick being clever enough to assume something that leads to a solution!) and then justify when that assumption leads to a physically meaningful answer. Sometimes physics and engineering students are unhappy when first encountering this line of reasoning, but it's perfectly okay.

If we next substitute (3.4.1) into Laplace's equation of (3.3.2), we get

$$Y\frac{d^2X}{dx^2} + X\frac{d^2Y}{dy^2} = 0$$

where I've dropped the partial derivative notation, since X and Y are each a function of just a single variable, or

$$\frac{1}{X}\frac{d^2X}{dx^2} = -\frac{1}{Y}\frac{d^2Y}{dy^2}. \tag{3.4.2}$$

Now comes a very clever argument: the left-hand side of (3.4.2) is a function of only x, while the right-hand side is a function only of y. How can two functions, each depending on a different independent variable, always be equal? Well, there is a way: they are equal to the same constant! If we call this constant α, then we have two ordinary differential equations, in place of one partial differential equation:

$$\frac{1}{X}\frac{d^2 X}{dx^2} = \alpha \tag{3.4.3}$$

and

$$-\frac{1}{Y}\frac{d^2 Y}{dy^2} = \alpha. \tag{3.4.4}$$

We'll solve each of these two differential equations in turn, starting with (3.4.3).

Let's begin by assuming $\alpha > 0$. It will turn out that this leads to a dead end, and in fact the only way to get an answer that makes physical sense is to have $\alpha < 0$. But let's go through the $\alpha > 0$ analysis anyway, so you can see where and how it fails (after all, that's how it was first discovered that $\alpha > 0$ doesn't work). If we assume an exponential solution for X (this is "experience"), that is, if we write $X(x) = Ae^{cx}$, where A and c are arbitrary constants, then substitution into (3.4.3) gives

$$\frac{1}{Ae^{cx}} \times Ac^2 e^{cx} = \alpha,$$

or $c = \pm\sqrt{\alpha}$; remember, $\alpha > 0$, and so c is real. Thus, most generally we have (with A_1 and A_2 as to-be-determined constants)

$$X(x) = A_1 e^{x\sqrt{\alpha}} + A_2 e^{-x\sqrt{\alpha}}. \tag{3.4.5}$$

Since $u(x,y) = X(x)Y(y)$, and $u(0,y) = 0$ from (3.3.2a), we have

$$X(0)Y(y) = 0, \tag{3.4.6}$$

and since $u(1,y) = 0$ from (3.3.2b), we have

$$X(1)Y(y) = 0. \tag{3.4.7}$$

Together, (3.4.6) and (3.4.7) say $X(0) = X(1) = 0$; the alternative conclusion, that $Y(y) = 0$, leads to $u(x, y) = 0$ for all x and y, which is clearly (on physical grounds) wrong; mathematicians are usually less judgmental on this, simply calling $u(x, y) = 0$ a *trivial* solution.

Using $X(0) = X(1) = 0$ in (3.4.5) gives

$$A_1 + A_2 = 0 \tag{3.4.8}$$

and

$$A_1 e^{\sqrt{\alpha}} + A_2 e^{-\sqrt{\alpha}} = 0. \tag{3.4.9}$$

From (3.4.8) we have $A_2 = -A_1$, and then from (3.4.9) we have

$$A_1 e^{\sqrt{\alpha}} - A_1 \frac{1}{e^{\sqrt{\alpha}}} = 0,$$

or

$$e^{2\sqrt{\alpha}} - 1 = 0,$$

which means $\alpha = 0$. But (3.4.5) then says $X(x) = A_1 + A_2$, a constant, and since $X(0) = 0$, the conclusion is that $X(x) = 0$, and so we are right back to a trivial solution!

We got into this unhappy state by assuming $\alpha > 0$. So let's start over and assume $\alpha < 0$. In fact, I'll now write $\alpha = -\beta^2$, which for any real β is guaranteed to be negative. Now, since $c = \sqrt{-\beta^2} = \pm i\beta$ (where, of course, $i = \sqrt{-1}$), then most generally,

$$X(x) = A_1 e^{i\beta x} + A_2 e^{-i\beta x}. \tag{3.4.10}$$

We still have $X(0) = 0$, which from (3.4.10) says $A_2 = -A_1$, just as before. Or, dropping the now unnecessary subscripts and writing $A_1 = k$ and $A_2 = -k$, we have from Euler's famous identity that

$$X(x) = k e^{i\beta x} - k e^{-i\beta x} = k2i \sin(\beta x)$$

or, writing the constant $k2i$ as simply A, we have,

$$X(x) = A \sin(\beta x). \tag{3.4.11}$$

Now, as before, we have $X(1) = 0$, and so (3.4.11) says $A \sin(\beta) = 0$, or $\sin(\beta) = 0$. Thus, β is not just any constant but rather is *quantized*, that is, $\beta = n\pi, n = 1, 2, 3, \ldots$. Notice that $n = 0$ is not included, since that value for n gives us $X(x) = 0$, which we've already twice dismissed as trivial. So, with A_n as generally a different constant for each value of n, we have from (3.4.11) that

$$X_n(x) = A_n \sin(n\pi x), \quad n = 1, 2, 3, \ldots. \tag{3.4.12}$$

We are now half done. All we have left to do is the calculation of $Y(y)$.
    From (3.4.4) we have

$$\frac{d^2 Y}{dy^2} + \alpha Y = 0,$$

or, as $\alpha = -\beta^2 = -n^2 \pi^2$,

$$\frac{d^2 Y}{dy^2} - n^2 \pi^2 Y = 0, \tag{3.4.13}$$

which we can solve pretty much as we did for $X(x)$. Assume $Y(y) = Be^{cy}$, where B and c are constants. Substitution into (3.4.13) gives

$$Bc^2 e^{cy} - n^2 \pi^2 Be^{cy} = 0,$$

or $c = \pm n\pi$. Thus,

$$Y(y) = B_1 e^{n\pi y} + B_2 e^{-n\pi y}. \tag{3.4.14}$$

Again, $u(x, y) = X(x)Y(y)$, and since $u(x, 1) = 0$ from (3.3.2c), we have

$$X(x)Y(1) = 0. \tag{3.4.15}$$

We now know from our earlier work that $X(x) \neq 0$, and so (3.4.15) says $Y(1) = 0$. Combined with (3.4.14), this tells us that

$$B_1 e^{n\pi} + B_2 e^{-n\pi} = 0,$$

or

$$B_2 = -B_1 e^{2n\pi}.$$

Putting this into (3.4.14) gives

$$Y(y) = B_1 e^{n\pi y} - B_1 e^{2n\pi} e^{-n\pi y},$$

or, dropping the now superfluous subscript,

$$Y(y) = B \left[e^{n\pi y} - e^{2n\pi} e^{-n\pi y} \right]. \tag{3.4.16}$$

Now a little algebraic trick. If we multiply through the right-hand side of (3.4.16) by $e^{n\pi}/e^{n\pi}$—a fancy way of writing 1—we have

$$Y(y) = Be^{n\pi} \left[\frac{e^{n\pi y} - e^{2n\pi} e^{-n\pi y}}{e^{n\pi}} \right] = Be^{n\pi} \left[e^{n\pi y} e^{-n\pi} - e^{n\pi} e^{-n\pi y} \right]$$

$$= Be^{n\pi} \left[e^{n\pi(y-1)} - e^{-n\pi(y-1)} \right] = Be^{n\pi} \left[e^{-n\pi(1-y)} - e^{n\pi(1-y)} \right]$$

$$= -Be^{n\pi} \left[e^{n\pi(1-y)} - e^{-n\pi(1-y)} \right] = -2Be^{n\pi} \sinh \left\{ n\pi(1-y) \right\}.$$

Calling $-2Be^{n\pi} = B_n$, a different (in general) constant for each value of n, we have

$$Y_n(y) = B_n \sinh \left\{ n\pi(1-y) \right\}. \tag{3.4.17}$$

So, writing $u_n(x, y) = X_n(x)Y_n(y)$, we have, from (3.4.12) and (3.4.17),

$$u_n(x, y) = A_n \sin(n\pi x) B_n \sinh \left\{ n\pi(1-y) \right\},$$

or, if we write $A_n B_n = F_n$, a constant (generally different) for each n,

$$u_n(x, y) = F_n \sin(n\pi x) \sinh \left\{ n\pi(1-y) \right\}, n = 1, 2, 3, \ldots.$$

The most general solution is therefore

$$u(x, y) = \sum_{n=1}^{\infty} u_n(x, y) = \sum_{n=1}^{\infty} F_n \sin(n\pi x) \sinh \left\{ n\pi(1-y) \right\}.$$

$$\tag{3.4.18}$$

We are almost done. Our last question to answer is, what are the F_n?

We can answer that question by using our last remaining boundary condition, (3.3.2d), which says $u(x, 0) = T$. Applying this condition to (3.4.18) gives

$$\sum_{n=1}^{\infty} F_n \sin(n\pi x) \sinh\{n\pi\} = T. \tag{3.4.19}$$

We can calculate the F_n from (3.4.19) by multiplying through by $\sin(m\pi x)$, where m is some (any) *fixed* positive integer, and integrating with respect to x from 0 to 1. Then,

$$\sum_{n=1}^{\infty} F_n \sinh\{n\pi\} \left\{ \int_0^1 \sin(n\pi x)\sin(m\pi x)\, dx \right\} = T \int_0^1 \sin(m\pi x)\, dx. \tag{3.4.20}$$

What motivates doing this is the so-called *orthogonality* property of the sine functions, that is,

$$\int_0^1 \sin(n\pi x)\sin(m\pi x)\, dx = \begin{cases} 0, & \text{if } m \neq n \\ \dfrac{1}{2}, & \text{if } m = n \end{cases}.$$

Using this in the integral on the left-hand side of (3.4.20), we have all the terms in the infinite sum vanish *except* for the one term that survives when $n = m$, which gives

$$F_m \sinh(m\pi) \cdot \frac{1}{2} = T \left[\frac{-\cos(m\pi x)}{m\pi} \right]_0^1 = T \frac{1 - \cos(m\pi)}{m\pi}.$$

Since $\cos(m\pi) = -1$ if m is odd and $+1$ if m is even, this says

$$F_m = 2T \frac{1 - (-1)^m}{m\pi \sinh(m\pi)}, \qquad m = 1, 2, 3, \ldots,$$

or, since

$$1 - (-1)^m = \begin{cases} 0 & \text{if } m \text{ is even} \\ 2 & \text{if } m \text{ is odd,} \end{cases}$$

we can write (replacing m with n, which is a trivial change in notation)

$$F_n = \frac{4T}{n\pi \sinh(n\pi)}, \quad n = 1, 3, 5, \ldots.$$

And so, at last, putting this F_n into (3.4.18), we have our solution:

$$u(x, y) = \frac{4T}{\pi} \sum_{n=1,3,5,\ldots}^{\infty} \frac{1}{n} \cdot \frac{\sinh\{n\pi(1-y)\}}{\sinh(n\pi)} \sin(n\pi x). \qquad (3.4.21)$$

The $u(x, y)$ in (3.4.21) is the theoretically exact steady-state temperature in our square hot plate, with the bottom edge at temperature T and the other three edges at temperature zero. This is an easy expression to code for a computer, and when I ran that code for the middle point ($x = y = 0.5$), for $T = 100°$, the sum very quickly (only the first three terms were needed) converged to $25°$, just as we concluded from a symmetry argument in the last section. For any other values of x and y, of course, our symmetry argument wouldn't work, while (3.4.21) always works. For example, (3.4.21) tells us that the temperature in the middle of the upper-right quadrant of the hot plate is $u(0.75, 0.75) = 6.7972°$, a value we'd not likely arrive at by any simple argument.

3.5 Solving the Hot Plate Problem by Iteration

So far we've seen how to solve Laplace's equation, with our hot plate boundary conditions, using a symmetry argument (which actually works for only *one* point!), and more generally by a fairly lengthy sequence of purely mathematical deductions. In this section I'll show you a computer-intensive approach (called *iteration*) to solving Laplace's equation that is, at its core, nothing but elementary school arithmetic. Simple as it is, however, in some circumstances iteration may be the only practical way to "solve" a partial differential equation (if, for example, the region's boundary isn't a nice simple square like our hot plate, but rather has some arbitrarily complicated shape). To start the development of how to solve Laplace's equation (with given boundary conditions) on a computer, I'll begin by showing you how to numerically approximate $\frac{\partial^2 u}{\partial x^2}$ and $\frac{\partial^2 u}{\partial y^2}$. This will then give us an enormously useful *physical* interpretation of what is going on in what otherwise appears, on the surface, to be just a lot of math symbols.

By the very definition of the first partial derivative, we have

$$\frac{\partial u}{\partial x} = \lim_{\Delta h \to 0} \frac{u(x + \Delta h, y) - u(x,y)}{\Delta h}, \tag{3.5.1}$$

or, alternatively,

$$\frac{\partial u}{\partial x} = \lim_{\Delta h \to 0} \frac{u(x, y) - u(x - \Delta h, y)}{\Delta h}. \tag{3.5.2}$$

Adding (3.5.1) and (3.5.2), we have

$$2\frac{\partial u}{\partial x} = \lim_{\Delta h \to 0} \frac{u(x + \Delta h, y) - u(x - \Delta h, y)}{\Delta h},$$

or

$$\frac{\partial u}{\partial x} = \lim_{\Delta h \to 0} \frac{u(x + \Delta h, y) - u(x - \Delta h, y)}{2\Delta h}. \tag{3.5.3}$$

Now, since $\dfrac{\partial^2 u}{\partial x^2} = \dfrac{\partial}{\partial x}\left(\dfrac{\partial u}{\partial x}\right)$, we can apply (3.5.3) to itself and immediately write

$$\frac{\partial^2 u}{\partial x^2} = \lim_{\Delta h \to 0} \frac{\frac{\partial u}{\partial x}\big|_{x+\Delta h, y} - \frac{\partial u}{\partial x}\big|_{x-\Delta h, y}}{2\Delta h}$$

$$= \lim_{\Delta h \to 0} \frac{\frac{u(x + 2\Delta h, y) - u(x,y) - u(x,y) + u(x - 2\Delta h, y)}{2\Delta h}}{2\Delta h},$$

or

$$\frac{\partial^2 u}{\partial x^2} = \lim_{\Delta h \to 0} \frac{u(x + 2\Delta h, y) - 2u(x,y) + u(x - 2\Delta h, y)}{(2\Delta h)^2}. \tag{3.5.4}$$

Or, if we write $\Delta s = 2\Delta h$, then (3.5.4) becomes

$$\frac{\partial^2 u}{\partial x^2} = \lim_{\Delta s \to 0} \frac{u(x + \Delta s, y) - 2u(x,y) + u(x - \Delta s, y)}{(\Delta s)^2}. \tag{3.5.5}$$

In the same way,

$$\frac{\partial^2 u}{\partial y^2} = \lim_{\Delta s \to 0} \frac{u(x,y + \Delta s) - 2u(x,y) + u(x,y - \Delta s)}{(\Delta s)^2}. \tag{3.5.6}$$

Using (3.5.5) and (3.5.6) to write Laplace's equation gives

$$\frac{\partial^2 u}{\partial x^2} + \frac{\partial^2 u}{\partial y^2}$$

$$= \lim_{\Delta s \to 0} \frac{u(x + \Delta s, y) + u(x - \Delta s, y) + u(x, y + \Delta s) + u(x, y - \Delta s) - 4u(x,y)}{(\Delta s)^2},$$

or, solving for $u(x,y)$,

$$u(x,y) = \lim_{\Delta s \to 0} \frac{1}{4} \{ u(x + \Delta s, y) + u(x - \Delta s, y)$$

$$+ u(x, y + \Delta s) + u(x, y - \Delta s) \}. \tag{3.5.7}$$

In words, the steady-state temperature at the point (x, y) in our hot plate is just the *average* of the temperatures of the tiny region (four nearby points) surrounding the point. In retrospect, perhaps we should have simply guessed this perfectly reasonable physical result!

Here's how we are going to use (3.5.7). If you'll look at Figure 3.5.1 you'll see that I've overlaid our square hot plate with a two-dimensional grid, with the grid lines spaced both horizontally and vertically by $\Delta s = 0.01$. This value of Δs certainly isn't zero, but it isn't unreasonable to say it is small compared to the size of the hot plate. We could, of course, make Δs even smaller, but as you'll soon see, there is a direct connection between decreasing Δs and increasing computational time. Using $\Delta s = 0.01$ is a compromise between numerical accuracy and computer processing time. The intersection points of the vertical and horizontal grid lines can be thought of as the elements of a 101×101 matrix, which I'll call $U(r, c)$—r for *row* and c for *column*—where r and c are each positive integers running from 1 to 101. (Having r and c run through the integers 0 to 100 might be more natural, but MATLAB doesn't support zero indexing into matrices.) The steady-state temperature $u(x, y)$ will be the value of the matrix element $U(r, c)$ where

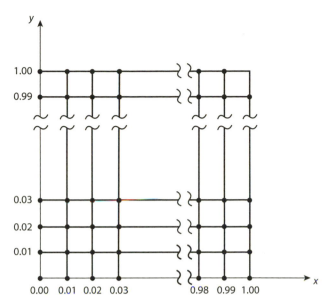

FIGURE 3.5.1. Grid approximation of the hot plate

$r = 100y + 1$ and $c = 100x + 1$. (*Caution:* When we talk of $u(x, y)$, the first index, x, moves us horizontally and the second index, y, moves us vertically, but if we talk of $U(r,c)$, then the first index, r, moves us vertically from row to row and the second index, c, moves us horizontally from column to column. Keep this distinction of the behavior between $u(x,y)$ and $U(r,c)$ in mind or you'll end up terribly confused!)

For example, along the entire left vertical edge of the hot plate we have $x = 0$ as y varies from 0 to 1, and so the values $U(r, 1)$, for $r = 1, 2, \ldots, 101$, are all set equal to zero because of the boundary condition (3.3.2a). And because of the boundary conditions (3.3.2b) and (3.3.2c) the values $U(r, 101)$, for $r = 1, 2, \ldots, 101$, and $U(101, c)$, for $c = 1, 2, \ldots, 101$, are also set equal to zero. By boundary condition (3.3.2d), however, the bottom horizontal edge, where $y = 0$, is fixed at temperature T, and so $U(1, c) = T$, for $c = 1, 2, \ldots, 101$. Notice that this last assignment sets $U(1, 1) = U(1, 101) = T$, which overwrites the earlier assignment $U(1, 1) = U(1, 101) = 0$, but that is simply a result of the "quantization noise" one runs into when transforming a spatially continuous problem into one discrete in space.

Once we are done solving our problem with these choices for $U(1, 1)$ and $U(1, 101)$ we could go back and make different choices (perhaps $U(1, 1) = U(1, 101) = 0$, or $U(1, 1) = 0$ and $U(1, 101) = T$) and see if it makes much difference. (It won't.)

Now, what about the values of the other elements of the U-matrix, that is, what are the $99 \times 99 = 9,801$ *interior* point temperatures of our quantized hot plate? Well, this is where (3.5.7) comes into play. First, we'll initially set all those element values to some arbitrary values (usually all the same value, typically the average of the individual boundary temperatures, which for our hot plate is $T/4$), and then iterate. That is, we'll run through each of the $9,801$ interior values of the U-matrix and, using (3.5.7), replace each one with the average of the values of its four nearest neighbors.[5] And then we'll do that again. And again. And again. And.... Eventually the numbers in the U-matrix will stop changing and the entire process will have *converged* to the steady-state temperatures of the hot plate at each of its $9,801$ interior points. I am not going to prove the convergence here, although such proofs exist.[6] My philosophical position on this is that of the Princeton computer scientist Forman Acton: "I have not usually tried to prove convergence of any iterative process. It is a commonplace that numerical processes that are efficient usually cannot be proven to converge while those amenable to proof are inefficient.... The best demonstration of convergence is convergence itself."[7]

Iteration of a matrix with thousands of values is an undeniably mind-numbing, error-prone task for a human, but it's duck soup for a computer. The MATLAB code **jacobi.m** does the job, running through *five thousand* iterations (millions of arithmetic operations) on my little home computer in less than seven seconds. This code, which implements Jacobi iteration (see note 5 again) is not very efficient compared to other possible iteration schemes, but it is very easy to understand. Figure 3.5.2 shows that the code requires thousands of iterations to converge to the temperature $u(0.75, 0.75)$—(3.4.21) told us that value is $6.7972°$, while after five thousand iterations **jacobi.m** tells us that $U(76, 76) = 6.8457°$ (another five thousand iterations, for a total of ten thousand, improves the code estimate to $6.7979°$)—but of course (3.4.21) has all sorts of high-powered math behind it while **jacobi.m** is nothing but lots and lots of grade-school arithmetic.

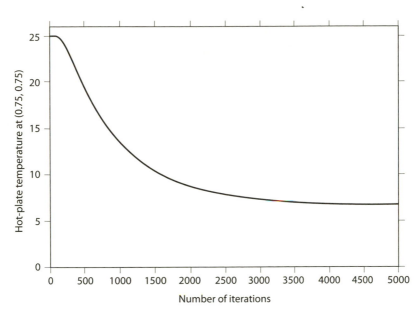

FIGURE 3.5.2. Jacobi iteration convergence can be slow!

jacobi.m

```
01      T=input('What is T?');
02      U=zeros(101,101);
03      for c=1:101
04          U(1,c)=T;
05      end
06      A=U;
07      for loop=1:5000
08          for r=2:100
09              for c=2:100
10                  A(r,c)=(U(r-1,c)+U(r+1,c)+U(r,c-1)
                            +U(r,c+1))/4;
11              end
12          end
13          U=A;
14      end
```

3.6 Solving the Hot Plate Problem with the Monte Carlo Technique

Let's solve the hot plate problem one more time, still using (3.5.7), but now with a *probabilistic* interpretation. Instead of using (3.5.7) to iterate, which simultaneously solves for the temperature $u(x, y)$ at many points over the entire plate, our new interpretation of (3.5.7) will be the theoretical basis for performing numerous so-called *random walks* that will estimate $u(x, y)$ at a given, *single* point location in the grid of Figure 3.5.1. And again, as with iteration, this new approach is feasible only because we have a high-speed computer available to us. Before starting the random walking that we'll soon be doing, I'll label the individual edges of the hot plate's boundary. This can be done in a variety of ways, but let's suppose we do it as follows: we'll call the four edges of the plate edges 1, 2, 3, and 4, with the bottom edge (the edge at temperature T) edge 1, and proceed around the boundary in a clockwise sense. Thus, the left vertical edge is edge 2, the top edge is edge 3, and the right vertical edge is edge 4.

Now, what is a random walk? Imagine you are at hot plate grid location (X, Y), and that (X, Y) is not on the boundary (I'll come back to that case in just a bit). Then, with probability 1/4, you move to the grid point directly to your right, with probability 1/4 to the grid point directly to your left, with probability 1/4 to the grid point directly above you, and with probability 1/4 to the grid point directly below you. (Since we are no longer using X and Y to index into a matrix, as we did in the last section with h and v into U when we performed Jacobi iteration, we no longer have to observe the MATLAB requirement that forced h and v to be positive integers—that is, starting at 1—but rather can imagine both X and Y running from 0 to 100, with integer steps representing movements of 0.01.) Then, after making that first move from (X, Y), you move again, and again, and so on, until eventually you move to a *boundary* point. That event terminates the walk, with your last task of the walk being the recording of the label of the edge you've landed on.

Then you go back to (X, Y) and do the whole process all over again. And again, and again, and so on, say, for a total of five thousand walks, with each one terminating somewhere on a boundary edge. When you're all done, you'll have a list of numbers, five thousand of them, each of which is a 1, 2, 3, or 4. From that list it is easy to then calculate the fraction of the walks that terminated on each edge. For example, suppose that your list of five thousand numbers has five hundred 1s, fifteen hundred 2s, one thousand 3s, and two thousand 4s. If we call f_i the fraction of the walks that terminate on boundary edge i, then $f_1 = 0.1$, $f_2 = 0.3$, $f_3 = 0.2$, and $f_4 = 0.4$. For the special case of (X, Y) being a boundary point, you should see that a walk starting at such a point immediately terminates right there on that boundary edge; if (X, Y) is on boundary edge e, $e = 1, 2, 3$, or 4, then $f_{i=e} = 1$ and $f_{i \neq e} = 0$.

Our formula for estimating $u(X, Y)$, the temperature of the hot plate at (X, Y), is the wonderfully simple (it is trivially obvious if (X, Y) is a boundary point)

$$u(X, Y) = \sum_{i=1}^{4} f_i T_i, \tag{3.6.1}$$

where T_i is the fixed temperature of boundary edge i. Why?

To understand where (3.6.1) comes from, suppose we call the sum in (3.6.1) the *value* of the walk starting from the point (X, Y), which I'll call $V(X, Y)$. $V(X, Y)$ is, of course, identical to $u(X, Y)$, but it avoids the psychological baggage of being called a *temperature*. $V(X, Y)$ is simply a value. (This is just a bit of a mind-game, and if you don't need it, fine, but it helps *me*!) Now, once we make the first move, the value of the walk from (X, Y) is the value of the walk starting from the new point *weighted by the probability of being at that new point*. That is,

$$V(X, Y) = \frac{1}{4}V(X - 1, Y) + \frac{1}{4}V(X + 1, Y) + \frac{1}{4}V(X, Y + 1)$$
$$+ \frac{1}{4}V(X, Y - 1),$$

or, since we know the value V is actually the temperature u, we have

$$u(x, y) = \frac{1}{4}\{u(x - 1, y) + u(x + 1, y) + u(x, y + 1) + u(x, y - 1)\}$$

with $\Delta s = 1$. But this, of course, is precisely (3.5.7).

The MATLAB code **walk.m** performs this entire process for our square hot plate. In the code I've taken $T_1 = 100$, and $T_2 = T_3 = T_4 = 0$. A natural presumption would be that as the number of walks increases we'll get better and better estimates for $u(X, Y)$, but we can experimentally check that presumption with the code. Before discussing what the code is doing, remember that we have two special cases for which we already know the answers: If $T = 100°$, then $u(0.5, 0.5) = 25°$ and $u(0.75, 0.75) = 6.7972°$. We got the first special result from a symmetry argument back in Section 3.3, but our new probabilistic interpretation of (3.5.7) now makes it obvious by inspection: if we start a random walk in the center of the hot plate, then clearly we'll end up on any particular one of the four edges with probability $1/4$. Thus, we hit the $100°$ edge with probability $1/4$ and a $0°$ edge with probability $3/4$. So, from (3.6.1) we immediately have $u(0.5, 0.5) = 25°$. Notice, too, that the Monte Carlo approach immediately tells us that $u(0.5, 0.5) = 50°$ or $75°$, respectively, if *any two*, or *any three*, edges are fixed at $100°$ with all other edges fixed at zero.

---

walk.m

```
01    N=input('How many walks?');
02    X=input('X=?');Y=input('Y=?');
03    X1=X;Y1=Y;
04    fraction=zeros(1,4);
05    for loop=1:N
06        X=X1;Y=Y1;
07        while X>0&&X<100&&Y>0&&Y<100
08            xdir=rand;ydir=rand;
09            if xdir<0.5
10                X=X +1;
11            else
12                X=X-1;
13            end
14            if ydir<0.5
15                Y=Y +1;
16            else
```

(Continued)

(Continued)

```
17                          Y=Y-1;
18                    end
19              end
20              if Y==0
21                    i=1;
22              elseif Y==100
23                    i=3;
24              elseif X==0
25                    i=2;
26              else
27                    i=4;
28              end
29              fraction(i)=fraction(i) +1;
30        end
31        fraction=fraction/N;
32        fraction,100*fraction(1)
```

Now, here's what **walk.m** does. Lines 01 and 02 ask for the values of X, Y, and N (the number of random walks to be performed from the starting point (X, Y)). Line 03 remembers X and Y (as $X1$ and $Y1$, respectively) so that each new random walk can start again from (X, Y); we need $X1$ and $Y1$ because, as the code executes, both X and Y will change (lines 10, 12, 15, and 17) and their values will have to be reset before starting the next walk. Line 04 creates the four element row vector *fraction*, where *fraction(i)* will be the value of f_i when the code finishes (after executing line 31). Lines 05 and 30 establish a *for/end* loop that will be executed N times, once for each of the N random walks. Using $X1$ and $Y1$, line 06 sets X and Y to a new walk's starting point. Lines 07 through 19 are a loop that runs for an indefinite number of cycles, exiting into line 20 only when the walk has reached a boundary edge. This loop is done via a *while* command, which checks for *all* the conditions that must be satisfied for the walk *not* having reached a boundary edge, that is, for $0 < X < 100$ *and* $0 < Y < 100$. While the code remains in the *while* loop it generates, for each new motion of the walk, two random numbers (*xdir* and *ydir*) uniform from 0 to 1 (line 08) and then uses them to "move" either by incrementing or decrementing X or Y by one (lines 09 through 18). Once the current walk terminates, lines 20 through 28 determine on which boundary edge the termination occurred, and then line 29 increments the appropriate element of *fraction* to update the current count of many walks have terminated on each of the four boundary edges. Once all N of the random walks have been completed, line 31 converts *fraction* to fractional form. Line 32 prints that converted form of *fraction*, and the value of $u(X, Y)$ for our square hot plate as computed from (3.6.1).

walk.m will produce somewhat different estimates for $u(X, Y)$ each time it executes because of the random nature of the walks. To "experimentally see" how stable the code's estimates are, the following table shows what **walk.m** produced for two separate runs, for each of the two starting points $(0.5, 0.5)$ and $(0.75, 0.75)$, for $N = 100$, $1,000$, $10,000$, and $100,00$ random walks. Remember, the theoretical values are $u(0.5, 0.5) = 25$ and $u(0.75, 0.75) = 6.7972$.

N	$u(0.5, 0.5)$	$u(0.75, 0.75)$
10^2	21	8
	31	7
10^3	26.6	6
	25.2	7.1
10^4	24.33	6.51
	25.25	6.98
10^5	25.230	6.850
	25.041	6.914

Finally, when out of curiosity I ran **walk.m** once for a million walks ($N = 10^6$) the code produced—after 183 seconds!—the estimate $u(0.75, 0.75) = 6.8328$.

The most obvious conclusion that I think comes from these results is that Monte Carlo is okay on our hot plate problem for a "fairly good" estimate (at a single point), but it is certainly inferior in terms of accuracy to iteration (which solves the entire hot plate all at once, to boot).

But there *is* something in **walk.m** that still makes Monte Carlo worth considering, and that is *fraction*. Once you have *fraction*, then (3.6.1) can be used over and over, with various different values for the T_i, *without having to perform any new random walks*. This is because *fraction* is completely determined only by the geometry of the boundary and by the location of the common starting point for each random walk.

3.7 ENIAC and MANIAC-I: The Electronic Computer Arrives

One of the great technological achievements in all of human history occurred in the mid-1940s with the construction (at the University of Pennsylvania's Moore School of Electrical Engineering) of the world's first electronic computer, the Electronic Numerical Integrator and Computer (ENIAC). There had been automatic computers before ENIAC, but they were very slow electromechanical gadgets that used relays, not electronics. The Harvard-IBM Mark I, for example, built in 1944, took six seconds to multiply two numbers together, and twice as long to do a division. ENIAC, on the other hand, used many thousands of vacuum tubes for its logical circuitry and was a thousand times faster. ENIAC was constructed under a U.S. Army contract and in 1946 was installed in the Ballistic Research Laboratory at the Aberdeen Proving Grounds in Maryland. It was used there to perform the vast numerical drudgery of computing ballistic trajectories for World War II heavy artillery guns firing huge shells through a realistic variable density atmosphere covering a rotating Earth. ENIAC had to be "programmed" for each new problem by an approach that would leave a modern student of computer science in shock: a rat's nest of banana-plug cables, along with numerous switches, all of which had to be manually adjusted. These cables and switches allowed different parts of ENIAC's hardware to electrically interact, as well as serving as the storage units of a problem's input data. It was about as far removed from a modern computer as a slingshot is from the Death Star, the planet-vaporizing weapon in *Star Wars*.

As astonishing an invention as was ENIAC, in its original form it would have been next to useless to a modern physicist. What would *almost* transform ENIAC into a machine that could be fairly called a modern electronic computer was a man who could see beyond ENIAC's conceptual design limitations, the mathematics genius John von Neumann (1903–1957), who was a consultant at Aberdeen as well as at the Los Alamos Scientific Laboratory. The Hungarian-born American physicist Edward Teller (1908–2003) wrote of von Neumann that he was one of the rare mathematicians who could descend to the level of a physicist.[9] Von Neumann quickly appreciated the crippling limitations of cable plug-board-and-switches "programming," and in fact had already conceived the revolutionary idea of not only storing a problem's input data inside the machine but also so storing the program itself that would manipulate that data. There would be no external rat's nest of plug cables for von Neumann! This is the so-called stored program concept, utilized today in every modern computer.

While ENIAC's basic architecture didn't allow much alteration, it was still possible to partially implement the stored program idea. In 1948 this altered configuration of ENIAC was actually achieved; with von Neumann's support the Los Alamos scientist Nicholas Metropolis (1915–1999) was able to set up and run the first ever Monte Carlo simulations on an electronic computer.[10] Meanwhile, von Neumann had been deeply involved in the development of the first stored-program computer, the EDVAC (Electronic Discrete VAriable Computer),[11] and its successor, the IAS machine (Institute for Advanced Study, Princeton). The IAS machine was the origin of some controversy at the institute, appearing to many of the deep thinkers there to be "merely engineering" and so more appropriately done elsewhere. Such was von Neumann's clout, however (the support of the institute's director, J. Robert Oppenheimer, also didn't hurt), that he got what he wanted. The IAS machine became operational in 1952 and served as the inspiration for several additional, quite similar machines constructed around the country, one of which was the famous MANIAC-I at Los Almos. MANIAC-I was built under the direction of Metropolis, who had spent a year at Princeton with von Neumann during the design of the IAS machine. Metropolis's computer got its name from Mathematical Analyzer, Numerical Integrator, and Computer, but the physicist George Gamow (1904–1968) playfully dubbed it "Metropolis and von Neumann Install Awful Computer."

MANIAC-I was directly programmed in low-level machine language (there was no BASIC, PASCAL, FORTRAN, or MATLAB in those days), had an instruction set of thirty-seven commands (such as "add number in memory location M to cleared accumulator" and "record number in quotient register to magnetic tape"), had programmable interfaces to a paper tape reader (input) and paper tape punch (output), magnetic tape drive (input/output), and teletype (output), and both "high-speed" memory (electrostatic storage tube with one thousand words) and "slow-speed" memory (rotating magnetic drum with ten thousand words).[12] As a measure of MANIAC-I's speed, it could add two numbers in ninety microseconds and multiply or divide two numbers in one millisecond. By today's standards it was slow, clunky, stupid, and huge (128 cubic feet of hardware dissipating 55,000 watts with the aid of a ten-ton air conditioner). By the standards of *its* day, however, it was a science fiction fantasy come true, a stupendously wonderful gadget that sent the genius Enrico Fermi into the highest clouds of ecstasy.[13]

When MANIAC-I became available to Los Alamos scientists in 1952 they immediately began searching for projects worthy of its capabilities. One of the fun ones was the development of the first realistic computer chess code, which implemented many of the ideas found in today's chess machines. That code looked two moves ahead, for both black and white, taking about twelve minutes per move.[14] The very first game the code played was against itself, something that a human can't really do. For a dispassionate machine, however, being privy to the "thinking processes" of both players is not psychologically illogical, as it would be for a single human playing against himself. The code's designers nevertheless found it hard to completely think of their creation as merely a big pile of nonhuman circuitry, but rather as more like them; at one point, they wrote, "the machine seemed to have a mortal fear of checks, since its freedom [to choose among many possible moves] after check was nearly nil and it tended to sacrifice material to avoid checks." Overall, it was felt that the code could be compared to a human player "who has average aptitude for the game and experience amounting to 20 or so full games played."

Well, fun it was, but it wasn't serious physics, a situation that the physicist Enrico Fermi and the mathematician Stanislaw Ulam were soon to address. There were several such serious projects[15] that eventually were tackled by MANIAC-I, but the most famous is the so-called Fermi-Pasta-Ulam (or FPU) experiment.[16] The late Russian mathematician Vladimir Arnol'd once observed that "The only computer experiments worth doing are those that yield a surprise," and, to quote an American physicist writing on the FPU computer experiment, "If generations of students have taught me anything, it is that few things fascinate them more than a scientific mystery—and the Fermi-Pasta-Ulam paradox is a cracker-jack mystery."[17]

3.8 The Fermi-Pasta-Ulam Computer Experiment

As the senior scientists involved, Ulam and Fermi decided to use MANIAC-I in a study of what is called "thermalization and mixing" in a one-dimensional crystal. Briefly, the problem was to study how the vibrational energy of the crystal, if initially located in just *one* mode of oscillation of the crystal's atoms, would over time spread itself throughout *all* the possible modes of oscillation (the meaning of *mode of oscillation* will be explained soon). This mode spreading of the energy is called the *equipartition of energy* in statistical physics, and it was believed at the time of the FPU experiment to occur in all systems described by nonlinear equations. This belief is what the FPU experiment was designed to test. The one-dimensional model was a compromise between a simulated system sufficiently complicated to have some connection to reality and yet simple enough to be within the computational capability of MANIAC-I.

To truly appreciate how impressive was the execution of the FPU experiment, it is important to take a little time up front to understand the mathematics behind the physics. Consider Figure 3.8.1, which shows a horizontal chain of N unit masses (the "atoms") connected one to another by identical, relaxed, massless springs.[18] The springs mimic the interatomic forces in the crystal. If these springs are linear (we'll make them *non*-linear in just a bit), that is, if they are springs that obey Hooke's law, and if we assume (as did FPU) that the spring constant is unity, then applying Newton's second law of motion to each of the unit masses gives the following set of second-order, linear

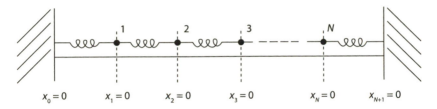

FIGURE 3.8.1. The FPU one-dimensional 'crystal' in its equilibrium state (all masses are unit masses, and all spring constants are unity)

differential equations, where x_j represents the displacement of the jth mass, $1 \leq j \leq N$, from its equilibrium position of $x_j = 0$:

$$\frac{d^2 x_1}{dt^2} = -(x_1 - x_0) + (x_2 - x_1) = x_2 - 2x_1 + x_0,$$

$$\frac{d^2 x_2}{dt^2} = -(x_2 - x_1) + (x_3 - x_2) = x_3 - 2x_2 + x_1,$$

$$\frac{d^2 x_3}{dt^2} = -(x_3 - x_2) + (x_4 - x_3) = x_4 - 2x_3 + x_2,$$

$$\cdots \tag{3.8.1}$$

$$\frac{d^2 x_{N-1}}{dt^2} = -(x_{N-1} - x_{N-2}) + (x_N - x_{N-1}) = x_N - 2x_{N-1} + x_{N-2},$$

$$\frac{d^2 x_N}{dt^2} = -(x_N - x_{N-1}) + (x_{N+1} - x_N) = x_{N+1} - 2x_N + x_{N-1},$$

with the boundary conditions of $x_0 = x_{N+1} = 0$, *always*.

Next, because we know from experience that masses connected to springs in a chain vibrate back and forth along the chain axis (remember, the masses move only along the horizontal, and have no vertical motion), we look for solutions to these equations in which all the masses oscillate sinusoidally at the same frequency, ω, but (most generally) with different amplitudes. That is, let's assume solutions of the form

$$x_j(t) = a_j \cos(\omega t + \theta), \ 1 \leq j \leq N, \theta \text{ arbitrary}, a_0 = a_{N+1} = 0. \tag{3.8.2}$$

I've included the arbitrary phase angle θ for generality, but as you'll see it will soon disappear and play no further role in the analysis. With

(3.8.2), the equations of (3.8.1) become

$$-\omega^2 a_1 \cos(\omega t + \theta) = a_2 \cos(\omega t + \theta) - 2a_1 \cos(\omega t + \theta)$$
$$+ a_0 \cos(\omega t + \theta)$$
$$-\omega^2 a_2 \cos(\omega t + \theta) = a_3 \cos(\omega t + \theta) - 2a_2 \cos(\omega t + \theta)$$
$$+ a_1 \cos(\omega t + \theta)$$
$$-\omega^2 a_3 \cos(\omega t + \theta) = a_4 \cos(\omega t + \theta) - 2a_3 \cos(\omega t + \theta)$$
$$+ a_2 \cos(\omega t + \theta)$$

$$\cdots$$

$$-\omega^2 a_{N-1} \cos(\omega t + \theta) = a_N \cos(\omega t + \theta) - 2a_{N-1} \cos(\omega t + \theta)$$
$$+ a_{N-2} \cos(\omega t + \theta)$$
$$-\omega^2 a_N \cos(\omega t + \theta) = a_{N+1} \cos(\omega t + \theta) - 2a_N \cos(\omega t + \theta)$$
$$+ a_{N-1} \cos(\omega t + \theta)$$

or, canceling all the common $\cos(\omega t + \theta)$ factors (and so vanishes θ)—this massive cancelation is precisely why our assumed solution in (3.8.1) is a *good* assumption!—we get the much simpler-looking system

$$(2 - \omega^2)a_1 - a_0 = a_2,$$
$$(2 - \omega^2)a_2 - a_1 = a_3,$$
$$(2 - \omega^2)a_3 - a_2 = a_4,$$
$$\cdots \qquad\qquad\qquad\qquad (3.8.3)$$
$$(2 - \omega^2)a_{N-1} - a_{N-2} = a_N,$$
$$(2 - \omega^2)a_N - a_{N-1} = a_{N+1},$$

where, again, $a_0 = a_{N+1} = 0$, *always*.

As you'll soon see, it will be of great help for us to write the system of equations in (3.8.3) in matrix form, as follows. If we define the 2×2 matrix **A** as

$$\mathbf{A} = \begin{bmatrix} 2 - \omega^2 & -1 \\ 1 & 0 \end{bmatrix} \qquad\qquad (3.8.4)$$

and if we define the $N + 1$ column vectors

$$\mathbf{u}_1 = \begin{bmatrix} a_1 \\ a_0 \end{bmatrix}, \quad \mathbf{u}_2 = \begin{bmatrix} a_2 \\ a_1 \end{bmatrix}, \quad \mathbf{u}_3 = \begin{bmatrix} a_3 \\ a_2 \end{bmatrix}, \ldots,$$

$$\mathbf{u}_N = \begin{bmatrix} a_N \\ a_{N-1} \end{bmatrix}, \quad \mathbf{u}_{N+1} = \begin{bmatrix} a_{N+1} \\ a_N \end{bmatrix}, \tag{3.8.5}$$

then (3.8.3) becomes

$$\mathbf{u}_2 = \mathbf{A}\mathbf{u}_1,$$

$$\mathbf{u}_3 = \mathbf{A}\mathbf{u}_2,$$

$$\mathbf{u}_4 = \mathbf{A}\mathbf{u}_3,$$

$$\ldots \tag{3.8.6}$$

$$\mathbf{u}_N = \mathbf{A}\mathbf{u}_{N-1},$$

$$\mathbf{u}_{N+1} = \mathbf{A}\mathbf{u}_N.$$

By successively substituting the equations of (3.8.6) into themselves, we can write

$$\mathbf{u}_3 = \mathbf{A}\mathbf{u}_2 = \mathbf{A}(\mathbf{A}\mathbf{u}_1) = \mathbf{A}^2\mathbf{u}_1,$$

$$\mathbf{u}_4 = \mathbf{A}\mathbf{u}_3 = \mathbf{A}(\mathbf{A}^2\mathbf{u}_1) = \mathbf{A}^3\mathbf{u}_1,$$

and so on, which leads immediately to the general result

$$\mathbf{u}_j = \mathbf{A}^{j-1}\mathbf{u}_1, \quad 1 \le j \le N + 1. \tag{3.8.7}$$

The crucial next step is that of calculating $\mathbf{A}^{j-1}$, which can be done by invoking a wonderful result from matrix theory called the Cayley-Hamilton theorem. You can find the complete details on how this is done in one of my previous books[19]—it isn't hard, but it does take more than just a few lines, so I won't repeat it here—with the final result being (you'll see why the + and − subscripts in just a moment)

$$\mathbf{A}^{j-1} = \frac{\lambda_+^{j-1} - \lambda_-^{j-1}}{\lambda_+ - \lambda_-}\mathbf{A} + \frac{\lambda_+\lambda_-^{j-1} - \lambda_-\lambda_+^{j-1}}{\lambda_+ - \lambda_-}\mathbf{I}, \tag{3.8.8}$$

where $\mathbf{I}$ is the 2×2 identity matrix, and the solutions to the determinant equation $\det[\mathbf{A} - \lambda\mathbf{I}] = 0$ (which is a quadratic equation, since $\mathbf{A}$ and $\mathbf{I}$ are 2×2) are λ_+ and λ_-. That is,

$$\mathbf{I} = \begin{bmatrix} 1 & 0 \\ 0 & 1 \end{bmatrix}$$

and λ_+ and λ_- are the solutions to

$$\begin{vmatrix} 2 - \omega^2 - \lambda & -1 \\ 1 & -\lambda \end{vmatrix} = -\lambda(2 - \omega^2 - \lambda) + 1 = 0,$$

or

$$\lambda^2 - (2 - \omega^2)\lambda + 1 = 0.$$

So, from the quadratic formula we have

$$\lambda_+ = \frac{(2 - \omega^2) + \sqrt{(2 - \omega^2)^2 - 4}}{2}$$

$$\lambda_- = \frac{(2 - \omega^2) - \sqrt{(2 - \omega^2)^2 - 4}}{2}. \tag{3.8.9}$$

Notice in particular that by direct multiplication we have (you should confirm this)

$$\lambda_+\lambda_- = 1, \tag{3.8.10}$$

and by addition (this should be obvious by inspection)

$$\lambda_+ + \lambda_- = 2 - \omega^2. \tag{3.8.11}$$

Putting (3.8.11) into (3.8.4), we get

$$\mathbf{A} = \begin{bmatrix} \lambda_+ + \lambda_- & -1 \\ 1 & 0 \end{bmatrix}, \tag{3.8.12}$$

and then, using (3.8.12) in (3.8.8), we arrive at

$$
\mathbf{A}^{j-1} = \frac{1}{\lambda_+ - \lambda_-} \left\{ (\lambda_+^{j-1} - \lambda_-^{j-1}) \begin{bmatrix} \lambda_+ + \lambda_- & -1 \\ 1 & 0 \end{bmatrix} \right.
$$

$$
\left. + (\lambda_+ \lambda_-^{j-1} - \lambda_- \lambda_+^{j-1}) \begin{bmatrix} 1 & 0 \\ 0 & 1 \end{bmatrix} \right\}
$$

$$
= \frac{1}{\lambda_+ - \lambda_-} \left\{ \begin{bmatrix} \lambda_+^j + \lambda_- \lambda_+^{j-1} - \lambda_+ \lambda_-^{j-1} - \lambda_-^j & \lambda_-^{j-1} - \lambda_+^{j-1} \\ \lambda_+^{j-1} - \lambda_-^{j-1} & 0 \end{bmatrix} \right.
$$

$$
\left. + \begin{bmatrix} \lambda_+ \lambda_-^{j-1} - \lambda_- \lambda_+^{j-1} & 0 \\ 0 & \lambda_+ \lambda_-^{j-1} - \lambda_- \lambda_+^{j-1} \end{bmatrix} \right\}
$$

$$
= \frac{1}{\lambda_+ - \lambda_-} \begin{bmatrix} \lambda_+^j - \lambda_-^j & \lambda_-^{j-1} - \lambda_+^{j-1} \\ \lambda_+^{j-1} - \lambda_-^{j-1} & \lambda_+ \lambda_- (\lambda_-^{j-2} - \lambda_+^{j-2}) \end{bmatrix},
$$

or, at last, using (3.8.10) in the lower right corner element of the last matrix, we have this pretty result for $\mathbf{A}^{j-1}$:

$$
\mathbf{A}^{j-1} = \frac{1}{\lambda_+ - \lambda_-} \begin{bmatrix} \lambda_+^j - \lambda_-^j & \lambda_-^{j-1} - \lambda_+^{j-1} \\ \lambda_+^{j-1} - \lambda_-^{j-1} & \lambda_-^{j-2} - \lambda_+^{j-2} \end{bmatrix}. \tag{3.8.13}
$$

Now, focus your attention on the conditions at the *ends* of the chain of masses. From (3.8.7), with $j = N + 1$, we have

$$
\mathbf{u}_{N+1} = \mathbf{A}^N \mathbf{u}_1,
$$

and from (3.8.5) we have $\mathbf{u}_1$ and $\mathbf{u}_{N+1}$, so

$$
\mathbf{u}_{N+1} = \begin{bmatrix} a_{N+1} \\ a_N \end{bmatrix} = \mathbf{A}^N \begin{bmatrix} a_1 \\ a_0 \end{bmatrix},
$$

or, in detail, with $j = N + 1$ in (3.8.13) and recalling that $a_0 = a_{N+1} = 0$,

$$
\begin{bmatrix} 0 \\ a_N \end{bmatrix} = \frac{1}{\lambda_+ - \lambda_-} \begin{bmatrix} \lambda_+^{N+1} - \lambda_-^{N+1} & \lambda_-^N - \lambda_+^N \\ \lambda_+^N - \lambda_-^N & \lambda_-^{N-1} - \lambda_+^{N-1} \end{bmatrix} \begin{bmatrix} a_1 \\ 0 \end{bmatrix}. \tag{3.8.14}
$$

Writing out the two equations contained in (3.8.14),

$$0 = a_1 \frac{\lambda_+^{N+1} - \lambda_-^{N+1}}{\lambda_+ - \lambda_-}$$

$$a_N = a_1 \frac{\lambda_+^{N} - \lambda_-^{N}}{\lambda_+ - \lambda_-}. \tag{3.8.15}$$

The first equation of (3.8.15), at first glance, seems to say $a_1 = 0$, which then makes $a_N = 0$. In fact, if you look back at (3.8.3) you'll see that $a_1 = 0$ (along with $a_0 = 0$) says that *every one* of the $a_j = 0$. That is, none of the masses move! This is what mathematicians call a *trivial* solution, and what physicists (who, of course, are far more vulgar) call ... well, never mind what kind of solution *those* guys call it!

There is a way out of this quandry, though, because instead of concluding from (3.8.15) that $a_1 = 0$, what if we say it is the *other* factor that is zero? That is, let's say

$$\frac{\lambda_+^{N+1} - \lambda_-^{N+1}}{\lambda_+ - \lambda_-} = 0. \tag{3.8.16}$$

If we multiply through (3.8.16) by $\dfrac{\lambda_+^{N+1}}{\lambda_+}$ we get

$$\frac{\lambda_+^{2N+2} - \lambda_+^{N+1}\lambda_-^{N+1}}{\lambda_+^{2} - \lambda_+\lambda_-} = 0,$$

or, using (3.8.10),

$$\frac{\lambda_+^{2N+2} - 1}{\lambda_+^{2} - 1} = 0. \tag{3.8.17}$$

Of course, if I instead had multiplied through (3.8.16) by $\dfrac{\lambda_-^{N+1}}{\lambda_-}$, we would have arrived at

$$\frac{\lambda_+^{N+1}\lambda_-^{N+1} - \lambda_-^{2N+2}}{\lambda_-\lambda_+ - \lambda_-^{2}} = 0 = \frac{1 - \lambda_-^{2N+2}}{1 - \lambda_-^{2}} = \frac{\lambda_-^{2N+2} - 1}{\lambda_-^{2} - 1},$$

which is precisely the same equation as (3.8.17). What this means is that λ_+ and λ_- are each solutions of the following equation (without the $+$ or $-$ subscripts):

$$\frac{\lambda^{2N+2} - 1}{\lambda^2 - 1} = 0. \tag{3.8.18}$$

The λ_+ and λ_- solutions to (3.8.18) come as *pairs*, because of (3.8.10).

We can solve (3.8.18) quite easily by noticing that the numerator can be factored as follows:

$$\lambda^{2N+2} - 1 = (\lambda^2 - 1)(\lambda^{2N} + \lambda^{2N-2} + \lambda^{2N-4} + \cdots + \lambda^2 + 1).$$

That is, since the denominator of (3.8.18) cancels the first factor on the right, the solutions to (3.8.18) are simply the solutions to

$$\lambda^{2N} + \lambda^{2N-2} + \lambda^{2N-4} + \cdots + \lambda^2 + 1 = 0,$$

a polynomial of degree $2N$, that is, there are $2N$ solutions, N of them for λ_+ and N of them for λ_-. As mentioned above, these solutions are not independent, with each λ_+ paired with a λ_- through (3.8.10). The solutions to (3.8.18) are therefore the solutions to $\lambda^{2N+2} - 1 = 0$ *except* for the obvious $\lambda = +1$ and $\lambda = -1$, which are canceled by the $\lambda^2 - 1$ in the denominator of (3.8.18). These solutions are easy to write down by inspection, as they are just the $2N + 2$ roots of unity (with two exceptions); that is, the solutions to (3.8.18) are

$$\lambda = 1^{1/(2N+2)}$$

except for $\lambda = \pm 1$.

Using Euler's famous identity to write $1 = e^{i2\pi r}$, where r is any integer and $i = \sqrt{-1}$, we thus have our solutions to (3.8.18) as

$$\lambda = \left(e^{i2\pi r}\right)^{1/(2N+2)} = e^{i\left(\frac{2\pi r}{2N+2}\right)} = e^{i\left(\frac{\pi}{N+1}\right)r},$$

where r runs from 1 to N, and from $N + 2$ to $2N + 1$. We exclude the two cases of $r = 0$ ($\lambda = +1$) and $r = N + 1$ ($\lambda = -1$), as explained earlier. All other integer values for r simply repeat these solutions. To write all this in a convenient way, let's define

$$\phi_r = r\frac{\pi}{N+1}, \tag{3.8.19}$$

and then we can write our $2N$ solutions to (3.8.18) as

$$\lambda_{+,r} = e^{i\phi_r}, \quad \lambda_{-,r} = e^{-i\phi_r}, \quad 1 \le r \le N, \tag{3.8.20}$$

and it is clear from (3.8.20) that (3.8.10) is satisfied. The $\lambda_{-,r}$ solution is simply the *complex conjugate* of the $\lambda_{+,r}$ solution.

Okay, now we are ready to answer the big question: what do we do with all this? First, looking back at (3.8.11), we see that we can write

$$\omega_r^2 = 2 - (\lambda_{+,r} + \lambda_{-,r}) = 2 - (e^{i\phi_r} + e^{-i\phi_r}) = 2 - 2\cos(\phi_r)$$
$$= 2[1 - \cos(\phi_r)] = 2\left[2\sin^2\left(\frac{1}{2}\phi_r\right)\right] = 4\sin^2\left(\frac{1}{2}\phi_r\right),$$

or

$$\boxed{\omega_r = 2\sin\left(\frac{\pi r}{2(N+1)}\right), \quad 1 \le r \le N.} \tag{3.8.21}$$

These N values of ω_r are called the *normal mode frequencies* of the crystal. For our N-mass crystal chain there are N normal modes, with each having its signature frequency ω_r. These frequencies increase, as given by (3.8.21), with increasing mode number, that is,

$$\omega_1 < \omega_2 < \omega_3 < \cdots < \omega_N.$$

Second, we can now also calculate the *relative amplitudes* of oscillation for each of the N masses, which will be different for each mode. So, instead of writing just a_j, I'll write $a_{j,r}$, which will denote the amplitude of oscillation for the jth mass when oscillating at the rth normal mode frequency. Looking back at (3.8.5), we see that we can write

$$\mathbf{u}_{j,r} = \begin{bmatrix} a_{j,r} \\ a_{j-1,r} \end{bmatrix},$$

and from (3.8.7) we have

$$\mathbf{u}_{j,r} = \mathbf{A}^{j-1}\mathbf{u}_{1,r},$$

and finally, using (3.8.13),

$$\mathbf{u}_{j,r} = \begin{bmatrix} a_{j,r} \\ a_{j-1,r} \end{bmatrix}$$

$$= \frac{1}{\lambda_{+,r} - \lambda_{-,r}} \begin{bmatrix} \lambda_{+,r}^{j} - \lambda_{-,r}^{j} & \lambda_{-,r}^{j-1} - \lambda_{+,r}^{j-1} \\ \lambda_{+,r}^{j-1} - \lambda_{-,r}^{j-1} & \lambda_{-,r}^{j-2} - \lambda_{+,r}^{j-2} \end{bmatrix} \begin{bmatrix} a_{1,r} \\ a_{0,r} \end{bmatrix}.$$

Since $a_{0,r} = 0$, this matrix equation says

$$a_{j,r} = \frac{\lambda_{+,r}^{j} - \lambda_{-,r}^{j}}{\lambda_{+,r} - \lambda_{-,r}} a_{1,r},$$

or, using (3.8.20),

$$a_{j,r} = \frac{e^{ij\phi_r} - e^{-ij\phi_r}}{e^{i\phi_r} - e^{-i\phi_r}} a_{1,r} = \frac{i2\sin(j\phi_r)}{i2\sin(\phi_r)} a_{1,r},$$

and so, finally, the normal mode amplitudes, for each mass in each mode are

$$\boxed{a_{j,r} = \frac{\sin\left(j\frac{\pi r}{N+1}\right)}{\sin\left(\frac{\pi r}{N+1}\right)} a_{1,r}, \quad r = 1, 2, \ldots, N \text{ and } j = 1, 2, \ldots, N.}$$

(3.8.22)

Notice that the oscillation amplitudes of all the masses are relative to the amplitude of mass 1.

Now, to illustrate how our two boxed equations, (3.8.21) and (3.8.22), work, consider the special case of $N = 2$. This, in fact, is almost surely the example you'll find discussed in just about any physics text that treats the topic of normal modes. From (3.8.21) we have the two normal mode frequencies $\omega_1 = 2\sin(\pi/6) = 1$ and

$\omega_2 = 2\sin(\pi/3) = \sqrt{3}$. From (3.8.22) we have the mode 1 oscillation amplitude for $a_{2,1}$, relative to $a_{1,1}$, as

$$a_{2,1} = \frac{\sin(2\pi/3)}{\sin(\pi/3)}a_{1,1} = a_{1,1}$$

because $\sin(2\pi/3) = \sin(\pi/3)$, and the mode 2 oscillation amplitude for $a_{2,2}$, relative to $a_{1,2}$, as

$$a_{2,2} = \frac{\sin(4\pi/3)}{\sin(2\pi/3)}a_{1,2} = -a_{1,2}$$

because $\sin(4\pi/3) = -\sin(2\pi/3)$. (Both $a_{1,1}$ and $a_{1,2}$ are arbitrary.) Figure 3.8.2 shows these two modes; when in mode 1 the two masses oscillate *in phase* (they always move in the same direction), and when in mode 2 they oscillate *out of phase* (they always move in opposite directions).

What makes the normal modes so interesting to physicists is that we can express any arbitrary oscillation of the N masses as some linear combination of the N normal mode oscillations. In this sense, the normal modes are like the pure sinusoids (with harmonically related

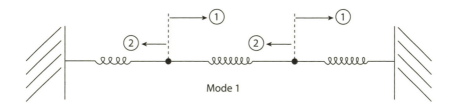

Mode 1

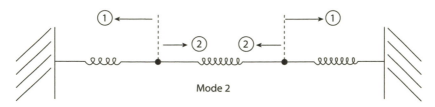

Mode 2

FIGURE 3.8.2. Normal modes for $N = 2$

frequencies) that make up the Fourier series expansion of an arbitrary periodic time signal. Indeed, FPU use the phrase "Fourier modes" synonymously with "normal modes."[20] Physically, the normal modes are *independent* of each other, in that if one starts the N-mass chain oscillating in just *one* mode, with some initial total energy (the kinetic energy of motion in the masses, plus the potential energy stored in the springs), then that energy remains in that mode and does not spread out into any of the other modes. That is, if E_r is the total energy in the rth normal mode, then

$$\frac{dE_r}{dt} = 0, 1 \leq r \leq N.$$

This is easy to show by direct calculation. For example, suppose our $N = 2$ chain has energy in mode 1. Then the total energy in that mode is

$$E_1 = \frac{1}{2}\left(\frac{dx_{1,1}}{dt}\right)^2 + \frac{1}{2}\left(\frac{dx_{2,1}}{dt}\right)^2 + \frac{1}{2}x_{1,1}^2 + \frac{1}{2}\left(x_{2,1} - x_{1,1}\right)^2$$
$$+ \frac{1}{2}x_{2,1}^2,$$

where the first two terms on the right are the kinetic energies of the masses, the third term is the potential energy in the leftmost spring, the fourth term is the potential energy in the middle spring, and the fifth term is the potential energy in the rightmost spring. So,

$$\frac{dE_1}{dt} = \frac{dx_{1,1}}{dt} \cdot \frac{d^2x_{1,1}}{dt^2} + \frac{dx_{2,1}}{dt} \cdot \frac{d^2x_{2,1}}{dt^2} + x_{1,1}\frac{dx_{1,1}}{dt}$$
$$+ \left(x_{2,1} - x_{1,1}\right)\left(\frac{dx_{2,1}}{dt} - \frac{dx_{1,1}}{dt}\right) + x_{2,1}\frac{dx_{2,1}}{dt}.$$

With just a bit of easy algebra this becomes

$$\frac{dE_1}{dt} = \frac{dx_{1,1}}{dt}\left(\frac{d^2x_{1,1}}{dt^2} + x_{1,1}\right) + \frac{dx_{2,1}}{dt}\left(\frac{d^2x_{2,1}}{dt^2} + x_{2,1}\right)$$
$$+ x_{2,1}\frac{dx_{2,1}}{dt} - x_{2,1}\frac{dx_{1,1}}{dt} - x_{1,1}\frac{dx_{2,1}}{dt} + x_{1,1}\frac{dx_{1,1}}{dt}$$

$$= \frac{dx_{1,1}}{dt} \left(\frac{d^2 x_{1,1}}{dt^2} + 2x_{1,1} - x_{2,1} \right)$$

$$+ \frac{dx_{2,1}}{dt} \left(\frac{d^2 x_{2,1}}{dt^2} + 2x_{2,1} - x_{1,1} \right).$$

Since $x_{1,1} = a_{1,1} \cos(\omega_1 t)$ and $x_{2,1} = a_{1,1} \cos(\omega_1 t)$, and since $\omega_1 = 1$, then $x_{1,1} = x_{2,1} = x = a_{1,1} \cos(t)$. Thus,

$$\frac{dE_1}{dt} = 2 \frac{dx}{dt} \left(\frac{d^2 x}{dt^2} + x \right) = 0$$

because $\dfrac{d^2 x}{dt^2} = -a_{1,1} \cos(t) = -x$. A similar calculation shows that if E_2 is the total energy in mode 2, then $\dfrac{dE_2}{dt} = 0$, too (see Challenge Problem 3.4).

And now you can see where the FPU computer experiment comes in: the *nonspreading* of energy initially in a single mode into the other modes is because of the linearity of our N-mass system. But what if, asked FPU, we make the equations of (3.8.1) very weakly *nonlinear*? "Very weakly" so as not to disturb the individual normal modes, but still not perfectly linear? Might we then, FPU wondered, see crystal energy initially in one mode, say mode 1, "leak" over time into the higher modes? And, after a long time ("long" means many multiples of the largest normal mode period), would the initial energy end up uniformly distributed over all the normal modes? That was the commonly held belief in the 1950s and, with the new MANIAC-I available to FPU, matters could at last be put to a controlled, scientific test by computer.

FPU performed many different experiments on MANIAC-I, but what I'll describe next is the classic version. To introduce a "weak" nonlinear coupling among the normal modes, FPU rewrote the equations of (3.8.1) to replace the original spring forces (which vary linearly with the mass displacements) with forces that also include *squared* terms; the equation for the jth mass became

$$\frac{d^2 x_j}{dt^2} = (x_{j+1} - 2x_j + x_{j-1}) + \alpha[(x_{j+1} - x_j)^2 + (x_j - x_{j-1})^2],$$

$$(3.8.23)$$

where α is "small" (FPU used $\alpha = 0.25$). The initial boundary conditions were $x_0 = x_N = 0$, and the chain was started from rest, that is,

$$\frac{dx_j}{dt}\bigg|_{t=0} = 0, \quad 1 \leq j \leq N. \tag{3.8.24}$$

The initial values of the x_j displacements of the masses from their equilibrium positions (FPU called these displacements, at any time, the *shape* of the chain) were determined by (3.8.22), and the total chain energy was initially in mode 1 ($r = 1$). The value of N was stated by FPU to be 32, although there is some variation on this in later papers (for example, some authors repeat the value of $N = 32$ in their texts, but appear to specifically write the normal mode *equations* for the case of $N - 1 = 31$ moving masses.[21])

The next step was to code the equations (3.8.23) and (3.8.24) into MANIAC-I. This was a *huge* task in those days—remember, it all had to be done in machine language![22]—and it was done brilliantly, not by the FPU senior scientists but instead by a young lady mathematician on the Los Alamos staff.[23] As one writer expressed this, perhaps just a bit flippantly, "Like all good senior scientists sporting a brand new idea, FPU [looked] for someone to do the actual work. Here they were extremely fortunate to find Mary Tsingou [born 1928], who programmed the dynamics, ensured its accuracy, and provided graphs of the results." [24] What her code did, over and over, was calculate the values of x_j, $\frac{dx_j}{dt}$, and $\frac{d^2x_j}{dt^2}$ from numerical difference equation approximations as the chain's oscillations evolved through time. Periodically, the current shape of the chain was subjected to a Fourier analysis to determine which normal modes were then present, and how much energy was in each mode. The accuracy of the numerical calculations was checked by confirming that the total energy remained constant.

So, what happened? I can't do better in answering that than by directly quoting from FPU themselves:

> Let us say here that the results of our computations show features which were, from the beginning, surprising to us. Instead of a gradual, continuous flow of energy from the first mode to the higher modes, all [of our simulations] show an entirely different behavior. Starting ... with a quadratic force and a pure sine

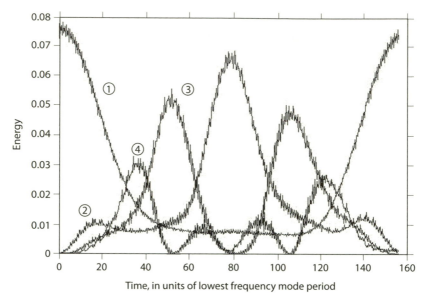

FIGURE 3.8.3. The 'classic' FPU MANIAC-I experiment

wave as the initial position of the [chain], we indeed observe initially a gradual increase of energy in the higher modes. . . . Mode
2 starts increasing first, followed by mode 3, and so on. Later on,
however, this gradual sharing of energy among successive modes
ceases. Instead, it is one or the other mode that predominates. For
example, mode 2 decides, as it were, to increase rather rapidly at
the cost of all the others put together! Then mode 3 undertakes
this role. It is only the first few modes which exchange energy
among themselves and they do this in a rather regular fashion.
Finally, at a later time mode 1 comes back to within one percent
of its initial energy. . . .[25]

Figure 3.8.3 reproduces the first four modes as given in Figure 1 of the
FPU Los Alamos report, and shows what prompted the above words.
As our semiflippant physicist so graphically expressed it, "Energy is
exchanged primarily only among modes 1 through 6 with all higher
modes writhing about in the noise gasping for energy."[26]

The explanation for this failure of the expected nonlinearity of
the FPU equations to exhibit energy equipartition took some years
to develop, but today it is well understood.[27] At the time, however, it

perplexed; after Fermi's death Ulam wrote that the MANIAC-I calculations were "interesting and quite surprising to Fermi," and that Fermi
believed "they really constituted a little discovery."[28] One desperate
"explanation" for the surprising result was that it all must simply be a
failure of the MANIAC-I computer and/or the programming, and not
actually true physics. This may well be the first occurrence of "blame
the computer," an excuse we still see often today!

3.9 Challenge Problems

CP 3.1. Derive (3.2.1).

CP 3.2. Imagine a uniformly thick (in the z-direction) plate that is
infinitely large in one of the other dimensions. Specifically, it has
the shape of a *strip*, with a finite width of π ($0 \leq x \leq \pi$), while
its length is infinite ($0 \leq y < \infty$). Its two faces are thermally insulated, and so the steady-state temperature $u(x, y, z) = u(x, y)$. The
temperatures of the infinitely long edges (parallel to the y-axis), at
$x = 0$ and $x = \pi$, are fixed at zero, while the temperature of the
x-axis edge is fixed at T. Find the steady-state temperature anywhere in the plate. That is, solve Laplace's equation with the
boundary conditions $u(0, y) = 0$, $u(\pi, y) = 0$, and $u(x, 0) = T$. *Hint:*
because the plate is infinitely long there isn't a fourth edge, but we
still need a fourth condition on $u(x, y)$. To get it, use the *physical*
requirement that $\lim_{y \to \infty} u(x, y) = 0$, that is, as we move ever farther away from the hot edge of the plate, the plate's temperature
approaches zero.

CP 3.3. Returning to CP 3.2, if T_0 is a given, constant temperature,
then the curve $y = y(x)$ along which $u(x, y) = T_0$ is called an *isotherm*.
Plot the isotherms of the infinite plate strip for $\frac{T_0}{T} = 0.9$, 0.8, and
0.7, for $0 \leq x \leq \pi$. *Hint:* if you solved CP 3.2 and got an infinite
series for $u(x, y)$, then you may find that the form of that series is
strongly reminiscent of the power series expansion

$$\frac{1}{2} \ln \left[\frac{1+w}{1-w} \right] = w + \frac{1}{3}w^3 + \frac{1}{5}w^5 + \cdots,$$

which converges for all complex w such that $|w| < 1$. So, writing $w = re^{i\theta}$ and invoking Euler's identity may lead you to discovering a simple, *closed-form* expression for $u(x, y)$ from which the equation for the isotherms immediately follows.

CP 3.4. For the case of $N = 2$ of the one-dimensional linear model of a crystal, show that all the energy E_2 initially in normal mode 2 remains in that mode. That is, show that $\dfrac{dE_2}{dt} = 0$.

3.10 Notes and References

1. *A Quantum Legacy: Seminal Papers of Julian Schwinger*, ed. Kimball A. Milton, vol. 26, World Scientific Series in 20th-Century Physics (Singapore: World Scientific, 2000, p. 1). The reference to Rabi "discovering" Schwinger refers to the astonishing fact that Schwinger had already entered the City College of New York in 1933 by age fifteen, and, after hearing of the wunderkind, Rabi got him out of CCNY and into Columbia University. By 1937 Schwinger had already finished his PhD dissertation, but he had to wait until 1939 (when twenty-one) to actually receive it. Although a world-class theoretician, Schwinger did not spend the war years on the atomic bomb project at Los Alamos as did many of his contemporaries, but rather at the MIT Radiation Laboratory developing radar technology.

2. See my books *Duelling Idiots* (2000, 2002) and *Digital Dice* (2008), both published by Princeton University Press.

3. H. L. Anderson, "Scientific Uses of the MANIAC" (*Journal of Statistical Physics*, June 1986, pp. 731–748). The connection between calculation and truth was made long ago by the English mathematician Augustus De Morgan in the introduction to his witty book, *A Budget of Paradoxes*. (We encountered De Morgan in Chapter 2.) There he writes,

> During the last two centuries and a half, physical knowledge has been gradually made to rest upon a basis which it had not before. It has become mathematical. The question now is, not whether this or that hypothesis is better or worse to the pure thought, but whether it [the asserted hypothesis] accords with observed phenomena in those consequences which can be shown necessarily to follow from it, if it be true.

4. See, for example, Paul W. Berg and James L. McGregor, *Elementary Partial Differential Equations* (San Francisco: Holden-Day, 1966, pp. 29–31). Professors Berg (1925–1990) and McGregor (1921–1988) were both on the math faculty

at Stanford when I was an undergraduate there, and I was fortunate enough to take my first course in ordinary differential equations (Math 130) in my sophomore year (1959) from Berg, and the first course in partial differential equations (Math 131) the next year from McGregor. Multilithed class handouts were the basis for their eventual excellent book, and from those notes I learned the standard methods for solving several linear types of partial differential equations important in mathematical physics, including Laplace's equation.

5. This replacement operation will be done in the following specific way: we'll create a new matrix A, with its elements receiving the average values created as we run through the U-matrix. Then we'll copy A back into U, and repeat. This is called *Jacobi iteration*, after the German mathematician Carl Jacobi (1804–1851): see the MATLAB code **jacobi.m** in the text, and in particular line 06, which initially creates A; line 10, which implements (3.5.7); and line 13, which copies A back into U.

6. If you're interested in looking further into the theory of convergence, a good place to start is G. D. Smith, *Numerical Solution of Partial Differential Equations: Finite Difference Methods* (Oxford: Clarendon Press 1985, esp. pp. 268–271).

7. This quotation is from Acton's terrific book, *Numerical Methods That Work* (New York: Harper & Row, 1970, p. vx). Acton (born 1920) holds a PhD in applied math but received his undergraduate degree in chemical engineering. He is decidedly a practical analyst whose approach to problems is not purely theoretical and abstract but rather is the single-minded pursuit of the question, *what works and produces an answer?* After serving for a while in the electrical engineering department at Princeton he evolved quickly into a computer scientist there, and his 1970 book is a "must read" for anybody who uses a computer for serious computation, as is his more recent *Real Computing Made Real* (Princeton, N.J.: Princeton University Press, 1996).

8. See, for example, the very nice history of the electronic computer by Herman H. Goldstine, *The Computer from Pascal to von Neumann* (Princeton, N.J.: Princeton University Press, 1972). Also quite interesting reading is Alice R. Burks and Arthur W. Burks, *The First Electronic Computer: the Atanasoff Story* (Ann Anbor: University of Michigan Press, 1988).

9. Edward Teller, "The Work of Many People" (*Science*, February 25, 1955, pp. 267–275). This paper is Teller's memory of how the "super" (the American H-bomb) came to be. ENIAC gets some mention.

10. The Monte Carlo concept first appeared in the open literature just a year later, in a paper written jointly by the mathematician Stanislaw

Ulam (1909–1984) and Metropolis: "The Monte Carlo Method" (*Journal of the American Statistical Association*, September 1949, pp. 335–341). See also Cuthbert C. Hurd, "A Note On Early Monte Carlo Computations and Scientific Meetings" (*Annals of the History of Computing*, April 1985, pp. 141–155), which reprints a Monte Carlo code (written by von Neumann) for the ENIAC, to simulate the history of 100 neutrons involved in a chain reaction process. There are many historians of computer science who balk at giving von Neumann as much credit in pioneering the modern computer as I do here. Rather, they claim, credit for the stored program concept should go to the English mathematician Alan Turing (1912–1954), the creator (on paper) in 1936 of the first machines that could change the contents of their memories. These machines are famous today as *Turing machines*. Turing and von Neumann first met in 1935 when von Neumann gave a lecture course at Cambridge, and again during Turing's stay at Princeton (1936–1939), where he earned a PhD in mathematics. There is no question that von Neumann was very much aware of the stored program nature of Turing's machines. It is a common undergraduate homework assignment today to simulate Turing machines on a digital computer (there is one such simulation, written in MATLAB, in my forthcoming Princeton book *Electric Logic*), but no such machine has ever actually been constructed in hardware. Von Neumann, on the other hand, was deeply involved in the actual construction of operational computers. You can read more about Turing in *The Essential Turing: The Ideas That Gave Birth to the Computer Age*, ed. B. Jack Copeland (Oxford: Oxford University Press, 2004). Turing appears again in this book in Chapter 9.

11. The EDVAC was conceived by the same people who designed ENIAC and, when completed in 1950, it followed ENIAC to Aberdeen. A sorting code written for EDVAC by von Neumann in early 1945 is discussed in Donald E. Knuth, "Von Neumann's First Computer Program" (*Computing Surveys*, December 1970, pp. 247–260). The EDVAC used sonic delay lines for its memory, a technological innovation used later in UNIVAC I (UNIVersal Automatic Computer), the first commercial computer.

12. Eugene H. Herbst, N. Metropolis, and Mark B. Wells, "Analysis of Problem Codes on the Maniac" (*Mathematical Tables and Other Aids to Computation*, January 1955, pp. 14–20).

13. I can personally identify with Fermi's emotional relationship with the MANIAC-I. After I graduated from Caltech in 1963 with an MSEE, I was hired by the (now defunct) Systems Division of Beckman Instruments, in Fullerton, California, to do the logic design of the stored program telemetry simulator for the Gemini phase of the American Moon Program. That machine had fifteen commands, was programmed in machine language, had a punched paper tape reader input/printer output interface (direct memory access was

FIGURE 3.10.1. The author at the console of the machine language programmable Gemini telemetry simulator (May 1964)

also possible, via control panel switches, for the quick entry of programming error fixes—see note 22), and a magnetic core memory of 256, 22-bit words. Constructed more than ten years after MANIAC-I, the simulator was perhaps just a bit faster, with a memory cycle time of ten microseconds (the technology was discrete solid state, that is, individual transistors, versus MANIAC-I's vacuum tubes), but it had no computational capability. For me, the Gemini machine was a fantastic gadget, and I spent many happy hours at its console (see Figure 3.10.1) writing all the final acceptance test programs. I couldn't believe people would actually *pay* me to have such fun!

14. J. Kisten, P. Stein, S. Ulam, W. Walden, and M. Wells, "Experiments in Chess" (*Journal of the Association for Computing Machinery*, April 1957, pp. 174–177, reprinted in Stanislaw Ulam, *Sets, Numbers, and Universes* [Cambridge, Mass.: MIT Press, 1974, pp. 346–349]). To make the decision-time-per-move reasonably short, Los Alamos chess was played on a 6 × 6 board, and not on the traditional 8 × 8 board. That was done by removing the bishops,

and so Los Alamos chess was known to its creators as "anticlerical" chess! As I write, the most powerful chess program in the world is Rybka 3, available for about $100, which runs on any home computer using the Windows operating system. Rybka 3 routinely defeats players at the grandmaster level, and in play against a world champion it would be a toss-up as to who would win any given game.

15. One of the MANIAC-I Monte Carlo applications Metropolis was working on in the early 1950s was that of determining the so-called *equation of state* of a substance. The equation of state is the relationship between the pressure (P), volume (V), and temperature (T) of the substance. (The ideal gas law, $PV = nRT$, where n is the number of moles of gas and R is the universal gas constant—known to all high school chemistry students—is such a relationship.) Metropolis and his Los Alamos colleagues (including Edward Teller) reported on this work in "Equation of State Calculations by Fast Computing Machines" (*Journal of Chemical Physics*, June 1953, 1087–1092). This paper is very interesting reading, not least because in one footnote it is stated that the random numbers used were produced by what is called a "middle-square" generator. That is, if r_n is the nth random number, m digits long, then r_{n+1} is produced by squaring r_n (thus producing a number with $2m$ digits) and then using the middle m digits. This type of generator is avoided today (for why, see my book *Duelling Idiots*, pp. 177–178). As another example, von Neumann had, since the late 1940s, been drawn into the use of electronic computers (including ENIAC and the MANIAC-I precursor, the IAS machine) for long-range weather forecasting; see Kristine C. Harper, *Weather by the Numbers: The Genesis of Modern Meteorology* (Cambridge, Mass.: MIT Press, 2008). And George Gamow, the man who "named" MANIAC—despite the initial negative impression his joke name may give, Gamow was as fascinated by MANIAC-I as was everybody else, and he introduced it to a general audience in one of his famous popular science books, *Mr. Tompkins Learns the Facts of Life*, first published in 1953—reported on one planned use of MANIAC-I, a study of stellar evolution, in his paper "Will Our Sun Ever Explode?" (*Proceedings of the American Academy of Arts and Sciences*, July 1951, pp. 291–294). It concludes with the hopeful words,

> It is too early to say whether or not [this paper has presented] a correct explanation of the dynamical states of aging stars. The study of the hydrodynamical equations involved in the problem is extremely difficult and can be done only by means of modern electronic computers. Work in this direction is now being done by the author and his colleagues ... with the hope of having solutions run on the new electronic computer ("Maniac") under construction at the Los Alamos Scientific Laboratory. When the Maniac starts "thinking" he will certainly turn out the answer to that problem in no time. Then, and only then, will we really know

how violent the explosions of smaller stars can be. Then, and only then we will know whether the explosion of our Sun will warm up, melt, or evaporate our Earth.

16. Fermi and Ulam have already appeared in this book, but the Pasta in the FPU experiment has not. He was John Pasta (1918–1981), who joined the Los Alamos staff in August 1951 as MANIAC-I was being built. He had an unusual background, having once worked as a New York City street cop. He eventually earned a PhD in physics, served as head of the Mathematics and Computer Branch of the Atomic Energy Commission, as head of the Computer Science Department at the University of Illinois, Urbana, and finally as head of the Office of Computing Activities at the National Science Foundation. A quite interesting biographical essay on Pasta's unusual life, written by several of his former colleagues (including Metropolis), can be found in *Annals of the History of Computing* (July 1983, pp. 224–238).

17. Joseph Ford, "The Fermi-Pasta-Ulam Problem: Paradox Turns Discovery" (*Physics Reports*, May 1992, pp. 271–310).

18. A generalization of the normal modes derivation that I give here can be found in James D. Louck, "Exact Normal Modes of Oscillation of a Linear Chain of Identical Particles" (*American Journal of Physics*, August 1962, pp. 585–590). The masses in this paper are still identical, but are not restricted to being unity masses, and the spring constants are also not restricted to being unity. Indeed, the interior springs are still identical but their constants can be anything, and the first and final spring constants can each be different from both each other and the interior spring constant. It is nicely appropriate that this paper is by Louck, as he (as I write) is a retired Laboratory Fellow of the Los Alamos National Laboratory, where the FPU computer experiment was performed.

19. *Dr. Euler's Fabulous Formula*, Princeton 2006, pp. 19–22.

20. E. Fermi, J. Pasta, and S. Ulam, "Studies of Non Linear Problems" (Los Alamos report LA-1940, May 1955, reprinted in volume 2 of Fermi's *Collected Papers* [Chicago: University of Chicago Press, 1965, pp. 977–988]) (the reprint includes an interesting historical introduction by Ulam).

21. Thierry Dauxois, Michel Peyrard, and Stefano Ruffo, "The Fermi-Pasta-Ulam 'Numerical Experiment': History and Pedagogical Perspectives" (*European Journal of Physics*, September 2005, pp. S3–S11). See also Ford, "The Fermi-Pasta-Ulam Problem."

22. When programming in machine language one is, of course, limited to just the primitive commands designed directly into the control logic circuitry

of the computer. There are no *while*, *if/else*, or *plot* commands. But even more burdensome is that the addressing is done for absolute, actual locations in a real, physical memory. Once a code is written, and then is found to need "correcting"—which means new instructions need to be inserted or old instructions removed, or even (most likely of all) that both tasks need to be done—then one has to go through the entire code (perhaps hundreds or thousands of instructions) and, one after the next, update the addressing of instructions that reference other instructions (which are now, of course, not necessarily where they once were in physical memory). It's awful, tedious work (I *know*—I've *done* it!), subject at every step to error. The introduction of assemblers and compilers with symbolic addressing (and the structured nature of modern software like MATLAB, which doesn't even use explicit addressing in its codes) eliminated those problems, but in the early 1950s Mary Tsingou had to do it all the hard way. One early computer scientist put it this way:

> By June 1949 people had begun to realize that it was not so easy to get a program right as had at one time appeared. I well remember when this realization first came on me.... I was trying to get working my first non-trivial program, which was one for the numerical integration of Airy's differential equation. It was [then] the realization came over me with full force that a good part of the remainder of my life was going to be spent in finding errors in my own programs.

Maurice V. Wilkes, *Memoirs of a Computer Pioneer* (Cambridge, Mass.: MIT Press, 1985, p. 145). Wilkes (born 1913) was the force behind the EDSAC (Electronic Delay Storage Automatic Computer), constructed using a sonic delay line memory, at Cambridge University in 1949 and used until 1958. The EDSAC was inspired by Wilkes's reading of a preliminary 1946 draft proposal written by von Neumann for the EDVAC (see note 11 again).

23. Thierry Dauxois, "Fermi, Pasta, Ulam, and a Mysterious Lady" (*Physics Today*, January 2008, pp. 55–57). This paper includes a flow-chart of the FPU algorithm (in Tsingou's hand) which gives a suggestive hint of the complexity of the code. Ulam wrote (see note 20 again) that "During one summer Fermi learned very rapidly how to *program* problems for the electronic computers and he not only could plan the general outline and construct the so-called flow diagram but would work out himself the actual *coding* of the whole problem in detail." The reference in note 3 contains images, in Fermi's hand, of the flow diagram and MANIAC-I machine code for converting binary numbers in memory to decimal and then printing the results. Apparently, however, the much more difficult FPU problem was just a bit beyond what Fermi wanted to tackle, and it was Tsingou who did the coding.

24. Ford, "The Fermi-Pasta-Ulam Problem."

25. Fermi, Pasta, and Ulam, "Studies of Non-Linear Problems," 981.

26. Ford, "The Fermi-Pasta-Ulam Problem," 274.

27. Thierry Dauxois and Michel Peyrard, *Physics of Solitons* (Cambridge: Cambridge University Press, 2006). Figure 3.8.3 was generated from a remarkably short (30 lines) MATLAB code that, because I didn't write it, I haven't reproduced here with a walkthrough. I have, however, listed the published code *as I very slightly modified it to produce Figure 3.8.3*, in an appendix at the end of the book. You also can easily find it in its original publication (see note 21 again) or on the Web (for example, at http://www.scholarpedia. org/article/Fermi_Pasta_Ulam_nonlinear _lattice_oscillations). Although not long, this code extensively uses one of MATLAB's efficient differential equation solvers, as well as its powerful discrete Fourier transform command. Unlike Mary Tsingou's MANIAC-I machine language code, the MATLAB code runs very quickly; even on my quite ordinary home computer; Figure 3.8.3 was generated in less than thirteen seconds. If you don't have MATLAB available, you're still in luck; take a look at the applet FPU simulator at http://stp.clarku.edu/simulations/fpu/index.html.

28. Fermi, Pasta, and Ulam, "Studies of Non-Linear Problems," 977.

4

THE ASTONISHING PROBLEM OF THE HANGING MASSES

It is well known that simple *nonlinear* systems may show very irregular oscillations that are termed "chaotic." It is not so well known that some simple *linear* systems can also show very irregular oscillations that resemble chaotic oscillations although, strictly speaking, they are not chaotic.
— Kenneth S. Mendelson and Frank G. Karioris (see note 5)

He thought of his old professor's saying: A Chinaman sneezing in Shen-si may set men to shoveling snow in New York City.
— A meteorologist pondering the sensitivity of weather to initial conditions in George R. Stewart's 1941 novel *Storm*, long before the now famous "butterfly effect" in chaos theory became the popular image

4.1 Springs and Harmonic Motion

In the FPU experiment we had a sophisticated system of coupled harmonic oscillators, a system capable of vibrating at a number of different frequencies—most generally, in fact, at more than one of the frequencies and even, perhaps, at all of them simultaneously. The geometry of the FPU experiment was simple in one sense: it did not involve any gravitational potential energy considerations. That was because all the masses and springs in the system were at the same elevation (which we implicitly took as our zero reference level). The only potential energy in the FPU system was that stored in the springs.

In this chapter I want to show you how a simple variation of the FPU experiment can lead to another result almost as surprising as that experiment's failure to exhibit energy equipartition. On the one hand, this variation will form a system simpler than the FPU system because it will, in its most elementary form, contain only two masses and two springs. It will also be more complicated, though, because now the system *will* have mass components at different elevations, and so we will have to include gravitational potential energy in our analysis. To set the stage for this modified FPU system, I'll start with a quick review of elementary harmonic oscillator physics.

When mathematical physicists see the second-order linear differential equation

$$\frac{d^2x}{dt^2} + \alpha x = 0, \quad \alpha > 0 \tag{4.1.1}$$

appear in an analysis, they instantly think *harmonic behavior*. That is, they know $x(t)$ will be described by a *sinusoidal oscillation* at a frequency determined by the positive constant α. A classic freshman physics textbook example of how (4.1.1) can come about is shown in Figure 4.1.1, which has a mass m hanging from one end of a *linear* (obeying Hooke's law) spring with spring constant k, with the other end of the spring attached to a ceiling. I'll write the acceleration of gravity (downward)

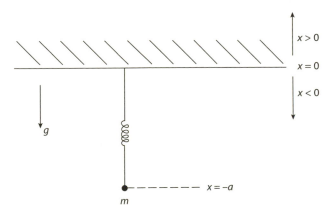

FIGURE 4.1.1. A mass hanging from a linear spring in a uniform gravitational field

as g, and the *relaxed* (neither stretched nor compressed) length of the spring as a. The undisturbed position of the mass, then, as it hangs motionless, is at distance $x < -a$ below the ceiling (I'm measuring x as positive upward, with $x = 0$ as the ceiling). If we now pull the mass downward and then let it go, we all know what will happen: the mass will bob up and down, with the spring alternating between stretching and compressing. What is happening during this oscillatory motion is a periodic exchange of the potential energy stored (either stretched *or* compressed) in the spring and the kinetic energy of the moving mass. There is no energy loss mechanism in (4.1.1)—energy loss mathematically appears as a *first* derivative term, and (4.1.1) has no such term[1]—and so energy is conserved; if $E(t)$ is the total system energy at any time t, then $\dfrac{dE(t)}{dt} = 0$.

From Newton's second law of motion we can write

$$m\frac{d^2x}{dt^2} = -k(a + x) - mg \tag{4.1.2}$$

because the left-hand side of (4.1.2) is the total upward force (mass times acceleration) acting on the mass, which the right-hand side says is the spring force (the first term, which you'll notice is indeed positive if $x < -a$, which means the spring is *stretched* and so is pulling the mass upward), and the gravitational force (the second term, which is always negative because it is always pulling the mass downward). Rewriting (4.1.2),

$$m\frac{d^2x}{dt^2} + kx = -ka - mg,$$

or

$$\frac{d^2x}{dt^2} + \frac{k}{m}x = -\frac{k}{m}a - g. \tag{4.1.3}$$

Now, (4.1.3) looks almost like (4.1.1), with $\alpha = \dfrac{k}{m}$, but the right-hand side isn't zero. Fortunately, that's a trivial complication, as I'll now show you.

One possible solution to (4.1.3) is a *constant*, that is, $x(t) = C$. This is because then $\dfrac{d^2x}{dt^2} = 0$, and so

$$\frac{k}{m}C = -\frac{k}{m}a - g,$$

and the constant is

$$C = -a - \frac{mg}{k}. \tag{4.1.4}$$

If $g = 0$ (no gravity), then $C = -a$, which makes physical sense: the mass has no weight and simply hangs at the end of the *unstretched* (massless) spring. The additional negative term in C, the $-\dfrac{mg}{k}$, is the stretching of the spring due to the weight of the mass.

Since (4.1.3) is a *linear* equation, we can write its general solution as the sum of the constant C and a time-varying component $p(t)$. So,

$$x(t) = C + p(t),$$

and substitution of this into (4.1.3) gives

$$\frac{d^2p}{dt^2} + \frac{k}{m}[p(t) + C] = -\frac{k}{m}a - g,$$

or, using (4.1.4),

$$\frac{d^2p}{dt^2} + \frac{k}{m}p = -\frac{k}{m}a - g - \frac{k}{m}C = -\frac{k}{m}a - g - \frac{k}{m}\left(-a - \frac{mg}{k}\right).$$

Thus,

$$\frac{d^2p}{dt^2} + \frac{k}{m}p = 0, \tag{4.1.5}$$

and (4.1.5) *does* look just like (4.1.1) with $\alpha = \dfrac{k}{m}$.

To solve (4.1.5), *assume* a solution of the form

$$p(t) = Ae^{st}, \tag{4.1.6}$$

where A and s are some (to-be-determined) constants. Then, substituting (4.1.6) into (4.1.5), we have

$$As^2e^{st} + \frac{k}{m}Ae^{st} = 0,$$

or, since the Ae^{st} in each term cancels away (this feature is, in fact, why our assumption is a *very good* assumption!), we have

$$s^2 + \frac{k}{m} = 0,$$

or

$$s = \pm\sqrt{-\frac{k}{m}} = \pm i\sqrt{\frac{k}{m}}, i = \sqrt{-1}.$$

Our most general solution for $p(t)$, then, is a sum of two terms, one for each value of s:

$$p(t) = A_1 e^{i\sqrt{\frac{k}{m}}t} + A_2 e^{-i\sqrt{\frac{k}{m}}t},$$

and so the general solution for the total motion of the hanging mass is

$$x(t) = p(t) + C,$$

or

$$x(t) = A_1 e^{i\sqrt{\frac{k}{m}}t} + A_2 e^{-i\sqrt{\frac{k}{m}}t} - a - \frac{mg}{k}. \tag{4.1.7}$$

To evaluate (4.1.7) further, that is, to determine A_1 and A_2, we need to introduce specific *initial conditions* on $x(t)$. For example, suppose we move the mass to $x = -2a$, twice the unstretched length of the spring, and then let the mass go *from rest*. Then we know that $x(0) = -2a$ and that $\left.\frac{dx}{dt}\right|_{t=0} = 0$. Since

$$\frac{dx}{dt} = i\sqrt{\frac{k}{m}}A_1 e^{i\sqrt{\frac{k}{m}}t} - i\sqrt{\frac{k}{m}}A_2 e^{-i\sqrt{\frac{k}{m}}t},$$

then the second condition says

$$0 = A_1 - A_2,$$

or

$$A_1 = A_2 = A.$$

Putting this into (4.1.7), and using Euler's wonderful identity, we have

$$x(t) = 2A \cos \left(\sqrt{\frac{k}{m}} t \right) - a - \frac{mg}{k}.$$

This says, along with the first initial condition, that

$$x(0) = 2A - a - \frac{mg}{k} = -2a.$$

Thus,

$$A = \frac{1}{2} \left(\frac{mg}{k} - a \right),$$

and so, at last,

$$x(t) = \left(\frac{mg}{k} - a \right) \cos \left(\sqrt{\frac{k}{m}} t \right) - a - \frac{mg}{k}, \tag{4.1.8}$$

which is, as claimed, a sinusoidal oscillation of the mass at angular frequency ω ($\omega t = \sqrt{\frac{k}{m}} t$, and so $\omega = \sqrt{\frac{k}{m}} = \sqrt{\alpha}$).

4.2 A Curious Oscillator

The previous section was (I hope!) pretty elementary and mostly familiar to you. But—and here is my reason for reviewing that material—it is very easy to fall into the trap of extending the nice, single-frequency result we got in (4.1.8) far beyond what is legitimate. For example, it is not true that all oscillating systems made of just masses and linear springs are sinusoidal oscillators. Perhaps even more surprising is that one does not need a spring to build an oscillator that *is* sinusoidal—it can be done with just masses! In this section and the next one I'll show you an example of both possibilities.

Figure 4.2.1 shows a mass m with a hole in it, through which is threaded a wire bent into the curve $y = x^2$. That is, the wire is a vertical parabola, and the mass can slide along the wire.[2] We'll take the sliding to occur without friction. Also attached to the mass is a linear, massless spring, which has its other end attached to a massless loop that can slide along the x-axis. That sliding is also taken to occur without friction. The spring constant is k, and the unstretched length of the spring is taken to be zero. This means the spring is almost always stretched, with a single exception occurring if the spring's x-axis loop happens to be at $x = 0$. Finally, we'll take the acceleration of gravity as uniformly constant everywhere (downward) and we'll write it as g.

At time $t = 0$ the mass is pulled to the right up along the wire so that its coordinates are (x_0, x_0^2), and then released *from rest*. What happens? The now initially stretched spring will pull the mass down along the parabolic wire and, since there is no energy loss mechanism present,

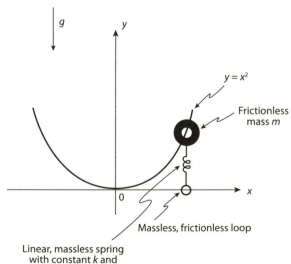

FIGURE 4.2.1. Oscillating on a parabolic wire

the mass will obviously oscillate back and forth along the wire, endlessly, between the coordinates $(-x_0, x_0^2)$ and (x_0, x_0^2). But what is the nature of these oscillations? To answer that, our starting point is the conservation of energy.

The energy $E(t)$ of the mass-spring system of Figure 4.2.1 is

$$E(t) = \frac{1}{2}m\left[\left(\frac{dx}{dt}\right)^2 + \left(\frac{dy}{dt}\right)^2\right] + \frac{1}{2}ky^2 + mgy, \qquad (4.2.1)$$

where the first term on the right-hand side, with the square brackets, is the kinetic energy of the mass, the second term is the potential energy stored in the spring, and the third term is the potential energy of the mass. We can rewrite (4.2.1) as follows:

$$E(t) = \frac{1}{2}m\left(\frac{dx}{dt}\right)^2\left[1 + \frac{\left(\frac{dy}{dt}\right)^2}{\left(\frac{dx}{dt}\right)^2}\right] + \frac{1}{2}ky^2 + mgy$$

$$= \frac{1}{2}m\left(\frac{dx}{dt}\right)^2\left[1 + \left(\frac{dy}{dx}\right)^2\right] + \frac{1}{2}ky^2 + mgy,$$

or, as $y = x^2$, then $\dfrac{dy}{dx} = 2x$, and so

$$E(t) = \frac{1}{2}m\left(\frac{dx}{dt}\right)^2\left[1 + 4x^2\right] + \frac{1}{2}ky^2 + mgy. \qquad (4.2.2)$$

Since energy is conserved we have

$$\frac{dE}{dt} = 0 = \frac{1}{2}m\left[2\left(\frac{dx}{dt}\right)\frac{d^2x}{dt^2}\left(1 + 4x^2\right) + \left(\frac{dx}{dt}\right)^2 8x\frac{dx}{dt}\right]$$

$$+ ky\frac{dy}{dt} + mg\frac{dy}{dt},$$

or, simplifying,

$$m\left(\frac{dx}{dt}\right)\frac{d^2x}{dt^2}\left(1 + 4x^2\right) + 4mx\left(\frac{dx}{dt}\right)^3 + \frac{dy}{dt}\left(ky + mg\right) = 0. \qquad (4.2.3)$$

Again, since $y = x^2$, then $\dfrac{dy}{dt} = 2x\dfrac{dx}{dt}$, and so (4.2.3) becomes

$$m\left(\frac{dx}{dt}\right)\frac{d^2x}{dt^2}(1 + 4x^2) + 4mx\left(\frac{dx}{dt}\right)^3 + 2x\frac{dx}{dt}(kx^2 + mg) = 0,$$

or, at last, dividing through by $m\left(\dfrac{dx}{dt}\right)$,

$$(1 + 4x^2)\frac{d^2x}{dt^2} + 4x\left(\frac{dx}{dt}\right)^2 + 2x\left(\frac{k}{m}x^2 + g\right) = 0, \qquad (4.2.4)$$

which is a pretty far cry from the elementary sinusoidal oscillator equation of (4.1.1). Remember, (4.2.4) is just a single mass sliding without friction on a parabolic wire while connected to the x-axis through a single linear spring that also slides without friction.

Perhaps almost as surprising as the complexity of (4.2.4) is that it is nonetheless still possible to get a pure sinusoidal solution for $x(t)$. To see this, suppose that somehow we do indeed observe

$$x(t) = x_0 \cos(\omega t). \qquad (4.2.5)$$

We can determine how this can occur by working backward. Thus,

$$\frac{dx}{dt} = -x_0\omega \sin(\omega t),$$

and so

$$\left(\frac{dx}{dt}\right)^2 = x_0^2\omega^2 \sin^2(\omega t) = x_0^2\omega^2\left[1 - \cos^2(\omega t)\right]$$

$$= \omega^2\left[x_0^2 - x_0^2\cos^2(\omega t)\right],$$

or

$$\left(\frac{dx}{dt}\right)^2 = \omega^2\left[x_0^2 - x^2\right]. \qquad (4.2.6)$$

Also,

$$\frac{d^2x}{dt^2} = -x_0\omega^2 \cos(\omega t),$$

and so

$$\frac{d^2x}{dt^2} = -\omega^2 x. \tag{4.2.7}$$

If we insert (4.2.6) and (4.2.7) into (4.2.4), we have

$$\left(-\omega^2 x\right)\left(1 + 4x^2\right) + 4x\left[\omega^2\left(x_0^2 - x^2\right)\right] + 2\frac{k}{m}x^3 + 2gx = 0,$$

or, collecting terms,

$$x\left(-\omega^2 + 4\omega^2 x_0^2 + 2g\right) + x^3\left(-8\omega^2 + 2\frac{k}{m}\right) = 0. \tag{4.2.8}$$

This can be true for *all* x only if each of the coefficients of x and x^3 in (4.2.8) separately vanish. That is, *if* (4.2.5) holds, *then* it is required that

$$-\omega^2 + 4\omega^2 x_0^2 + 2g = 0$$

and

$$-8\omega^2 + 2\frac{k}{m} = 0.$$

That is,

$$x_0 = \frac{\sqrt{\omega^2 - 2g}}{2\omega} = \sqrt{\frac{1}{4} - \frac{g}{2\omega^2}}$$

and

$$k = 4m\omega^2.$$

We can rewrite these equations in slightly more compact form by defining

$$c = \frac{k}{m}$$

and then

$$\omega = \frac{1}{2}\sqrt{\frac{k}{m}} = \frac{1}{2}\sqrt{c} \qquad (4.2.9)$$

and

$$x_0 = \sqrt{\frac{1}{4} - \frac{g}{2\frac{k}{4m}}} = \sqrt{\frac{1}{4} - \frac{2g}{\frac{k}{m}}} = \sqrt{\frac{1}{4} - \frac{2g}{c}}. \qquad (4.2.10)$$

Then, using Earth's gravity of $g = 9.81$ m/s^2 (remember, we are using the mks system of units), we have (4.2.5) as true if

$$x_0 = \sqrt{0.25 - \frac{19.62}{c}} \text{ meters,} \qquad (4.2.11)$$

and this will give sinusoidal oscillations at the angular frequency ω (in radians per second) given by (4.2.9), where c is *almost* an arbitrary positive constant (c isn't totally arbitrary since it obviously must be at least large enough to give a real value to x_0 in (4.2.11). For example, if $c = 100, 400$, or $1,600$, then ω is equal to 5, 10, or 20 (respectively) radians per second. Since $\omega = \frac{2\pi}{T}$, where T is the period of the resulting sinusoidal oscillation, then for these three values of c we have $T = \frac{2\pi}{\omega}$ seconds, or $\frac{2\pi}{5}, \frac{2\pi}{10}$, or $\frac{2\pi}{20}$ seconds (1.257 seconds, 0.628 seconds, and 0.314 seconds). In just a bit I'll show you the $x(t)$ solutions for these three particular values of c, and so if we plot each $x(t)$ over the interval 0 to 2 seconds, then even the lowest frequency $x(t)$ will display more than one full period.

To show you that (4.2.11) really does lead to sinusoidal oscillations at the frequency predicted by (4.2.9), I'll now turn to doing a direct computer solution of the system's differential equation, (4.2.4). MATLAB has a very powerful capability[3] for numerically solving differential equations, which starts by asking the analyst to write the system differential equation(s) as all *first*-order differential equations. If the system's differential equation is already just a first-order one, then, of course, the task is done, but in our case we have a second-order differential equation and so just a bit of preliminary work to do. I'll start by defining the two variables u_1 and u_2 as

$$u_1 = x \tag{4.2.12}$$

and

$$u_2 = \frac{dx}{dt}, \tag{4.2.13}$$

and then, since $\frac{du_2}{dt} = \frac{d^2x}{dt^2}$, we can rewrite (4.2.4) as, using (4.2.9), too,

$$(1 + 4u_1^2)\frac{du_2}{dt} + 4u_1u_2^2 + 2cu_1^3 + 19.62u_1 = 0,$$

or

$$\frac{du_2}{dt} = -\frac{4u_1u_2^2 + 2cu_1^3 + 19.62u_1}{1 + 4u_1^2}. \tag{4.2.14}$$

Now, this book is not intended in any way to be a MATLAB programming text, so what I next discuss will be a very brief overview of how to code our two first-order differential equations (4.2.13) and (4.2.14) in MATLAB 7 (specifically, MATLAB 7.3), in sufficient detail for you to understand what's going on (what *can* be done in MATLAB to solve differential equations far exceeds what I'll actually do here). It will be doubly useful to have this introductory discussion now because we'll be numerically solving differential equations with MATLAB in Chapter 5, too.

Our coding will consist of a so-called *function m-file* (which will, itself, contain a second, internal or *nested* function) that I'll name **osc.m** (for *oscillator*). When run, **osc.m** generates Figure 4.2.2, which shows plots of $x(t)$ for our three values of c, using the three associated values of x_0 calculated from (4.2.11), x_0 values that I claim will result in sinusoidal behavior for $x(t)$. The following listing shows the essentials of **osc.m**, except for the routine statements that add axis, title, and legend labels. The line numbers in **osc.m** are, as always, not part of MATLAB but rather serve as reference labels for the walkthrough that follows the listing.

```
osc.m
01    function osc
02    options=odeset('AbsTol',1e-7,'RelTol',1e-4);
03    tspan=[0 2];
04    c=100;
05    x0=sqrt(0.25-19.62/c);
06    uzero=[x0;0];
07    [ t,u]=ode45(@wire,tspan,uzero,options);
08    subplot(311)
09    plot(t,u(:,1),'-k')
10    c=400;
11    x0=sqrt(0.25-19.62/c);
12    uzero=[x0;0];
13    [ t,u]=ode45(@wire,tspan,uzero,options);
14    subplot(312)
15    plot(t,u(:,1),'-k')
16    c=1600;
17    x0=sqrt(0.25-19.62/c);
18    uzero=[x0;0];
19    [ t,u]=ode45(@wire,tspan,uzero,options);
20    subplot(313)
21    plot(t,u(:,1),'-k')
22        function xderivative=wire(t,u)
23        xderivative=[u(2);-(4*u(1)*u(2)^2 +2*c*u(1)^3
                +19.62*u(1))/(1 +4*u(1)^2)];
24        end
25    end
```

The three plots in Figure 4.2.2 certainly *look* sinusoidal (more on this point in the next section), with the periods we calculated earlier (immediately after (4.2.11)).

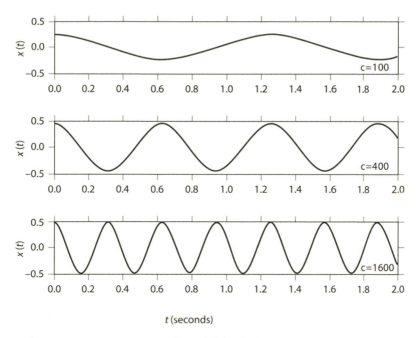

FIGURE 4.2.2. Sinusoidal solutions to (4.2.4)

Here's how the function m-file **osc.m** works, with line 01 defining the code as a function with the name *osc*. Lines 02 through 06 set the values of the parameters we'll pass to the MATLAB differential equation solver in line 07. Specifically, line 02 defines the absolute and relative accuracies we wish our solution to have (the default values are 10^{-6} and 10^{-3}, respectively, and I've arbitrarily made each ten times smaller just to show how *options* works), line 03 defines the time span over which we want the code to calculate $x(t)$—here we are looking for $x(t)$ in the interval 0 to 2 seconds, line 04 specifies the value of c, line 05 calculates x_0 for the given c, and line 06 sets the two row elements of the column vector *uzero* equal to the initial conditions of our problem (in this case, $x(0) = x_0$ and $\left. \dfrac{dx}{dt} \right|_{t=0} = 0$—remember, we are releasing the mass *from rest*). With all that done, line 07 then invokes the MATLAB differential equation solver *ode45*. On the right-hand side of line 07 we are asking MATLAB to visit the internal function *wire* (see lines 22 to 24) with the input arguments *tspan*, *uzero*, and *options* that we defined earlier in lines 02, 03, and 06. On the left-hand side of line 07 we are asking MATLAB to give back to us the two output arguments t and u.

Now, here's what happens when *wire* is visited. Line 22 defines the variable *xderivative* as synonymous with *wire*, and line 23 tells MATLAB that *xderivative* is a two-row column vector. The first element computed is the value of the first derivative $\frac{dx}{dt}$ (see (4.2.13)) and the second element computed is the value of the second derivative $\frac{d^2x}{dt^2}$ (see (4.2.14)). These row elements are calculated from the matrix *u*; *u* has two column vectors. To start, the first row of *u* is loaded with the initial conditions in *uzero* passed to *wire* in line 07. That is, the first row element of the first row of *u* is the value of x_0, and the second row element of the first row of *u* is the value of $\left.\frac{dx}{dt}\right|_{t=0}$. From that, MATLAB constructs all the rest of the rows of *u*, one row after the other; when finished the *u* matrix for Figure 4.2.2 had 113 rows. The *first column* of *u* is a column of *x* values and the *second column* of *u* is a column of $\frac{dx}{dt}$ values. Each *u* matrix *row* corresponds to a specific instant of time, and again, for Figure 4.2.2, the *t* column vector of increasing time values had 113 rows. The *t* and *u* row counts counts will always be equal, of course, but it is MATLAB (not you) that automatically determines what that row count is. The row count is a function of both the duration of time over which we ask MATLAB to produce a solution (that is, of *tspan*, the value of which was passed to *wire* as an argument in *ode45*) and how rapidly the solution changes with time in the solution interval.[4] Notice that the nested function *wire* is terminated by an *end* in line 24.

When *ode45* has completed its calculations it returns both the vector *t* and the matrix *u* as outputs, as requested in the left-hand side of line 07. Line 08 says to select subplot 1 (the top plot) from a column of three plots (311 means three rows, one column, plot one), and line 09 plots the first column vector of *u* (= *x*) on the vertical axis against the time vector *t* on the horizontal axis. Lines 10 through 15 do it all over again for *c* = 400, and lines 16 through 21 do it yet again for *c* = 1600.

4.3 Phase-Plane Portraits

As I mentioned before, the curves in Figure 4.2.2 at least *look* like the sinusoidal oscillations that our theoretical analysis in the last section predicted. A natural question to ask now is, what do the oscillations of $x(t)$ look like if, for a given c, we *don't* use the value for $x(0) = x_0$ given by (4.2.11)? That's an easy question to answer using **osc.m**, and Figure 4.3.1 shows, for example, $x(t)$ for $c = 400$ and $x(0) = 7x_0$, where x_0 is given by (4.2.11).

Well, I can speak only for myself on this point, but *I* think Figure 4.3.1 *still* looks sinusoidal! This probably shouldn't have been as surprising as it was for me—after all, we know, physically, that the mass just slides back and forth along the parabolic wire, and so $x(t)$ will *always* "look wavy" *no matter what* value of $x(0)$ we use. So, is there any other way we can distinguish between sinusoidal and nonsinusoidal oscillations, if "just looking" isn't enough? One possibility is to create what is called a *phase-plane portrait* of the system. To see what this means, recall the basic harmonic oscillator differential equation of (4.1.1), which I'll repeat here as

$$\frac{d^2x}{dt^2} + \alpha x = 0, \quad \alpha > 0. \tag{4.3.1}$$

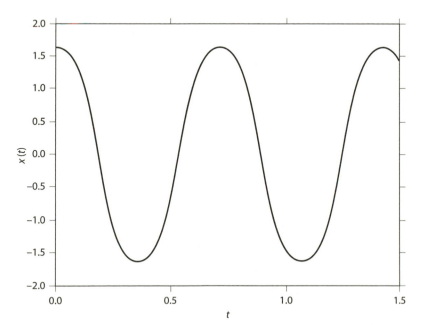

FIGURE 4.3.1. $x(t)$ when $x(0) \neq x_0$

If we reduce this second-order equation to two first-order equations, that is, if we write

$$y = \frac{dx}{dt} \tag{4.3.2}$$

and so

$$\frac{dy}{dt} = -\alpha x, \tag{4.3.3}$$

then

$$\frac{dy}{dx} = \frac{dy/dt}{dx/dt}$$

or

$$\frac{dy}{dx} = -\alpha \frac{x}{y}. \tag{4.3.4}$$

(4.3.4) is a *separable* differential equation and is easily integrated: $y \, dy = -\alpha x \, dx$, or, with $\frac{1}{2}C$ as the arbitrary constant of indefinite integration (which depends both on the amplitude of the oscillations and on their frequency),

$$\frac{1}{2}y^2 = -\frac{1}{2}\alpha x^2 + \frac{1}{2}C$$

or

$$y^2 + \alpha x^2 = C. \tag{4.3.5}$$

That is, a plot of $x(t)$ against its derivative $y(t)$, as given in (4.3.2), will (for a harmonic oscillator) generally result in an *ellipse* (a *circle*, in fact, if $\alpha = 1$). This plot of $x(t)$ versus $\frac{dx}{dt}$ is called the *phase-plane portrait* of the harmonic oscillator differential equation of (4.3.1). A phase-plane portrait tells you immediately, by just looking at it, two quite different things.

First, if the plot is a closed curve, then the system it describes has *periodic* oscillations. That is, the oscillations repeat after a finite time interval T (called the *period* or *orbit time*)—*orbit time* because that is the time it takes to travel once around the closed curve in the phase-plane portrait. This is because the solution to a second-order differential equation is completely determined by knowledge of both $x(t)$ and $\frac{dx}{dt}$ at time $t = 0$, and so, if at some later time $t = T$ the value of $x(T) = x(0)$ and $\left.\frac{dx}{dt}\right|_{t=T} = \left.\frac{dx}{dt}\right|_{t=0}$, then the system is back in its original state (electrical engineers, in fact, often call $x(t)$ and $\frac{dx}{dt}$ the *state variables* of a second-order system) and $x(t)$ simply starts afresh. (As an aside, note that if $x(t)$ is periodic with period T then it *automatically* is true that *all* derivatives of $x(t)$ are also periodic with the same period—just endlessly differentiate the very definition of periodicity, $x(t) = x(t + T)$.)

These periodic oscillations may or may not be sinusoidal. And that leads us to the second "at-a-glance" feature of the portrait. If the phase-plane portrait is (not) an ellipse, then the oscillations are (not) sinusoidal.

We can easily generate the phase-plane portrait of the mass-on-a-parabolic-wire oscillator with a single modification of **osc.m**. All we need to recall is that the second column of the u matrix holds the values of $\frac{dx}{dt}$. So, instead of plotting the first column of u (which holds the values of x) against t, let's plot the first column of u on the horizontal axis and the second column of u on the vertical axis. This modification is simply that of replacing line 21 of **osc.m** with *plot(u(:,1),u(:,2),'-k')*. When run, the modified code produces Figure 4.3.2, and it is obvious that the phase-plane portrait is *closed*, and so we have oscillations that are periodic, which is no surprise. Notice, too, that the plot is *concave*, and so is certainly *not* elliptical, and so the oscillations are *not* sinusoidal, which was not completely obvious from Figure 4.3.1. Figure 4.3.3, on the other hand, shows the phase-plane portrait for $c = 400$ when $x(0) = x_0$, and that portrait does indeed look elliptical.

4.4 Another (Even More?) Curious Oscillator

In Figure 4.4.1 we have two equal masses (m), one constrained to sliding along the x-axis and the other to sliding along the y-axis; the two

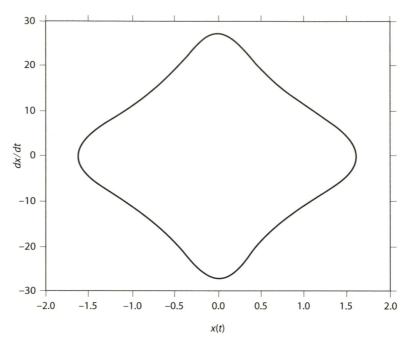

FIGURE 4.3.2. Phase-plane portrait of Figure 4.3.1

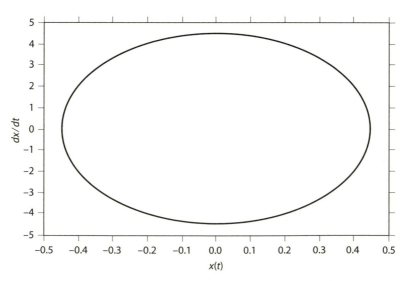

FIGURE 4.3.3. Elliptical phase-plane portrait for a sinusoidal oscillation

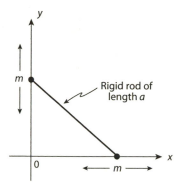

FIGURE 4.4.1. A two-mass, springless oscillator

masses are connected (via flexible hinges) by a rigid, massless rod of length a. There is no friction or any other mechanism of energy loss, and so the total energy of the system is constant. Finally, we imagine that the x-y plane is the horizontal plane, which means that the total gravitational potential energy is constant. And therefore we conclude that the total kinetic energy is constant. Now, notice carefully—no springs!

If we write the coordinates of the two masses as $(x(t), 0)$ and $(0, y(t))$ then, with C as some constant, we can write

$$x^2 + y^2 = a^2 \qquad (4.4.1)$$

and

$$\frac{1}{2}m\left(\frac{dx}{dt}\right)^2 + \frac{1}{2}m\left(\frac{dy}{dt}\right)^2 = C. \qquad (4.4.2)$$

The first equation is pure math (the Pythagorean theorem) and the second is pure physics (the conservation of energy). Now, it should be physically obvious that as the x-axis mass passes through the origin ($t = 0$), the y-axis mass will, at that instant, have zero speed, that is, zero kinetic energy. Thus,

$$\left(\frac{dy}{dt}\right)\Bigg|_{t=0} = 0,$$

which, from (4.4.2), says

$$\frac{1}{2}m\left(\frac{dx}{dt}\right)^2\bigg|_{t=0} = C. \tag{4.4.3}$$

Putting (4.4.3) into (4.4.2) we have

$$\frac{1}{2}m\left(\frac{dx}{dt}\right)^2 + \frac{1}{2}m\left(\frac{dy}{dt}\right)^2 = \frac{1}{2}m\left(\frac{dx}{dt}\right)^2\bigg|_{t=0},$$

or

$$\left(\frac{dx}{dt}\right)^2 + \left(\frac{dy}{dt}\right)^2 = \left(\frac{dx}{dt}\right)^2\bigg|_{t=0}. \tag{4.4.4}$$

From (4.4.1) we have $y = (a^2 - x^2)^{1/2}$, and so

$$\frac{dy}{dt} = -\frac{1}{2}(a^2 - x^2)^{-1/2}2x\frac{dx}{dt},$$

or

$$\left(\frac{dy}{dt}\right)^2 = \frac{x^2}{a^2 - x^2}\left(\frac{dx}{dt}\right)^2. \tag{4.4.5}$$

Combining (4.4.4) and (4.4.5),

$$\left(\frac{dx}{dt}\right)^2 + \frac{x^2}{a^2 - x^2}\left(\frac{dx}{dt}\right)^2 = \left(\frac{dx}{dt}\right)^2\bigg|_{t=0},$$

or

$$\left(\frac{dx}{dt}\right)^2\left[1 + \frac{x^2}{a^2 - x^2}\right] = \left(\frac{dx}{dt}\right)^2\bigg|_{t=0} = \left(\frac{dx}{dt}\right)^2\frac{a^2}{a^2 - x^2},$$

or

$$a^2\left(\frac{dx}{dt}\right)^2 = a^2\left(\frac{dx}{dt}\right)^2\bigg|_{t=0} - x^2\left(\frac{dx}{dt}\right)^2\bigg|_{t=0},$$

or

$$\left(\frac{dx}{dt}\right)^2 = \left(\frac{dx}{dt}\right)^2\Bigg|_{t=0} - x^2 \frac{\left(\frac{dx}{dt}\right)^2\Big|_{t=0}}{a^2},$$

or

$$\left(\frac{dx}{dt}\right)^2 + \frac{\left(\frac{dx}{dt}\right)^2\Big|_{t=0}}{a^2}x^2 = \left(\frac{dx}{dt}\right)^2\Bigg|_{t=0}. \tag{4.4.6}$$

If we differentiate (4.4.6) with respect to time, then

$$2\frac{dx}{dt} \cdot \frac{d^2x}{dt^2} + \frac{\left(\frac{dx}{dt}\right)^2\Big|_{t=0}}{a^2}2x\frac{dx}{dt} = 0,$$

or, making the obvious cancelation of the common $2\dfrac{dx}{dt}$ factor in each term,

$$\frac{d^2x}{dt^2} + \frac{\left(\frac{dx}{dt}\right)^2\Big|_{t=0}}{a^2}x = 0. \tag{4.4.7}$$

But (4.4.7) is simply (4.1.1), our starting point for sinusoidal oscillations at angular frequency $\omega = \sqrt{\alpha}$, with

$$\alpha = \frac{\left(\frac{dx}{dt}\right)^2\Big|_{t=0}}{a^2}, \tag{4.4.8}$$

where it is clear that $\alpha > 0$ (and so $\omega = \sqrt{\alpha}$ is real) is indeed the case.

The two rigidly connected masses of Figure 4.4.1, in which there is no spring, oscillate sinusoidally up and down the y-axis and back and forth along the x-axis at an angular frequency that is independent of m but does depend on both the length of the connecting rod and on the initial speed of the masses. These oscillations are not the result of an endless exchange of kinetic and potential energy but rather the result of an endless kinetic energy to kinetic energy exchange. Can you *really* say that (4.4.8) was obvious before we did the analysis? I don't think so!

4.5 Hanging Masses

Now, at last, after all the previous buildup, we come to the central problem of this chapter. Figure 4.5.1 shows N identical masses (each m), hanging in a chain one below the other from N identical mass-less linear springs. The common spring constant is k, the common unstretched length of each spring is a, and the constant, uniform gravitational acceleration is g. The uppermost spring is attached to a ceiling at $y = y_0 = 0$ ($y > 0$ is upward and $y < 0$ is downward). The whole arrangement looks something like the FPU experiment rotated into the vertical direction, but now with one end (the bottom end) free; you'll recall that in the FPU experiment *both* ends of a *horizontal* chain were *fixed*. Imagine that this chain of springs and masses is initially hanging motionless. If we then disturb the chain in some way—if, for example, we lift one of the masses upward and then let it go— we'd expect the chain to oscillate up and down in some manner. Are

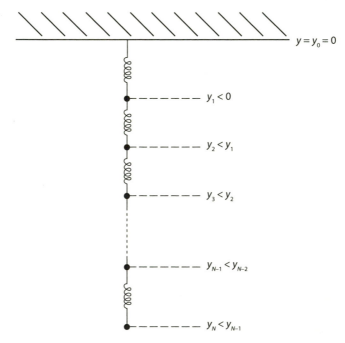

FIGURE 4.5.1. N hanging masses

those oscillations sinusoidal? Be careful with this, and don't answer too quickly!

One thing you should be able to see immediately is that, when the chain is hanging motionless, the topmost spring is stretched the most, the next spring down is stretched not so much, and so on until we get to the bottom spring, which is stretched the least. Let's write y_n as the position of the nth mass, $1 \leq n \leq N$, as measured from the ceiling, where obviously

$$y_N < y_{N-1} < \cdots < y_3 < y_2 < y_1 < y_0 = 0.$$

That is, all the ys are *negative*.

What I'll do now is use Newton's second law of motion to write the differential equations that describe the motions of the N masses. To do this, all we need to keep in mind is that if the upper spring to a mass is stretched, then that spring exerts an *upwardly* directed force on the mass, but if the lower spring attached to that mass is stretched it exerts a *downwardly* directed force on that mass. So, for each of the upper $N - 1$ masses, the ones that have both upper and lower springs attached to them, we can write

$$m\frac{d^2y_n}{dt^2} = k\left[(y_{n-1} - y_n) - a\right] - k\left[(y_n - y_{n+1}) - a\right] - mg,$$

$$1 \leq n \leq N - 1,$$

because the left-hand side is mass times acceleration, which is the total net upward force on the nth mass, while the first term on the right-hand side is the *upwardly* directed spring force on that mass, due to its top spring (notice that if $(y_{n-1} - y_n) - a > 0$, then that spring is *stretched*), and the second term on the right-hand side is the *downwardly* (hence the minus sign) directed spring force due to its bottom spring because if $(y_n - y_{n+1}) - a > 0$, then that spring is *stretched*. (Don't forget: all the y_n are *negative*, except for $y_0 = 0$). The third term is, of course, simply the weight of the mass, which is a force always directed downward. Thus,

$$m\frac{d^2y_n}{dt^2} = k(y_{n+1} + y_{n-1} - 2y_n) - mg, \quad 1 \leq n \leq N - 1. \quad (4.5.1)$$

For the bottom, $n = N$ mass, which doesn't have a lower spring, the equation of motion is simply

$$m\frac{d^2 y_N}{dt^2} = k(y_{N-1} - y_N - a) - mg. \tag{4.5.2}$$

Now, again imagine the chain of masses hanging motionless. If none of the springs stretched, then the equilibrium positions of the masses would be given by $Y_n = -na, 1 \leq n \leq N$. But of course, when hanging motionless, each of the N masses is actually positioned lower than given by this expression because the springs *do* stretch. To account for that stretching, let's write

$$Y_n = -na - b_n, \quad 1 \leq n \leq N, \tag{4.5.3}$$

where, of course, all the b_n will be positive. Since the masses are motionless, by assumption, $\dfrac{d^2 Y_n}{dt^2} = 0$, and so substituting (4.5.3) into (4.5.1) and (4.5.2) gives us

$$0 = k\left[-(n+1)a - b_{n+1} - (n-1)a - b_{n-1} - 2(-na - b_n)\right] - mg,$$
$$1 \leq n \leq N - 1$$

and

$$0 = k\left[-(N-1)a - b_{N-1} - (-Na - b_N) - a\right] - mg,$$

or, doing the easy algebraic simplifications with each expression,

$$b_{n+1} + b_{n-1} - 2b_n = -\frac{mg}{k}, \quad 1 \leq n \leq N - 1 \tag{4.5.4}$$

and

$$b_{N-1} - b_N = -\frac{mg}{k}. \tag{4.5.5}$$

We could now solve (4.5.4) and (4.5.5) to find an explicit expression for b_n as a function of N and n. For our purposes here, however—the exploration of the nature of the *oscillations* of the masses when the chain is disturbed—we don't need to do that.

What we *will* do is solve for the oscillations about the *equilibrium* positions of the masses. That is, let's write $x_n(t)$ as the position of the nth mass about Y_n as given in (4.5.3). That is, let's write

$$y_n(t) = Y_n + x_n(t),$$

or

$$y_n(t) = -na - b_n + x_n(t), \quad 1 \le n \le N. \tag{4.5.6}$$

If we substitute (4.5.6) into (4.5.1), then

$$m\frac{d^2y_n}{dt^2} = m\frac{d^2x_n}{dt^2}$$

$$= k\left[\begin{array}{c} -(n+1)a - b_{n+1} + x_{n+1}(t) - (n-1)a - b_{n-1} \\ + x_{n-1}(t) + 2na + 2b_n - 2x_n(t) \end{array}\right] - mg$$

$$= k\left[x_{n+1}(t) + x_{n-1}(t) - 2x_n(t) + 2b_n - b_{n+1} - b_{n-1}\right] - mg$$

$$= k\left[x_{n+1}(t) + x_{n-1}(t) - 2x_n(t)\right] - k\left[b_{n+1} + b_{n-1} - 2b_n\right] - mg,$$

or, using (4.5.4) in the second term,

$$\frac{d^2x_n}{dt^2} = \frac{k}{m}\left[x_{n+1}(t) + x_{n-1}(t) - 2x_n(t)\right] - \frac{k}{m}\left[-\frac{mg}{k}\right] - g,$$

or, as the second and third terms cancel, we have

$$\frac{d^2x_n}{dt^2} = \Omega^2\left[x_{n+1}(t) + x_{n-1}(t) - 2x_n(t)\right], \quad 1 \le n \le N-1, \tag{4.5.7}$$

where $x_0(t) = 0$ and

$$\Omega = \sqrt{\frac{k}{m}}. \tag{4.5.8}$$

For the bottom, $n = N$ mass, I'll let you substitute (4.5.6) into (4.5.2), which, remembering (4.5.5), should lead you to

$$\frac{d^2x_N}{dt^2} = \Omega^2\left[x_{N-1}(t) - x_N(t)\right]. \tag{4.5.9}$$

The N equations of (4.5.7) and (4.5.9) are linear, *coupled*, second-order differential equations with constant coefficients. How can we solve them for the x_n, $1 \le n \le N$? One way is to follow the same general approach used in the FPU experiment, in terms of the normal modes of the hanging chain,[5] but I want to do something different here. First, I'll solve the special $N = 2$ case analytically, and you'll see that, even in that very simplest form of the problem, we get a quite surprising result: despite the linearity of the system equations the oscillations are not sinusoidal and, in fact, are not even periodic! Then I'll show you how to set up the MATLAB computer solution for values of $N \ge 2$, and we'll use the analytical results for the $N = 2$ case to partially validate the computer code.

4.6 Two Hanging Masses and the Laplace Transform

For $N = 2$, (4.5.7) and (4.5.9) reduce to (remember, $x_0(t) = 0$ because that's the ceiling!)

$$\frac{d^2 x_1}{dt^2} = \Omega^2 (x_2 - 2x_1) \qquad (4.6.1)$$

and

$$\frac{d^2 x_2}{dt^2} = \Omega^2 (x_1 - x_2). \qquad (4.6.2)$$

To solve for $x_1(t)$ and $x_2(t)$, I'll first get (4.6.1) and (4.6.2) into a more convenient mathematical form. What I'll do is use the *Laplace transform* (see note 6 for a quick review of the transform), a technique that turns linear, constant coefficient *differential* equations ("hard" to solve) into *algebraic* equations ("easy" to solve). Writing the Laplace transform pairs $x_1(t) \leftrightarrow X_1(s)$ and $x_2(t) \leftrightarrow X_2(s)$, we have

$$s^2 X_1 - s x_1(0) - \frac{dx_1}{dt}\bigg|_{t=0} = \Omega^2 (X_2 - 2X_1) \qquad (4.6.3)$$

and

$$s^2 X_2 - s x_2(0) - \frac{dx_2}{dt}\bigg|_{t=0} = \Omega^2 (X_1 - X_2). \qquad (4.6.4)$$

We've already agreed that we going to start our chain oscillating *from rest*, which means that

$$\frac{dx_1}{dt}\bigg|_{t=0} = \frac{dx_2}{dt}\bigg|_{t=0} = 0.$$

But what about $x_1(0)$ and $x_2(0)$? There is a very nice mathematical and *physically easy to achieve* choice for them. Imagine taking a finger and *very slowly*, so as not to induce any vibrations, *gently* lifting the upper mass through distance A. The lower mass will then, of course, also move upward through distance A. Then—quickly!—remove your finger. Off goes the system into oscillation, with both masses starting from rest and with $x_1(0) = x_2(0) = A$. With these initial conditions, (4.6.3) and (4.6.4) reduce to

$$s^2 X_1 - As = \Omega^2 X_2 - 2\Omega^2 X_1$$

and

$$s^2 X_2 - As = \Omega^2 X_1 - \Omega^2 X_2,$$

or

$$\left(s^2 + 2\Omega^2\right) X_1 - \Omega^2 X_2 = As \tag{4.6.5}$$

and

$$-\Omega^2 X_1 + \left(s^2 + \Omega^2\right) X_2 = As. \tag{4.6.6}$$

We can solve (4.6.5) and (4.6.6) for $X_1(s)$ and $X_2(s)$ using Cramer's determinant rule for solving a system of simultaneous linear algebraic equations:

$$X_1(s) = \frac{\begin{vmatrix} As & -\Omega^2 \\ As & s^2 + \Omega^2 \end{vmatrix}}{\begin{vmatrix} s^2 + 2\Omega^2 & -\Omega^2 \\ -\Omega^2 & s^2 + \Omega^2 \end{vmatrix}} = \frac{As\left(s^2 + \Omega^2\right) + As\Omega^2}{\left(s^2 + 2\Omega^2\right)\left(s^2 + \Omega^2\right) - \Omega^4},$$

or

$$X_1(s) = \frac{As^3 + 2As\Omega^2}{s^4 + 3\Omega^2 s^2 + \Omega^4}, \tag{4.6.7}$$

as well as

$$X_2(s) = \frac{\begin{vmatrix} s^2 + 2\Omega^2 & As \\ -\Omega^2 & As \end{vmatrix}}{\begin{vmatrix} s^2 + 2\Omega^2 & -\Omega^2 \\ -\Omega^2 & s^2 + \Omega^2 \end{vmatrix}} = \frac{As\left(s^2 + 2\Omega^2\right) + As\Omega^2}{s^4 + 3\Omega^2 s^2 + \Omega^4},$$

or

$$X_2(s) = \frac{As^3 + 3As\Omega^2}{s^4 + 3\Omega^2 s^2 + \Omega^4}. \tag{4.6.8}$$

Now, since

$$s^4 + 3\Omega^2 s^2 + \Omega^4 = \left[s^2 + \left(\Omega \sqrt{\frac{3 - \sqrt{5}}{2}} \right)^2 \right]$$
$$\times \left[s^2 + \left(\Omega \sqrt{\frac{3 + \sqrt{5}}{2}} \right)^2 \right],$$

we can rewrite (4.6.7) and (4.6.8) as

$$X_1(s) = A \frac{s^3}{\left(s^2 + \omega_1^2\right)\left(s^2 + \omega_2^2\right)} + 2A\Omega^2 \frac{s}{\left(s^2 + \omega_1^2\right)\left(s^2 + \omega_2^2\right)} \tag{4.6.9}$$

and

$$X_2(s) = A \frac{s^3}{\left(s^2 + \omega_1^2\right)\left(s^2 + \omega_2^2\right)} + 3A\Omega^2 \frac{s}{\left(s^2 + \omega_1^2\right)\left(s^2 + \omega_2^2\right)}, \tag{4.6.10}$$

where

$$\omega_1 = \Omega \sqrt{\frac{3 - \sqrt{5}}{2}} \qquad (4.6.11)$$

and

$$\omega_2 = \Omega \sqrt{\frac{3 + \sqrt{5}}{2}}. \qquad (4.6.12)$$

From math tables we find the Laplace transform pair

$$\frac{s}{(s^2 + \omega_1^2)(s^2 + \omega_2^2)} \leftrightarrow \frac{\cos(\omega_1 t) - \cos(\omega_2 t)}{\omega_2^2 - \omega_1^2},$$

from which it follows[7] that

$$\frac{s^3}{(s^2 + \omega_1^2)(s^2 + \omega_2^2)} \leftrightarrow \frac{-\omega_1^2 \cos(\omega_1 t) + \omega_2^2 \cos(\omega_2 t)}{\omega_2^2 - \omega_1^2}.$$

Using these two transform pairs on (4.6.9) to return to the time domain,

$$\frac{x_1(t)}{A} = \frac{-\omega_1^2 \cos(\omega_1 t) + \omega_2^2 \cos(\omega_2 t)}{\omega_2^2 - \omega_1^2} + 2\Omega^2 \frac{\cos(\omega_1 t) - \cos(\omega_2 t)}{\omega_2^2 - \omega_1^2},$$

$$(4.6.13)$$

from which it immediately follows that

$$\frac{dx_1/dt}{A} = \frac{\omega_1^3 \sin(\omega_1 t) - \omega_2^3 \sin(\omega_2 t)}{\omega_2^2 - \omega_1^2}$$

$$+ 2\Omega^2 \frac{-\omega_1 \sin(\omega_1 t) + \omega_2 \sin(\omega_2 t)}{\omega_2^2 - \omega_1^2}. \qquad (4.6.14)$$

Notice that in (4.6.13), using the definitions of ω_1 and ω_2 given in (4.6.11) and (4.6.12), there is a complete cancelation of Ω except for its appearance in the cosine arguments where it multiplies t. That is, all Ω does is establish the *time scale*. So, if we redefine ω_1 and ω_2 (for ease in writing our equations) as before but *without* the Ω factor, that is, as

$$\omega_1 = \sqrt{\frac{3 - \sqrt{5}}{2}} \qquad (4.6.15)$$

and

$$\omega_2 = \sqrt{\frac{3 + \sqrt{5}}{2}}, \qquad (4.6.16)$$

and then explicitly include Ω in the cosine arguments as Ωt, we have the *amplitude-normalized, time-scaled* result

$$\frac{x_1(\Omega t)}{A} = \frac{-\omega_1^2 \cos(\omega_1 \Omega t) + \omega_2^2 \cos(\omega_2 \Omega t)}{\omega_2^2 - \omega_1^2}$$
$$+ 2 \frac{\cos(\omega_1 \Omega t) - \cos(\omega_2 \Omega t)}{\omega_2^2 - \omega_1^2},$$

or

$$\boxed{\frac{x_1(\Omega t)}{A} = \frac{\left(2 - \omega_1^2\right) \cos(\omega_1 \Omega t) - \left(2 - \omega_2^2\right) \cos(\omega_2 \Omega t)}{\omega_2^2 - \omega_1^2}.}$$

$$(4.6.17)$$

Further, with the new definitions of ω_1 and ω_2, we see that in (4.6.14) a Ω factor is left in every term on the right-hand side, and so we can write the amplitude-normalized, time-scaled result

$$\boxed{\begin{aligned} & \frac{dx_1(\Omega t)/d(\Omega t)}{A} \\ & = \frac{-\omega_1 \left(2 - \omega_1^2\right) \sin(\omega_1 \Omega t) + \omega_2 \left(2 - \omega_2^2\right) \sin(\omega_2 \Omega t)}{\omega_2^2 - \omega_1^2}. \end{aligned}}$$

$$(4.6.18)$$

If *you* repeat the above for (4.6.10) then *you* should be able to confirm that

$$\frac{x_2(\Omega t)}{A} = \frac{\left(3 - \omega_1^2\right) \cos(\omega_1 \Omega t) - \left(3 - \omega_2^2\right) \cos(\omega_2 \Omega t)}{\omega_2^2 - \omega_1^2}$$

(4.6.19)

and

$$\frac{dx_2(\Omega t)/d(\Omega t)}{A}$$
$$= \frac{-\omega_1 \left(3 - \omega_1^2\right) \sin(\omega_1 \Omega t) + \omega_2 \left(3 - \omega_2^2\right) \sin(\omega_2 \Omega t)}{\omega_2^2 - \omega_1^2}.$$

(4.6.20)

As a quick, partial check on the algebra, notice that (4.6.17) and (4.6.18) do indeed reduce, at $t = 0$, to $x_1(0) = A$ and $\left.\frac{dx_1}{dt}\right|_{t=0} = 0$, and similarly (4.6.19) and (4.6.20) reduce to $x_2(0) = A$ and $\left.\frac{dx_2}{dt}\right|_{t=0} = 0$.

Figure 4.6.1 shows what the oscillations of the upper mass (solid line) and the lower mass (dashed line) look like for $0 \leq \Omega t \leq 50$. Neither oscillation is sinusoidal, even though each is the sum of just two sinusoidal terms. Indeed, it *appears* from even the limited display in the figure that the oscillations may not even be periodic! That this is indeed the case is the result of the fact that the ratio $\frac{\omega_2}{\omega_1}$ is irrational.[8] Figure 4.6.2 shows the phase-plane portrait for the upper mass, and Figure 4.6.3 does the same for the lower mass. Both phase-plane plots make it even more evident than does Figure 4.6.1 that there is no periodicity in the motion of either mass. In both of the portrait plots $0 \leq \Omega t \leq 50$, and an **X** marks the $t = 0$ instant.

4.7 Hanging Masses and MATLAB

To try to solve the hanging masses problem for $N > 2$ using the Laplace transform proves to be an algebraic horror (try it!). To obtain

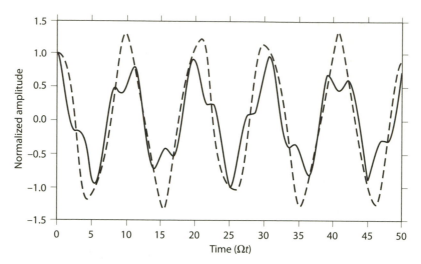

FIGURE 4.6.1. Oscillations of the upper (solid) and lower (dashed) masses

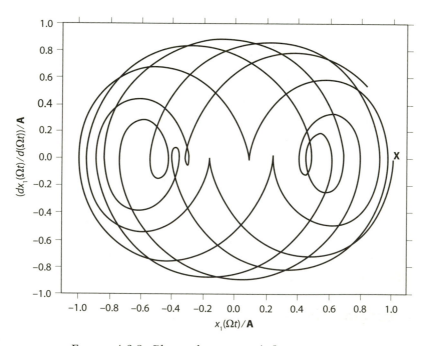

FIGURE 4.6.2. Phase-plane portrait for upper mass

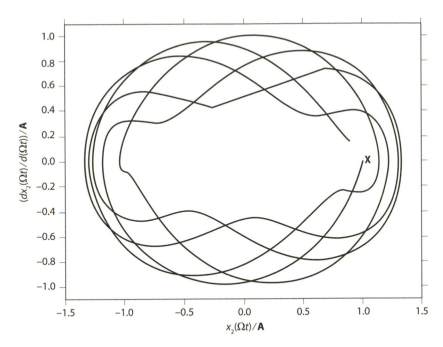

FIGURE 4.6.3. Phase-plane portrait for lower mass

plots of the oscillations of the individual masses, and of their phase-plane portraits, using a computer and MATLAB, however, is not at all difficult. It is, in fact, easy. What I'll do in this final section of the chapter is show you how to set up the MATLAB code for the $N = 2$ case, demonstrate that this code does indeed replicate our analytical solution from the previous section, and then show you how the $N = 2$ code *easily* extends to greater values of N (I'll use $N = 5$) for which the analytical Laplace transform solution would be *extremely* difficult.

Taking $\dfrac{k}{m} = 1$ (that is, $\Omega = 1$) and $N = 2$, our starting differential equations are (4.5.7) and (4.5.9), which I'll repeat here:

$$\frac{d^2x_1}{dt^2} = -2x_1 + x_2 \tag{4.7.1}$$

and

$$\frac{d^2x_2}{dt^2} = x_1 - x_2. \tag{4.7.2}$$

(Taking $\Omega = 1$ is really not a loss of generality, since all our analytical work in the previous section was normalized with respect to Ω, and to A, too, anyway.) Now, as usual with a MATLAB differential equation solver, we have to rewrite (4.7.1) and (4.7.2) in the form of first-order differential equations. So, define

$$u_1 = x_1 \tag{4.7.3}$$

and

$$u_2 = \frac{dx_1}{dt} = \frac{du_1}{dt}, \tag{4.7.4}$$

and so

$$\frac{du_2}{dt} = \frac{d^2x_1}{dt^2} = -2x_1 + x_2.$$

Also, define

$$u_3 = x_2 \tag{4.7.5}$$

and

$$u_4 = \frac{dx_2}{dt} = \frac{du_3}{dt}, \tag{4.7.6}$$

and so

$$\frac{du_4}{dt} = \frac{d^2x_2}{dt^2} = x_1 - x_2.$$

Thus, in summary we have

$$\frac{du_1}{dt} = u_2 = \frac{dx_1}{dt}, \tag{4.7.7a}$$

$$\frac{du_2}{dt} = -2u_1 + u_3 = \frac{d^2x_1}{dt^2}, \tag{4.7.7b}$$

$$\frac{du_3}{dt} = u_4 = \frac{dx_2}{dt}, \tag{4.7.7c}$$

$$\frac{du_4}{dt} = u_1 - u_3 = \frac{d^2x_2}{dt^2}. \tag{4.7.7d}$$

We can now write the MATLAB code **mass2.m** using our earlier **osc.m** from Section 4.2 as a guide. (As before, I have not included the routine labeling commands.) To understand what the code is doing, all you need to keep in mind is that the u matrix has four columns: $u(:, 1) = x_1$ (see (4.7.3)), $u(:, 2) = \dfrac{dx_1}{dt}$ (see (4.7.4)), $u(:, 3) = x_2$ (see (4.7.5)), and $u(:, 4) = \dfrac{dx_2}{dt}$ (see (4.7.6)). The nested function *xderivative* calculates the required derivatives to construct the u matrix, using the four equations of (4.7.7). To start the creation of the u matrix, that is, to form the first row of that matrix, *uzero* sets that row's values to $x_1(0) = 1$, $\left.\dfrac{dx_1}{dt}\right|_{t=0} = 0, x_2(0) = 1$, and $\left.\dfrac{dx_2}{dt}\right|_{t=0} = 0$, just as we did in the analytical Laplace transform solution of the previous section.

---

mass2.m

```
function mass2
options=odeset('AbsTol',1e-8,'RelTol',1e-5);
tspan=[0 50];
uzero=[1;0;1;0];
[ t,u]=ode45(@hanging,tspan,uzero,options);
figure(1)
plot(t,u(:,1),'-k'),hold on,plot(t,u(:,3),'-k')
figure(2)
plot(u(:,1),u(:,2),'-k')
figure(3)
plot(u(:,3),u(:,4),'-k')
        function xderivative=hanging(t,u)
        xderivative=[u(2);-2*u(1) +u(3);u(4);u(1)-u(3)];
        end
end
```

When run, **mass2.m** created Figures 4.7.1, 4.7.2, and 4.7.3, which are (to my eye, at least) *identical* to Figures 4.6.1, 4.6.2, and 4.6.3, respectively.

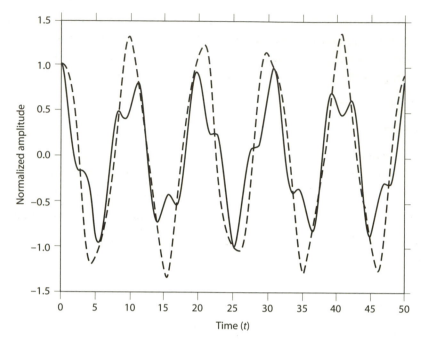

FIGURE 4.7.1. Oscillations of the upper (solid) and lower (dashed) masses

For $N = 5$ hanging masses, we simply repeat this process. Our defining differential equations are

$$\frac{d^2x_1}{dt^2} = x_2 - 2x_1, \quad \frac{d^2x_2}{dt^2} = x_3 + x_1 - 2x_2,$$

$$\frac{d^2x_3}{dt^2} = x_4 + x_2 - 2x_3, \quad \frac{d^2x_4}{dt^2} = x_5 + x_3 - 2x_4, \quad \frac{d^2x_5}{dt^2} = x_4 - x_5$$

that, with the definitions

$$u_1 = x_1, \quad u_2 = \frac{dx_1}{dt}, \quad u_3 = x_2, \quad u_4 = \frac{dx_2}{dt}, \quad u_5 = x_3,$$

$$u_6 = \frac{dx_3}{dt}, \quad u_7 = x_4, \quad u_8 = \frac{dx_4}{dt}, \quad u_9 = x_5, \quad u_{10} = \frac{dx_5}{dt},$$

result in the derivative equations

$$\frac{dx_1}{dt} = u_2, \quad \frac{d^2x_1}{dt^2} = u_3 - 2u_1, \quad \frac{dx_2}{dt} = u_4, \quad \frac{d^2x_2}{dt^2} = u_5 + u_1 - 2u_3,$$

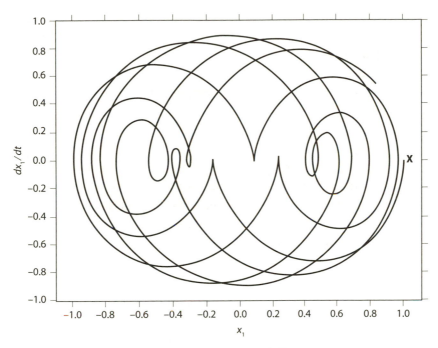

FIGURE 4.7.2. Phase-plane portrait for upper mass

$$\frac{dx_3}{dt} = u_6, \qquad \frac{d^2x_3}{dt^2} = u_7 + u_3 - 2u_5, \qquad \frac{dx_4}{dt} = u_8,$$

$$\frac{d^2x_4}{dt^2} = u_9 + u_5 - 2u_7, \qquad \frac{dx_5}{dt} = u_{10}, \qquad \frac{d^2x_5}{dt^2} = u_7 - u_9.$$

The MATLAB code **mass5.m** then creates Figure 4.7.4 (the oscillation amplitude versus time for the uppermost mass) and Figure 4.7.5 (the phase-plane portrait for that mass). (The triple dots in two of the lines of the nested function *xderivative* is the "line continuation" notation for MATLAB, forming one very long line.) As in earlier phase plane plots, an **X** marks the $t = 0$ instant in Figure 4.7.5. How to generate plots for the other four masses should now be obvious.

---

mass5.m

```
function mass5
options=odeset('AbsTol',1e-8,'RelTol',1e-5);
```

(Continued)

(Continued)

```
tspan=[0 50];
uzero=[1;0;1;0;1;0;1;0;1;0];
[ t,u]=ode45(@hanging,tspan,uzero,options);
figure(1)
plot(t,u(:,1),'-k')
figure(2)
plot(u(:,1),u(:,2),'-k')
        function xderivative=hanging(t,u)
        xderivative=[u(2);-2*u(1)+u(3);u(4);u(5)+u(1)-2*u(3);...
            u(6);u(7)+u(3)-2*u(5);u(8);u(9)+u(5)-2*u(7);...
            u(10);u(7)-u(9)];
        end
end
```

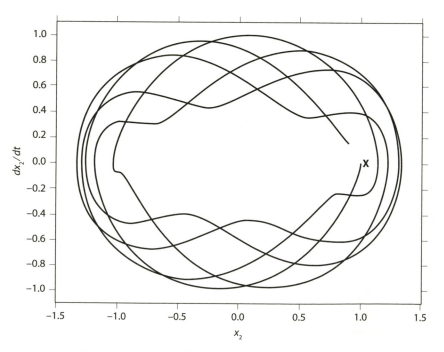

FIGURE 4.7.3. Phase-plane portrait for lower mass

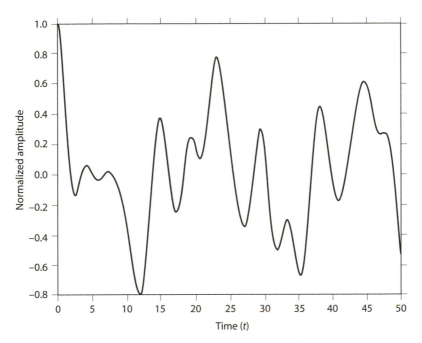

FIGURE 4.7.4. Oscillation of upper-most mass for $N = 5$

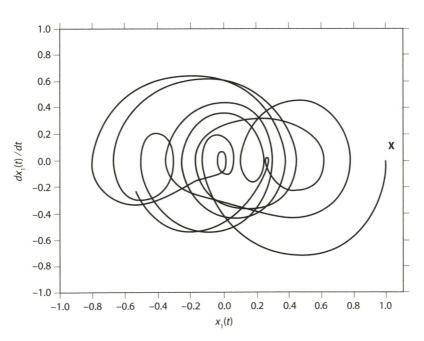

FIGURE 4.7.5. Phase plane portrait of upper-most mass for $N = 5$

Now, in conclusion, three comments. First, I believe the question of the nonperiodicity of the oscillations of the masses for the $N > 2$ case is still open. Proving the nonperiodicity for the $N = 2$ case (see note 8 again) was pretty straightforward, and it seems incredibly unlikely that the oscillations for *any* $N > 2$ case, for even a single one of the masses hanging in a chain, would be periodic, but I have not seen a *proof* of that. If a reader discovers such a proof, or comes across one in the published literature, please let me know!

Second, this nonperiodicity of the oscillations of a hanging chain revealed itself in the phase-plane portraits as trajectories that curve, twist, loop, and bend every which way. It can be shown, in fact, that the phase-plane portraits of nonperiodic oscillations are examples of what mathematicians call *space-filling curves*. What that means is, given any point inside the region encompassed by the furthest wanderings of the trajectory, the phase-plane trajectory eventually passes arbitrarily close to the given point. "Arbitrarily close" means that one can construct a tiny circle centered on the point, a circle with a vanishingly small radius, and the phase-plane trajectory will pass through that circle an *infinite* number of times during the interval $0 \leq t < \infty$![9] As a suggestive hint to this space-filling property, Figure 4.7.6 repeats Figure 4.7.2, the phase-plane portrait of the upper mass in the $N = 2$ case, except now the time duration has been increased by a factor of ten; that is, I've replaced the third line in **mass2.m** with *tspan=[0 500]*.

Third, so surprisingly odd and curious are space-filling curves that one might be tempted to call them *chaotic*. But that is not so. The signature feature of chaotic behavior is *extreme sensitivity to initial conditions* (take a look again at the second opening quotation to this chapter).[10] The phase-plane curves of our nonperiodic oscillations are, however, *not* sensitive to initial conditions, and this is easy to experimentally explore with our MATLAB computer codes. In Table 4.7.1, for example, I've listed the terminal values for the state variables x_1 and $\dfrac{dx_1}{dt}$ in the $N = 2$ case for two sets of initial conditions (the original *uzero=[1;0;1;0]* and *uzero* perturbed by 10% to *uzero=[0.9;0;0.9;0]*), for *tspan=[0 50]* and *tspan=[0 500]*.

Table 4.7.2 has the same information for x_2 and $\dfrac{dx_2}{dt}$. As you can see from the tables, with a 10% variation in the initial conditions there is

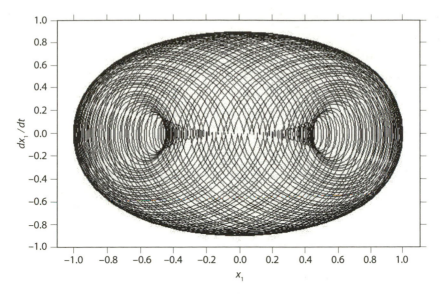

FIGURE 4.7.6. Does this suggest a space-filling curve to you?

TABLE 4.7.1

uzero	$x_1(50)$	$x_1(500)$	$\dfrac{dx_1}{dt}\Big|_{t=50}$	$\dfrac{dx_1}{dt}\Big|_{t=500}$
[1;0;1;0]	0.8226	0.3174	0.5343	0.0396
[0.9;0;0.9;0]	0.7439	0.2857	0.4809	0.0356

TABLE 4.7.2

uzero	$x_2(50)$	$x_2(500)$	$\dfrac{dx_2}{dt}\Big|_{t=50}$	$\dfrac{dx_2}{dt}\Big|_{t=500}$
[1;0;1;0]	0.8979	0.4782	0.1616	−0.9335
[0.9;0;0.9;0]	0.8081	0.4303	0.1455	−0.8401

a corresponding 10% variation in the ending values when we compute out to $t = 50$, and we still see the same 10% variation in the ending values when we go out ten times farther in time, to $t = 500$. With a chaotic phenomenon, on the other hand, we'd see the ending values of the state variables wildly *different* with a 10% change in the initial conditions.

Later in this book we will encounter chaotic behavior, and in a rather unexpected place, Newton's gravitational equations when calculating planetary orbits! But that's for later.

4.8 Challenge Problems

CP 4.1. Consider the two time functions $x(t)$ and $y(t)$ that are related through the linear, coupled, first-order differential equations

$$\frac{dx}{dt} = ax - by, \quad \frac{dy}{dt} = cx - ay,$$

where a, b, and c are constants. Show that both $x(t)$ and $y(t)$ are sinusoidal in time, with the same frequency, *if* a, b, and c satisfy a certain inequality. Also, what happens when the inequality is violated?

CP 4.2. In Section 4.3 it is stated that a *closed* phase-plane portrait means *periodic* behavior. As Figures 4.6.2 and 4.6.3 (and 4.7.2 and 4.7.3) show, however, the hanging masses problem for the $N = 2$ and $N = 5$ cases have phase-plane portraits that self-intersect at numerous places (that is, they appear to close), and yet we know the oscillations in those cases are not periodic. Explain why this apparent contradiction actually isn't one.

4.9 Notes and References

1. Here's a quick way to see this. Suppose that we have, for some non-negative constant R,

$$\frac{d^2x}{dt^2} + R\frac{dx}{dt} + \alpha x = 0, \quad R \geq 0,$$

rather than (4.1.1). This equation obviously does reduce to (4.1.1), however, if $R = 0$. Multiplying through by $\frac{dx}{dt}$, we have

$$\frac{dx}{dt} \cdot \frac{d^2x}{dt^2} + R\left(\frac{dx}{dt}\right)^2 + \alpha x\frac{dx}{dt} = 0,$$

or

$$\frac{d}{dt}\left[\frac{1}{2}\left(\frac{dx}{dt}\right)^2 + \frac{1}{2}\alpha x^2\right] + R\left(\frac{dx}{dt}\right)^2 = 0.$$

The first and second terms in the square brackets on the left are immediately identified with the kinetic energy of a moving mass and the potential energy of a spring, respectively. That is, if $E(t)$ is the total system energy, then

$$\frac{dE(t)}{dt} + R\left(\frac{dx}{dt}\right)^2 = 0,$$

or

$$\frac{dE(t)}{dt} = -R\left(\frac{dx}{dt}\right)^2,$$

and we see that the rate at which system energy is lost (notice that $\frac{dE(t)}{dt} < 0$ since R is non-negative) is proportional to the square (and so always non-negative) of the first derivative. But *if* $R = 0$ (and so the first derivative term vanishes), *then* $\frac{dE(t)}{dt} = 0$, and our system conserves energy.

2. The relationship $y = x^2$ has a hidden assumption built into it, one all too easy to miss. It can be true only in a specific set of units! If it holds when distance is measured in meters, for example, it fails when distance is measured in feet. Try it and see. I'm using the mks system of units, and so distances are measured in meters.

3. Of course, if you are using non-MATLAB software, you won't find my discussion very helpful, but also of course I have to use *some* software to generate solutions, and I do have MATLAB 7. If you use something else, you can at least check your solutions against mine (independent of software, the solutions had better be the same!). Now, if you do have MATLAB 7, then in my opinion, *the* book to have handy is by Desmond J. Higham and Nicholas J. Higham, *MATLAB Guide*, 2nd ed. (SIAM, 2005), which is stuffed with interesting programming examples, neat MATLAB tricks, and many amusing (as well as informative) musings.

4. The differential equation solver *ode45* uses an adaptive Runge-Kutta algorithm (named after the German mathematicians Carle Runge (1856–1927) and Martin Kutta (1867–1944)) that adjusts the spacing between adjacent time instants so that, as the solution changes more rapidly, that spacing decreases, and vice versa. The t vector and u matrix row counts also depend on the accuracy with which we ask MATLAB to calculate a solution. For example, when

the values in *options* were each made even smaller than they are in **osc.m** by a factor of ten, the row counts increased from 113 to 177.

5. The normal modes solution for the hanging masses problem can be found in the paper (which was the inspiration for this chapter) by Kenneth S. Mendelson and Frank G. Karioris, "Chaoticlike Motion of a *Linear* Dynamical System" (*American Journal of Physics*, March 1991, pp. 221–224).

6. If $f(t)$ is a time function such that $f(t) = 0$ for $t < 0$, then the *Laplace transform* of $f(t)$, written as $\pounds\{f(t)\}$, is defined as

$$\pounds\{f(t)\} \triangleq \int_0^\infty e^{-st}f(t)dt = F(s), \tag{1}$$

where s is the *transform variable* (s is complex, usually written as $\sigma + i\omega$, $i = \sqrt{-1}$ and $\sigma > 0$ to ensure convergence of the integral). It doesn't really matter if $f(t) = 0$ for $t < 0$, as the defining integral ignores what $f(t)$ is doing for negative time, but the restriction does give a unique, one-to-one correspondence between $f(t)$ and $F(s)$. From a "practical" point of view, any $f(t)$ that represents a physical entity (a voltage or a force, for example) in any system we could actually build *is* zero if we go back far enough in time (such as before the system is built!), and that time shift defines the moment we will call $t = 0$. What makes the definition in (1) so useful is the following property of the transform:

$$\pounds\left\{\frac{df(t)}{dt}\right\} = \int_0^\infty e^{-st}\frac{df(t)}{dt}dt = sF(s) - f(0). \tag{2}$$

We can establish (2) with integration by parts. That is, in the traditional notation used in all calculus texts,

$$\int_0^\infty u\,dv = (uv)\,|_0^\infty - \int_0^\infty v\,du,$$

where I'll write $dv = \dfrac{df(t)}{dt}dt = df$ (and so $v = f$) and $u = e^{-st}$ (and so $du = -se^{-st}dt$). Thus,

$$\pounds\left\{\frac{df(t)}{dt}\right\} = \{f(t)e^{-st}\}\,|_0^\infty - \int_0^\infty f(t)\{-se^{-st}\}dt = \lim_{t\to\infty}\{f(t)e^{-st} - f(0)\}$$

$$+ s\int_0^\infty e^{-st}f(t)dt = -f(0) + sF(s),$$

where I've assumed that $\lim_{t\to\infty} f(t)\,e^{-st} = \lim_{t\to\infty} f(t)\,e^{-(\sigma+i\omega)t} = 0$. Since $\sigma > 0$, and for any physical system we can actually build it is not unreasonable to expect that $f(t)$ will always be *finite*, then our assumption is not outrageous. So, as claimed, we have (2). You can repeat this entire argument to show that

$$\pounds\left\{\frac{d^2f(t)}{dt^2}\right\} = s^2F(s) - sf(0) - \frac{df}{dt}\bigg|_{t=0}. \tag{3}$$

One other fundamental property of the Laplace transform, essential to transforming linear, constant coefficient differential equations into algebraic equations, is its *linearity*. That is, if $g(t)$ and $f(t)$ are two time functions, then, with c_1 and c_2 as arbitrary constants,

$$\pounds\{c_1 g(t) + c_2 f(t)\} = c_1 G(s) + c_2 F(s), \tag{4}$$

and (4) follows immediately from (1) because the integral of a sum is the sum of the integrals. By convention, we'll use lower case letters to denote time functions and uppercase letters for the associated transform, and we'll write the resulting *transform pair* as $f(t) \longleftrightarrow F(s)$ (or $F(s) \longleftrightarrow f(t)$).

7. If we write

$$F(s) = \frac{s}{(s^2 + \omega_1^2)(s^2 + \omega_2^2)} \longleftrightarrow \frac{\cos(\omega_1 t) - \cos(\omega_2 t)}{\omega_2^2 - \omega_1^2} = f(t),$$

then we obviously have $f(0) = 0$, and so, using (2) from the previous note, it follows that

$$\pounds\left\{\frac{df(t)}{dt}\right\} = \pounds\left\{\frac{-\omega_1 \sin(\omega_1 t) + \omega_2 \sin(\omega_2 t)}{\omega_2^2 - \omega_1^2}\right\} = sF(s).$$

Thus, if we write $g(t) = \dfrac{df(t)}{dt}$, then $g(t) \longleftrightarrow G(s) = sF(s)$. Now, $g(0) = 0$, and so, differentiating one more time,

$$\pounds\left\{\frac{dg(t)}{dt}\right\} = sG(s) = \pounds\left\{\frac{-\omega_1^2 \cos(\omega_1 t) + \omega_2^2 \cos(\omega_2 t)}{\omega_2^2 - \omega_1^2}\right\} = s^2 F(s)$$

$$= \frac{s^3}{(s^2 + \omega_1^2)(s^2 + \omega_2^2)},$$

as claimed.

8. The irrationality of $\dfrac{\omega_2}{\omega_1}$ is easily demonstrated by writing

$$\frac{\omega_2}{\omega_1} = \sqrt{\frac{3 + \sqrt{5}}{3 - \sqrt{5}}} = \sqrt{\frac{(3 + \sqrt{5})(3 + \sqrt{5})}{(3 - \sqrt{5})(3 + \sqrt{5})}} = \sqrt{\frac{(3 + \sqrt{5})^2}{9 - 5}}$$

$$= \sqrt{\frac{(3 + \sqrt{5})^2}{4}} = \frac{1}{2}(3 + \sqrt{5}),$$

which is irrational because $\sqrt{5}$ is irrational (indeed, the square roots of *all* the non-square integers are irrational). The nonperiodicity of $A \sin(\omega_1 t) + B \sin(\omega_2 t)$ can be shown by assuming this sum *is* periodic with period T and then drawing a contradiction, as follows. The period of $A\sin(\omega_1 t)$ is $2\pi/\omega_1$, and the period of $B\sin(\omega_2 t)$ is $2\pi/\omega_2$. The period T of the sum must be an integer multiple of each of these periods, that is, $T = p2\pi/\omega_1 = q2\pi/\omega_2$. Thus, $\dfrac{\omega_2}{\omega_1} = \dfrac{q2\pi}{p2\pi} = \dfrac{q}{p}$, which says $\dfrac{\omega_2}{\omega_1}$ is rational, in contradiction to the given irrationality of $\dfrac{\omega_2}{\omega_1}$. Thus, our *assumption* that there exists a period T must be wrong. The same argument obviously holds for $A\sin(\omega_1 t) + B\cos(\omega_2 t)$ and $A\cos(\omega_1 t) + B\cos(\omega_2 t)$, as well. When $\dfrac{\omega_2}{\omega_1}$ is irrational we have nonperiodic oscillations, but still, the behavior of those oscillations does look "sort of" repetitive. The oscillations of the hanging chain belong, in fact, to a class of functions called *almost periodic*, a relatively new subject invented almost single-handedly in the early 1920s by the Danish mathematician Harald Bohr (1887–1951). Bohr was the younger brother of the 1922 Nobel Prize winner, physicist Niels Bohr (1885–1961).

9. The mathematics required to show the space-filling property of the phase-plane trajectory of a nonperiodic oscillation is surprisingly straightforward, depending mostly on an elementary use of continued fractions. See, for example, Harold T. Davis, *Introduction to Nonlinear Differential and Integral Equations* (New York: Dover, 1962, pp. 284–291). The theory of continued fractions is not a topic routinely studied by physicists or engineers, but continued fractions do occur in numerous surprising applications in physics and mathematics. You'll recall how useful they were back in Chapter 2. As another example of their value, consider the equation $\alpha = 1 + \dfrac{1}{1 + \alpha}$. Multiplying through by $1 + \alpha$, we get $\alpha + \alpha^2 = 1 + \alpha + 1 = \alpha + 2$, or $\alpha^2 = 2$, and so $\alpha = \sqrt{2}$. Putting this back into the original equation, we have $\sqrt{2} = 1 + \dfrac{1}{1 + \sqrt{2}} = 1 + \dfrac{1}{1 + \left(1 + \frac{1}{1+\sqrt{2}}\right)} = 1 + \dfrac{1}{2 + \frac{1}{1+\sqrt{2}}}$. If you repeat this self-substitution over and over, you arrive at a beautiful continued fraction expression for $\sqrt{2}$, discovered in 1572 by the Italian engineer-architect Rafael Bombelli (1526–1573);

$$\sqrt{2} = 1 + \cfrac{1}{2 + \cfrac{1}{2 + \cfrac{1}{2 + \cfrac{1}{2 + \cfrac{1}{2 + \cdots}}}}}.$$

This expression provides an amazingly simple way to calculate the numerical value of $\sqrt{2}$. We simply truncate the infinite continued fraction at any finite "level"; the "deeper" we go in the fraction, the more accurate will be our result. (The "number of levels" is the number of division bars we retain.) For example, if we truncate at level 4, we have $\sqrt{2} \approx 1 + \cfrac{1}{2 + \cfrac{1}{2 + \cfrac{1}{2 + \frac{1}{2}}}}$. A MATLAB code that first evaluates just the staircase of divisions from the bottom up, and then adds one, for any specified level ≥ 2, is the following (where I've set the level to 4):

```
f=2.5;level=4;
for loop=1:level-2
    f=2+1/f;
end
root2=1+1/f
```

When run, the result was $root2 = 1.413793103\ldots$, compared to the true value of $1.414213562\ldots$. The continued fraction converges quite rapidly; when run at level 10, for example, the result was $root2 = 1.414213551\ldots$. Just to be complete here, there is an entirely different way to calculate $\sqrt{2}$ from the continued fraction expansion that goes in the opposite direction, that is, from the top down. If we define $f_k = \cfrac{1}{2+\cfrac{1}{2+\cfrac{1}{2+\cfrac{1}{2+\cdots}}}}$, where k is the number of division bars retained, then we can write the *difference equation* $f_k = \dfrac{1}{1+2f_{k-1}}$ and so $\sqrt{2} = \lim_{k\to\infty}(1+f_k)$, where $f_1 = \dfrac{1}{2}$ (we retain one division bar). A MATLAB code that implements this scheme is

```
f=0.5;level=4;
for loop=1:level-1
    f=1/(2+f);
end
root2=1+f
```

When run at level 4 the result was $root2 = 1.416666666\ldots$, while at level 10 the result was $root2 = 1.414213551\ldots$.

10. Two classic examples of chaos are the pool table and the weather. In the case of the pool table (billiards), the slightest variation in the direction at which the ball comes off the stick makes it geometrically impossible to predict where the ball will be after only a very few collisions with, and rebounds off, the edges of the pool table. And that is with the ideal assumptions of a perfectly smooth, frictionless tabletop and zero ball spin. Put *those* variables into the picture and the impossibility of the long-term prediction of "where the ball will be" becomes an obvious certainty! See, for example, Charles Ruhla, *The Physics of Chance* (Oxford: Oxford University Press, 1992, esp. Chapter 6, "Poincaré, or Deterministic Chaos (Sensitivity to Initial Conditions)," pp. 126–149). The

differential equations of meteorology are extraordinarily sensitive to initial conditions, and that explains the failure, even with today's superfast computers, to be able to accurately and reliably predict the weather out beyond a small number of days (three or four). The inputs to the weather equations are from a relatively small number of physical locations over the Earth's surface (with huge unrepresented areas, such as remote deserts and vast regions of the oceans), and the measurements we do have are not known with an accuracy better than a few percent. The evolution of faster and faster computers means only that we can more and more quickly, *but unsuccessfully,* predict the weather farther and farther out in time (a dubious achievement), and it is my opinion that the inherent mathematical chaos of the weather equations means we will never, with any confidence, be able to predict next week's weather, much less next month's. This idea has been around for a long time, too, predating by decades both Edward Lorentz's now famous butterfly (the title of a 1972 talk he gave to the American Association for the Advancement of Science was "Does the Flap of a Butterfly's Wings in Brazil Set off a Tornado in Texas?") and the 1941 Chinese sneeze in Shen-si that opens this chapter. In a review of a book written by the French physicist and philosopher Pierre Duhem (1861–1914), *Traité elémentaire de méchanique fondée sur la thermodynamique* (1897), the Lehigh University physics professor William Suddards Franklin (1863–1930) wrote (*Physical Review*, March 1, 1898, pp. 170–175),

> Long range detailed weather prediction is, therefore, impossible, and the only detailed prediction which is possible is the inference of the ultimate trend and character of a storm from observations of its early stages; and the accuracy of this prediction is subject to the condition that the flight of a grasshopper in Montana may turn a storm aside from Philadelphia to New York!

For an interesting history on these matters, see the paper by Robert C. Hilborn, "Sea Gulls, Butterflies, and Grasshoppers: A Brief History of the Butterfly Effect in Nonlinear Dynamics" (*American Journal of Physics*, April 2004, pp. 425–427).

5
THE THREE-BODY PROBLEM AND COMPUTERS

The three-body problem arises in many different contexts in nature.... It is an old problem and logically follows from the two-body problem which was solved by Newton in his *Principia* in 1687. Newton also considered the three-body problem in connection with the motion of the Moon under the influences of the Sun and the Earth, the consequences of which included a headache.
   —Mauri Valtonen and Hannu Karttunen (see note 29)

Newton next turned to the problem of describing the orbit of the moon. This is a harder problem [than describing the orbit of Mars] since the first approximation should be a three-body problem—the earth, moon, and sun. The problem he encountered caused him to remark to the astronomer John Machin that *"... his head never ached but with his studies on the moon."*
   —Kenneth R. Meyer (see note 32)

To consider simultaneously all these causes of [planetary] motion and to define these motions by exact laws admitting of easy calculation exceeds, if I am not mistaken, the force of any human mind.
   —Isaac Newton writing (in late 1684) in his *De Motu Corporum in Gyrum* (*On the Motion of Revolving Bodies*), the precursor to the *Principia*, 260 years before the invention of the electronic digital computer, a development that would surely have caused Newton to rethink his words

5.1 Newton's Theory of Gravity

Isaac Newton (1642–1727) is a man who almost certainly needs no introduction to readers of this book. Newton's *Principia* transported gravity from a topic about which just about everybody had an opinion, based on how they personally imagined God had constructed the world, to one unambiguously described in dispassionate mathematics.[1] This momentous intellectual achievement allowed some people (those possessing sufficient analytical ability) to actually *derive without speculation* the observable effects of gravity. This is not to say, however, that that is necessarily an easy thing to do. As long as one talks about the Sun and the Earth alone, or the Earth and the Moon alone, and in general about any two massive bodies alone, the mathematics is indeed doable. But as the opening quotation implies, as soon as we take the very next step, to considering systems that have three massive bodies, then matters generally seem to suddenly become difficult. Dramatically and extraordinarily difficult, in fact. The study of how three massive bodies gravitationally interact is called the three-body problem, and its general solution eluded even the genius of Newton.

But, of course, that didn't mean it was necessarily an *impossible* problem, just that it was a very, very difficult problem. Perhaps, it was hoped, with advancing technical knowledge and the appearance of an even greater intellect than Newton's, the three-body problem could be solved. That lure, of being the one that finally did what the great Newton could not, attracted some of the most brilliant analysts over the next 250 years. Such luminaries as the Swiss Leonhard Euler (1707–1783), the French Joseph Lagrange (1736–1813), and the German Carl Jacobi (1804–1851), for example, all tried their hands at it, but except for limited partial successes made under special, restrictive assumptions, the general solution of the three-body problem defeated every one of them.

Now, what does it mean to "solve" the three-body problem? In general, to completely specify a three-body problem we first have to give the size of each of the three masses. Second, we have to specify where in space (in some given coordinate system) each of the three bodies is at some specified instant of time (usually called $t = 0$), and since space is three-dimensional, it takes nine more numbers (three for each mass) to do that. And finally, at that same specific instant of time ($t = 0$), we have to give the velocity component along each space dimension, for each body, which means yet another nine more numbers. All together, then, to define the initial state of an arbitrary three-body problem requires twenty-one numbers. To solve the three-body problem means to completely describe the resulting motion of each body (its *orbit*) as a function of time ($t \geq 0$) and those twenty-one numbers. With this in mind, perhaps it's a bit easier to understand why the general three-body problem is so hard to solve in the traditional way of writing down formulas for the orbits.

What has lulled so many down the decades is that the underlying mathematical physics of the three-body problem doesn't actually seem that hard, at least not at first look. All that is involved are the following:

$$\mathbf{F} = m\mathbf{a}, \tag{5.1.1}$$

Newton's famous second law of motion, which relates the instantaneous force $\mathbf{F}$ (a vector, hence in bold) acting on a body with mass m to the instantaneous acceleration ($\mathbf{a}$, another vector) of the body,[2] and

$$|\mathbf{F}| = G\frac{m_1 m_2}{r^2}, \tag{5.1.2}$$

Newton's famous inverse-square law for the mutual gravitational force that two *point*[3] bodies (with masses m_1 and m_2) distance r apart exert on each other, where (in the physicist's mks system of units) $G = 6.67 \cdot 10^{-11} \frac{\text{meters}^3}{\text{kilograms·seconds}^2}$ is the so-called *universal constant of gravitation*; (5.1.2) comes with the caveat that while each body "feels" the same magnitude of force as does the other, as given by (5.1.2), those two forces are in opposite directions. That is, the force on m_1 due to m_2 is directed toward m_2, and the force on m_2 due to m_1 is directed toward m_1. Such a force, one that lies along the line connecting the two interacting bodies, is called a *central force*.[4] That's it! How hard can all that be? Well, *you'll see....*

It *is* hard to see just how astonishingly subtle the above three statements are, to see how complicated can be the behaviors that are built into them. So, before going any farther with the three-body problem, let me give you a dramatic, historical example of such a complicated behavior when we deal with just *two* bodies.[5] The Danish astronomer Tycho Brahe (1546–1601) for many years made careful observations of the motion of Mars across the night sky, and those observations were continued after his death by his German assistant Johannes Kepler (1571–1630). Kepler was a follower of the revolutionary claim of the Polish astronomer Nicolaus Copernicus (1473–1543) that all the planets of the Solar System orbit the Sun (the so-called *heliocentric view*), contrary to the longstanding *geocentric* view of the ancient (second century A.D.) Greek astronomer Ptolemy that the Sun and all the other planets orbit the Earth, each traveling along its own "perfect" circular path. What Kepler deduced from all the observational data available to him, however, was in direct contradiction to Ptolemy. Mars actually appears at times to stop its motion across the night sky and *to go backward*, that is, to reverse the direction of its progress and to exhibit what is called *retrograde motion*.

People were understandably puzzled by this. It can be explained as a purely *observational effect*, however, one that arises from looking at Mars from a *rotating reference system* (centered on the moving Earth), *if* one accepts Copernicus's view that it is the Sun, not the Earth, that is at the center of the Solar System. Here's how.

Suppose we have a planet with mass m, in a circular orbit of radius r around the Sun (which has mass M). Kepler established from his observational data that the planetary orbits are actually *elliptical*, but in fact the planetary ellipses are almost circular and for our purposes here it's a lot easier to use a circular orbit. The orbital speed is v, and the orbital period is T. The inwardly directed centripetal acceleration of the planet moving along its curved orbit is provided by the Sun's gravitational pull on the planet, and so we have

$$G\frac{Mm}{r^2} = \frac{mv^2}{r}. \tag{5.1.3}$$

Thus,

$$GM = v^2 r, \tag{5.1.4}$$

and since $T = 2\pi r/v$ and $v = \sqrt{GM/r}$, we have

$$T = \frac{2\pi r}{\sqrt{\frac{GM}{r}}} = \frac{2\pi r^{3/2}}{\sqrt{GM}},$$

or

$$T^2 = \frac{4\pi^2}{GM} r^3. \tag{5.1.5}$$

This is another of Kepler's laws (usually called the third law) of planetary motion.

Next, we do what all astrophysicists do at this point, to keep the numerical values in our equations "nice" (in the usual mks system of physicists the numbers are *not* nice, with G small and M huge)—we *scale* (5.1.5). That is, let's define the unit of distance as the Earth's orbital radius (the so-called *astronomical unit*, or AU, which is about 93 million miles), and let's also define the unit of time as Earth's orbital period (one year). So, $T = 1$ when $r = 1$, and therefore, in this system of units, we have

$$GM = 4\pi^2, \tag{5.1.6}$$

and so, as far as computing numerical values go, we have the wonderfully simple

$$T^2 = r^3. \tag{5.1.7}$$

In Figure 5.1.1 I've drawn the Earth and a more distant planet (that is, one of the so-called superior planets—Mars, Jupiter, Saturn, Neptune, Uranus, and, once upon a time, Pluto) in circular orbits around the Sun (which is at the origin of an x,y-coordinate system). At every instant of time the Sun, Earth, and the planet define the vertices of a vector triangle. The vector from the Sun to the Earth is $\mathbf{z}_E$, the vector from the Sun to the other planet is $\mathbf{z}_P$, and the vector from the Earth to the other planet is $\mathbf{z}_{EP}$. The first two vectors are called the *position vectors* of the Earth and of the other planet, respectively. I am making the assumption that the orbits of both the Earth and the other planet are in the same plane. This is not strictly true, but it is almost true. All of the other planets in the Solar System do, in fact, have their orbital planes inclined only slightly with respect to the Earth's. With the exception of Pluto (which has actually lost its planetary status), the various orbital plane inclinations do not exceed Mercury's 7°.

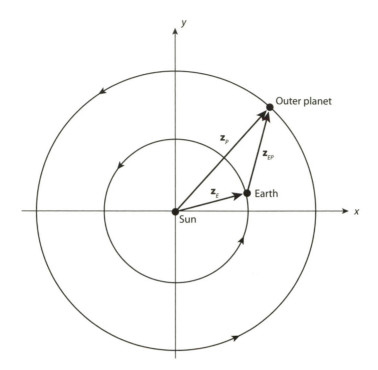

FIGURE 5.1.1. The Sun, Earth, and outer planet vector triangle

By elementary vector algebra we have

$$\mathbf{z}_P = \mathbf{z}_E + \mathbf{z}_{EP}. \tag{5.1.8}$$

We can represent $\mathbf{z}_E$ as a *rotating* vector by using complex exponentials, that is, we have (with $i = \sqrt{-1}$)

$$\mathbf{z}_E = e^{i2\pi t}, \tag{5.1.9}$$

which simply says that $|\mathbf{z}_E| = 1$ and that the Earth's position vector rotates once, through 2π radians, as time goes from $t = 0$ to $t = 1$ (which is Earth's orbital period). We can do the same thing for the other planet, too, recognizing of course that its rotating position vector spins once around, through 2π radians, as time varies through one of that planet's orbital periods, that is, from $t = 0$ to $t = T_p$, where, from (5.1.7), we have $r_p = |\mathbf{z}_P| = T_p^{2/3}$. That is,

$$\mathbf{z}_P = r_p e^{i2\pi(t/T_P)} = T_p^{2/3} e^{i2\pi(t/T_P)}. \tag{5.1.10}$$

Notice that at $t = 0$ the Sun, the Earth, and the other planet are colinear (all three are on the x-axis), and that this represents no loss in generality because that event is bound to occur at some time (infinitely often, in fact) and, at one of those instants, we'll simply define the time to be $t = 0$.

Combining (5.1.9) and (5.1.10) with (5.1.8), we have

$$\mathbf{z}_{EP} = \mathbf{z}_P - \mathbf{z}_E = T_p^{2/3} e^{i2\pi(t/T_P)} - e^{i2\pi t},$$

or, since $\mathbf{z}_{EP}$ is a complex quantity with real and imaginary parts, that is,

$$\mathbf{z}_{EP} = x(t) + iy(t),$$

we can write the location of the other planet *as seen from the Earth* as follows (using Euler's identity to expand the complex exponentials);

$$x(t) = T_p^{2/3} \cos\left(2\pi \frac{t}{T_P}\right) - \cos(2\pi t) \tag{5.1.11}$$

and

$$y(t) = T_p^{2/3} \sin\left(2\pi \frac{t}{T_P}\right) - \sin(2\pi t). \tag{5.1.12}$$

Figure 5.1.2 shows the observed motion resulting from (5.1.11) and (5.1.12) when the other planet is Mars (Mars has an orbital period of 687 Earth days), and Figure 5.1.3 does the same for Jupiter (which has an orbital period of $4,332$ Earth days) (the asterisk locates the Earth, and the cross is the start of the other planet's path at $t = 0$). In both plots we see the signature "looping" that gives the illusion of retrograde motion. The *precessing* of the orbits is due to the fact that the orbital periods of both Mars and Jupiter are a *noninteger* number of Earth's orbital period.

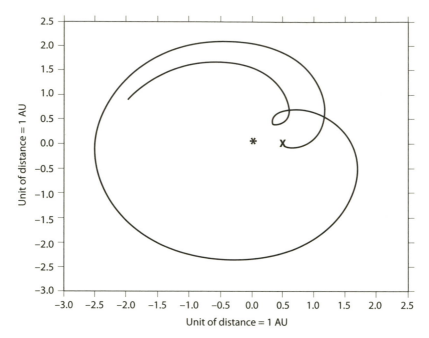

FIGURE 5.1.2. Retrograde motion of Mars

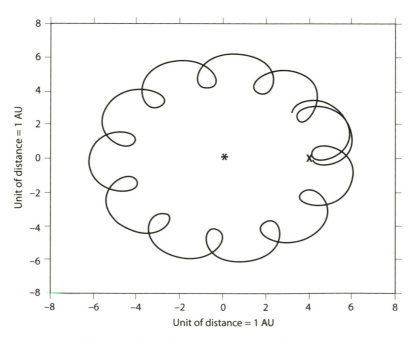

FIGURE 5.1.3. Retrograde motion of Jupiter

5.2 Newton's Two-Body Solution

My central goal in this chapter is to illustrate how computers can help illuminate what happens in the three-body problem, but it will be quite useful to have the analytical solution for the two-body problem available. That's because we can then use that solution to at least partially check our three-body computer code; after all, if we have a code that allows us to arbitrarily specify *three* masses m_1, m_2, and m_3, then the code's results had better be in agreement with Newton if we set (any) one of those three masses equal to zero. When Newton wrote the *Principia* he didn't have modern vector algebra and vector calculus as tools, but we do.[6] And with them we can derive many of his results quickly and almost without pain (not totally, but almost).

Imagine that we have two masses, M *fixed* at the origin of an x,y, z-coordinate system, and m free to move about in the x,y-plane of that coordinate system *with its velocity vector entirely in that plane*. For example, M could be the massive Sun, and m could be the Earth. As shown in Figure 5.2.1, the *position* and *velocity* vectors of m are $\mathbf{r}$ and $\mathbf{v}$, respectively, where I'll write

$$\mathbf{r} = \left[x, y, 0\right] \tag{5.2.1}$$

and

$$\mathbf{v} = \frac{d\mathbf{r}}{dt} = \left[\frac{dx}{dt}, \frac{dy}{dt}, 0\right]. \tag{5.2.2}$$

I'll justify that zero component of $\mathbf{r}$ (and so of $\mathbf{v}$, too) in just a moment. My convention here will be to write the magnitude of a (bold) vector as nonbold, that is, $r = |\mathbf{r}|$ and $v = |\mathbf{v}|$. It will also be convenient at times to speak of the location of m in terms of the polar coordinates r and θ (see Figure 5.2.1 again), where

$$\mathbf{r} = [r\cos(\theta), r\sin(\theta), 0]. \tag{5.2.3}$$

We can write Newton's second law of motion for m as

$$m\frac{d^2\mathbf{r}}{dt} = -\frac{GMm}{r^2}\left(\frac{\mathbf{r}}{r}\right)$$

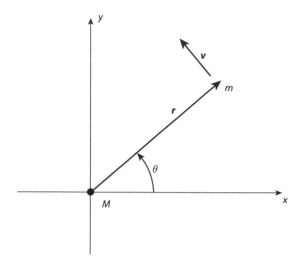

FIGURE 5.2.1. Two-body motion in a plane

where $\dfrac{\mathbf{r}}{r}$ is the *unit* vector from the origin to m, and so

$$\frac{d^2\mathbf{r}}{dt^2} = \frac{d\mathbf{v}}{dt} = -\frac{GM}{r^3}\mathbf{r}. \qquad (5.2.4)$$

Taking a vector cross-product of (5.2.4) gives

$$\mathbf{r} \times \frac{d\mathbf{v}}{dt} = -\frac{GM}{r^3}\mathbf{r} \times \mathbf{r} = 0, \qquad (5.2.5)$$

as any vector crossed with itself gives zero. Since

$$\frac{d}{dt}(\mathbf{r} \times \mathbf{v}) = \left\{ \mathbf{r} \times \frac{d\mathbf{v}}{dt} \right\} + \left\{ \frac{d\mathbf{r}}{dt} \times \mathbf{v} \right\},$$

and using the definition of $\mathbf{v}$ from (5.2.2) in the second term on the right, we have

$$\frac{d}{dt}(\mathbf{r} \times \mathbf{v}) = \left\{ \mathbf{r} \times \frac{d\mathbf{v}}{dt} \right\} + \left\{ \mathbf{v} \times \mathbf{v} \right\},$$

or, using (5.2.5) for the first term on the right, and again remembering that any vector crossed with itself gives zero, we have

$$\frac{d}{dt}(\mathbf{r} \times \mathbf{v}) = 0. \tag{5.2.6}$$

This last result says $\mathbf{r} \times \mathbf{v}$ must be a *constant* vector, which I'll simply call **c**, that is,

$$\mathbf{r} \times \mathbf{v} = \mathbf{c}. \tag{5.2.7}$$

This is, in fact, our first major theoretical result: the cross-product of the position and velocity vectors of m in Newtonian gravity is a constant. Physicists call $m\mathbf{c}$ the *angular momentum* of m, and since both vectors are time-varying and yet always give the same cross-product, then (5.2.7) is the mathematical statement of the *conservation of angular momentum* in Newtonian gravity. If we assume $\mathbf{c} \neq \mathbf{0}$, then (5.2.7) says that $\mathbf{r}$ and $\mathbf{v}$ are each perpendicular to $\mathbf{c}$, and so the motion of m (which we assumed to be in the x,y-plane at the start of this analysis) *stays* in the x,y-plane, and that explains the zero z-component of $\mathbf{r}$ in (5.2.3).[7] The angular momentum vector of m is always perpendicular to the x,y-plane.

Taking a vector dot product with (5.2.4) gives

$$\mathbf{v} \cdot \frac{d\mathbf{v}}{dt} = \mathbf{v} \cdot \left(-\frac{GM}{r^3}\mathbf{r}\right) = -\frac{GM}{r^3}\mathbf{v} \cdot \mathbf{r} = -\frac{GM}{r^3}\frac{d\mathbf{r}}{dt} \cdot \mathbf{r}.$$

But since, from (4) of note 6,

$$\frac{d\mathbf{r}}{dt} \cdot \mathbf{r} = r\frac{dr}{dt}$$

and

$$\mathbf{v} \cdot \frac{d\mathbf{v}}{dt} = v\frac{dv}{dt},$$

then

$$v\frac{dv}{dt} = -\frac{GM}{r^2}\left(\frac{dr}{dt}\right),$$

or

$$v\,dv = -GM\left(\frac{dr}{r^2}\right).$$ (5.2.8)

This is easily integrated to give (with W some constant)

$$\frac{1}{2}v^2 = \frac{GM}{r} + W.$$

Solving for W and then multiplying through by m, we have

$$Wm = \frac{1}{2}mv^2 - \frac{GMm}{r}.$$

Since m is a constant, Wm is also a constant, which I'll call E, and so

$$E = \frac{1}{2}mv^2 - \frac{GMm}{r},$$ (5.2.9)

and now you know why I used the literal E; (5.2.9) is the mathematical statement of the *conservation of total energy* in Newtonian gravity, where the first term on the right is the *kinetic* energy of m and the second term on the right (including the minus sign) is the *potential* energy of m.

The potential energy is a *shared* property of both M and m, not just of m, while the kinetic energy is entirely the property of m alone. Later, in the three-body problem, this will be important to keep in mind; the total gravitational potential energy of three interacting bodies is

$$G\left(-\frac{m_1m_2}{r_{12}} - \frac{m_1m_3}{r_{13}} - \frac{m_2m_3}{r_{23}}\right),$$

where $r_{ij} = r_{ji}$ is the distance between masses m_i and m_j, and *not* twice this. That is, one does not, for example, argue that m_1 has gravitational potential energy $-\dfrac{Gm_1m_2}{r_{12}}$ due to m_2, and m_2 has gravitational potential energy $-\dfrac{Gm_1m_2}{r_{12}}$ due to m_1. *Together*, the *total* gravitational potential energy of m_1 and m_2 is $-\dfrac{Gm_1m_2}{r_{12}}$. The same goes for the gravitational potential energies of the remaining two mass-pairs m_1 and m_3, and m_2 and m_3.

Since

$$\frac{d}{dt}(\mathbf{r} \cdot \mathbf{r}) = \frac{d(r^2)}{dt} = 2r\frac{dr}{dt},$$

and since

$$\frac{d}{dt}(\mathbf{r} \cdot \mathbf{r}) = 2\mathbf{r} \cdot \frac{d\mathbf{r}}{dt} = 2\mathbf{r} \cdot \mathbf{v},$$

then

$$r\frac{dr}{dt} = \mathbf{r} \cdot \mathbf{v}. \tag{5.2.10}$$

Differentiating (5.2.10), we have

$$\frac{d}{dt}\left(r\frac{dr}{dt}\right) = \frac{d}{dt}(\mathbf{r} \cdot \mathbf{v}),$$

and so, expanding both sides,

$$r\frac{d^2r}{dt^2} + \left(\frac{dr}{dt}\right)^2 = \mathbf{r} \cdot \frac{d\mathbf{v}}{dt} + \frac{d\mathbf{r}}{dt} \cdot \mathbf{v} = \mathbf{r} \cdot \frac{d\mathbf{v}}{dt} + \mathbf{v} \cdot \mathbf{v},$$

or

$$r\frac{d^2r}{dt^2} + \left(\frac{dr}{dt}\right)^2 = \mathbf{r} \cdot \frac{d\mathbf{v}}{dt} + v^2. \tag{5.2.11}$$

It is not hard to show (see Challenge Problem 5.3) that

$$v^2 = \left(\frac{dr}{dt}\right)^2 + \frac{c^2}{r^2},$$

where $c = |\mathbf{c}|$ is the magnitude of the angular momentum vector $\mathbf{c}$ defined in (5.2.7). Thus, (5.2.11) becomes

$$r\frac{d^2r}{dt^2} + \left(\frac{dr}{dt}\right)^2 = \mathbf{r} \cdot \frac{d\mathbf{v}}{dt} + \left(\frac{dr}{dt}\right)^2 + \frac{c^2}{r^2},$$

or

$$r\frac{d^2r}{dt^2} = \mathbf{r} \cdot \frac{d\mathbf{v}}{dt} + \frac{c^2}{r^2}.$$

(5.2.12)

Using (5.2.4) for $\dfrac{d\mathbf{v}}{dt}$, (5.2.12) becomes

$$r\frac{d^2r}{dt^2} = \mathbf{r} \cdot \left(-\frac{GM}{r^3}\mathbf{r} \right) + \frac{c^2}{r^2} = -\frac{GM}{r^3}\mathbf{r} \cdot \mathbf{r} + \frac{c^2}{r^2} = -\frac{GM}{r} + \frac{c^2}{r^2},$$

or

$$\frac{d^2r}{dt^2} - \frac{c^2}{r^3} = -\frac{GM}{r^2}.$$

(5.2.13)

With (5.2.13) we can now solve for the *shape* of m's orbit about M. That is, we'll find r as a function of θ (which is not the same as finding r as a function of time, which is a *much* more difficult problem).[8] To do this, we'll eliminate time from (5.2.13). Returning to (5.2.7)—the statement of the conservation of angular momentum—and using the polar coordinates for $\mathbf{r}$ given in (5.2.3), that is

$$\mathbf{r} = [r\cos(\theta), r\sin(\theta), 0],$$

we have

$$\mathbf{v} = \frac{d\mathbf{r}}{dt} = \left[-r\sin(\theta)\frac{d\theta}{dt} + \frac{dr}{dt}\cos(\theta), r\cos(\theta)\frac{d\theta}{dt} + \frac{dr}{dt}\sin(\theta), 0 \right].$$

We'll use this $\mathbf{v}$ to expand the left-hand side of (5.2.7), that is, $\mathbf{r} \times \mathbf{v}$, using the determinant formula for the vector cross-product (see (2) of note 6):

$$\mathbf{r} \times \mathbf{v} = \begin{vmatrix} \mathbf{i} & \mathbf{j} & \mathbf{k} \\ r\cos(\theta) & r\sin(\theta) & 0 \\ \left\{-r\sin(\theta)\dfrac{d\theta}{dt} + \dfrac{dr}{dt}\cos(\theta)\right\} & \left\{r\cos(\theta)\dfrac{d\theta}{dt} + \dfrac{dr}{dt}\sin(\theta)\right\} & 0 \end{vmatrix}$$

$$= \left\{ r^2\cos^2(\theta)\frac{d\theta}{dt} + r\frac{dr}{dt}\cos(\theta)\sin(\theta) + r^2\sin^2(\theta)\frac{d\theta}{dt} \right.$$

$$\left. - r\frac{dr}{dt}\cos(\theta)\sin(\theta) \right\} \mathbf{k},$$

or

$$\mathbf{r} \times \mathbf{v} = r^2 \frac{d\theta}{dt} \mathbf{k}. \tag{5.2.14}$$

As you'll recall from earlier (right after we derived (5.2.7)), we already know that $\mathbf{r} \times \mathbf{v} = \mathbf{c}$, a constant vector (the angular momentum vector) that is *perpendicular* to the x,y-plane. That is, $\mathbf{c} = c\mathbf{k}$. So, (5.2.14) says that

$$r^2 \frac{d\theta}{dt} = c, \tag{5.2.15}$$

where c is the constant magnitude of the angular momentum vector.[9]

Now, let's make the change of variable

$$r = \frac{1}{\rho}. \tag{5.2.16}$$

Then, using the chain rule from calculus,

$$\frac{dr}{dt} = -\frac{1}{\rho^2} \frac{d\rho}{dt} = -\frac{1}{\rho^2} \left(\frac{d\rho}{d\theta} \right) \left(\frac{d\theta}{dt} \right),$$

or, using (5.2.15),

$$\frac{dr}{dt} = -\frac{1}{\rho^2} \left(\frac{d\rho}{d\theta} \right) \frac{c}{r^2} = -r^2 \left(\frac{d\rho}{d\theta} \right) \frac{c}{r^2},$$

and so

$$\frac{dr}{dt} = -c \frac{d\rho}{d\theta}. \tag{5.2.17}$$

Differentiating (5.2.17) with respect to t, and again using the chain rule,

$$\frac{d^2 r}{dt^2} = -c \frac{d}{dt} \left(\frac{d\rho}{d\theta} \right) = -c \frac{d}{d\theta} \left(\frac{d\rho}{d\theta} \right) \frac{d\theta}{dt},$$

or

$$\frac{d^2r}{dt^2} = -c\frac{d^2\rho}{d\theta^2}\left(\frac{d\theta}{dt}\right),$$

(5.2.18)

or, using (5.2.15) for the $\dfrac{d\theta}{dt}$ factor in (5.2.18),

$$\frac{d^2r}{dt^2} = -c\frac{d^2\rho}{d\theta^2}\left(\frac{c}{r^2}\right) = -\frac{c^2}{r^2}\left(\frac{d^2\rho}{d\theta^2}\right),$$

or, remembering (5.2.16),

$$\frac{d^2r}{dt^2} = -c^2\rho^2\frac{d^2\rho}{d\theta^2}.$$

(5.2.19)

If we then use (5.2.19) in (5.2.13), along with (5.2.16), we arrive at

$$-c^2\rho^2\frac{d^2\rho}{d\theta^2} - c^2\rho^3 = -GM\rho^2,$$

or, at last, we have the astonishingly familiar (I hope!)

$$\frac{d^2\rho}{d\theta^2} + \rho = \frac{GM}{c^2}.$$

(5.2.20)

I say "astonishingly familiar" because (5.2.20) is our old friend, the differential equation for the harmonic oscillator that we discussed in Chapter 4. Back there we found the solution to be sinusoidal, and you should now find it obvious that the general solution is

$$\rho = \frac{GM}{c^2} + A\cos(\theta - \theta_0),$$

where A and θ_0 are constants (*two* arbitrary constants because we have a *second*-order differential equation), and this tells us (because of (5.2.16)) that

$$r = \frac{\alpha}{1 + \epsilon\cos(\theta - \theta_0)}$$

(5.2.21)

where α, θ_0, and ϵ are constants. This is the polar equation of a *conic section* with r as the distance from one of the foci. If $0 \leq \epsilon < 1$, the conic section is an ellipse with *eccentricity* ϵ ($\epsilon = 0$ is the degenerate case of a circle), and if $\epsilon = 1$ we have a parabolic orbit, and if $\epsilon > 1$ we have a hyperbolic orbit. And so we have Kepler's first law of planetary motion: the orbit of any one of the planets around the Sun is an *ellipse*, with the Sun at one of the foci.

5.3 Euler's Restricted Three-Body Problem

There is much more that can be done with the two-body problem, but beyond that of establishing the conservation of energy in an inverse-square, central force field, and what the gravitational potential energies of multiple bodies are in such a field, my only other objective was to convince you that the math is certainly not trivial. My major goal was to make it credible when I say that to add a third body to the general gravitational motion problem pushes the math right to the edge, even *over* the edge into "beyond the pale." An enormous amount of analytical talent has been devoted to the three-body problem over the last three centuries, starting with Newton's Earth-Moon-Sun system[10] and continuing right up to modern times.[11] Even the most enthusiastic of the purely analytical persuasion would agree, however, that the three-body problem is so full of mathematical difficulties and physical surprises (you'll see a few of both before we are done!) that it is virtually certain to *forever* resist any sort of useful, general solution. The discontinuity in difficulty in going from the two-body to the three-body problem is a particular example of what the great Russian mathematician Andrei Nikolaevich Kolmogorov (1903–1987) had in mind when, in September 1943, he wrote in his diary, "There is only a thin layer separating what is trivial from what is impossibly difficult." That isn't to say that there aren't special cases of three-body motions that *can* be exactly solved—see Challenge Problems 5.4 and 5.5—but those solutions depend on quite special symmetry conditions to make them analytically "doable."

As a sort of compromise, then, analysts have defined what are called "restricted" problems that, while not full-blown, general three-body problems, do make simplifying assumptions that are not as severe as is the imposition of symmetry. The first such restricted problem was treated in 1760 by Euler. It takes two of the three bodies as each having any mass desired, but also *fixes them in space*. That is, these two massive bodies each produce a gravitational field throughout space, but neither of them moves because of the field of the other. The third body in the problem is taken as having a very small mass (in the limit, zero mass), and it *does* move according to the total gravitational force it "feels" from the first two bodies. A realistic version of such a restricted model would be, for example, the motion of a rocket under the gravitational influence of the Earth and the Moon. This special form of the three-body problem can, in fact, be treated analytically, but in support of my opening words to this section, it is not an "easy" solution; rather, it requires the appearance of elliptic integrals.[12] Doable, but definitely not "nice."

As an alternative approach to Euler's "two fixed bodies" problem, the use of a computer is an attractive one. Numerical approaches to solving the three-body problem can be traced back to the end of the nineteenth century (and in a later section I'll tell you more about that), but it was with the appearance of the electronic computer that such an approach really became popular. Beginning with the 1960s physics education literature, for example, one can find simple codes written in BASIC or FORTRAN that directly integrate Newton's gravitational equations to plot orbits.[13] Even programmable handheld calculators made an appearance as orbit integrators.[14] These early codes all used the simplest possible integration algorithm (also due, interestingly, to Euler—he is *everywhere* in mathematical physics!),[15] that of approximating a derivative with a finite difference (remember Jacobi iteration in the hot plate problem of Chapter 3?). That is, if one knows the value of x at time t, then at time $t + \Delta t$ one writes, $x(t + \Delta t) = x(t) + \Delta t \dfrac{dx}{dt}$, and starts the numerical work from *given* values of $x(0)$ and $\left. \dfrac{dx}{dt} \right|_{t=0}$, and the given time step Δt.

Three-body computer codes are not a panacea, and they have their problems, too. The loss of numerical accuracy over time with Euler's elementary integration algorithm, for example, was appreciated from the start, and it is common to read in those early papers comments to the effect that "of course, better results would be obtained by using a better integration algorithm." Usually it was suggested that something like "a fourth-order Runge-Kutta method" would be the proper tool. And, as you now know from our earlier use of MATLAB to solve differential equations, that is precisely the underlying algorithm of *ode45*. We'll make good use of that powerful command to "solve" three-body problems. To start, let's see how to do it with Euler's two fixed bodies problem.

In Figure 5.3.1 we see two massive bodies, M_1 and M_2, located respectively (with no loss in generality) on the x-axis at $x = 0$ and at $x = d$. These two bodies are fixed in place and will not move. (We'll remove, as did Euler, the fixed restriction in just a bit, but even then there will still be a restriction on just *how* M_1 and M_2 can move. It will be a far more realistic restriction than is *fixing* the two masses but still far short of allowing arbitrary motion.) Free to move under the influence of the combined gravitational forces due to M_1 and M_2 is the "small" mass m, which is located at the arbitrary (x,y). It should be clear that however m moves, it will remain in the x,y-plane if its initial velocity vector lies entirely in the x,y-plane.

We can write Newton's second law of motion for m, along each coordinate axis, as follows, in the notation of Figure 5.3.1. For motion

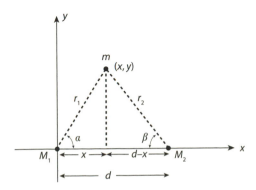

FIGURE 5.3.1. Geometry of Euler's *fixed* three-body problem

parallel to the x-axis, we have

$$m\frac{d^2x}{dt^2} = G\left[-\frac{mM_1}{r_1^2}\cos(\alpha) + \frac{mM_2}{r_2^2}\cos(\beta)\right],$$

where the algebraic signs of the two terms on the right are physically clear because, given the position of m relative to M_1 and M_2 in Figure 5.3.1, we see that M_1 will accelerate m along the x-axis to the *left*, while M_2 will accelerate m along the x-axis to the *right*. Since

$$\cos(\alpha) = \frac{x}{r_1}, \quad \cos(\beta) = \frac{d-x}{r_2},$$

then (after canceling all the m's) we have

$$\frac{d^2x}{dt^2} = G\left[-\frac{xM_1}{r_1^3} + \frac{(d-x)M_2}{r_2^3}\right],$$

or, as

$$r_1 = \left\{x^2 + y^2\right\}^{1/2}, \quad r_2 = \left\{(d-x)^2 + y^2\right\}^{1/2},$$

then

$$\frac{d^2x}{dt^2} = G\left[-\frac{xM_1}{\left\{x^2 + y^2\right\}^{3/2}} + \frac{(d-x)M_2}{\left\{(d-x)^2 + y^2\right\}^{3/2}}\right]. \tag{5.3.1}$$

Repeating for motion parallel to the y-axis, we have

$$m\frac{d^2x}{dt^2} = G\left[-\frac{mM_1}{r_1^2}\sin(\alpha) - \frac{mM_2}{r_2^2}\sin(\beta)\right],$$

where now both signs on the right are minus since both M_1 and M_2 will accelerate m *downward*. And, as

$$\sin(\alpha) = \frac{y}{r_1}, \quad \sin(\beta) = \frac{y}{r_2},$$

then (after canceling all the m's) we have

$$\frac{d^2y}{dt^2} = G\left[-\frac{yM_1}{r_1^3} - \frac{yM_2}{r_2^3}\right],$$

and so

$$\frac{d^2y}{dt^2} = G\left[-\frac{yM_1}{\{x^2+y^2\}^{3/2}} - \frac{yM_2}{\{(d-x)^2+y^2\}^{3/2}}\right]. \tag{5.3.2}$$

Notice that in both (5.3.1) and (5.3.2) the mass m does not appear and so our assumption that m is "small" may seem unnecessary. Actually, it *would* be unnecessary *if* we always imagine that M_1 and M_2 are somehow glued in place—which we are indeed free to imagine if we wish (and here we do)—but later we will allow M_1 and M_2 to move (in a restricted way, and only under the influence of each other) and so, to justify ignoring the influence of m on M_1 and M_2, we must assume m is "small." We of course need m to be *something* greater than zero, or else m simply wouldn't respond to M_1 and M_2 at all (after all, $m\frac{d^2x}{dt^2}$ and $m\frac{d^2y}{dt^2}$, the x- and y-component forces on m, are both *zero* if $m = 0$)!

Now, some final steps before we write computer code. First, as in the first section, let's take our units to be such that the unit of distance is one AU and the unit of time is one year. Then, writing $M_1 = M$, we have the result given in (5.1.6), that $GM = 4\pi^2$. Second, is it clear to you that in this system of units we can simply *define* the unit mass to be the mass of the Sun? If it isn't, think about this point for a bit and, if you still remain uncertain, take a look at the notes.[16] And finally, third, if we normalize M_2 to M_1 ($= M$) by writing $M_2 = \alpha M$, $\alpha \geq 0$, then (5.3.1) and (5.3.2) become

$$\frac{d^2x}{dt^2} = 4\pi^2 M\left[-\frac{x}{\{x^2+y^2\}^{3/2}} + \frac{\alpha(d-x)}{\{(d-x)^2+y^2\}^{3/2}}\right] \tag{5.3.3}$$

and

$$\frac{d^2y}{dt^2} = -4\pi^2 My\left[\frac{1}{\{x^2+y^2\}^{3/2}} + \frac{\alpha}{\{(d-x)^2+y^2\}^{3/2}}\right]. \tag{5.3.4}$$

As usual, to use a MATLAB differential equation solver, we must first write each of these second-order differential equations as two first-order ones. So, define

$$u_1 = x \tag{5.3.5a}$$

and

$$u_2 = \frac{dx}{dt}. \tag{5.3.5b}$$

Also, let's write

$$u_3 = y \tag{5.3.5c}$$

and

$$u_4 = \frac{dy}{dt}. \tag{5.3.5d}$$

Then

$$\frac{du_2}{dt} = \frac{d^2x}{dt^2}$$

and

$$\frac{du_4}{dt} = \frac{d^2y}{dt^2}.$$

Instead of (5.3.3) and (5.3.4), then, the equations we'll feed to MATLAB are:

$$\frac{du_1}{dt} = u_2, \tag{5.3.6a}$$

$$\frac{du_2}{dt} = 4\pi^2 \left[-\frac{u_1}{\{u_1^2 + u_3^2\}^{3/2}} + \frac{\alpha(d - u_1)}{\{(d - u_1)^2 + u_3^2\}^{3/2}} \right], \tag{5.3.6b}$$

$$\frac{du_3}{dt} = u_4, \tag{5.3.6c}$$

$$\frac{du_4}{dt} = -4\pi^2 u_3 \left[\frac{1}{\{u_1^2 + u_3^2\}^{3/2}} + \frac{\alpha}{\{(d - u_1)^2 + u_3^2\}^{3/2}} \right], \tag{5.3.6d}$$

where M is taken as equal to the solar mass ($M = 1$) and the values of d and α are constants that must be defined at the start of the code.

Okay, now we can write code. The MATLAB function m-file code **twobody.m** implements the equations of (5.3.6), where I've set the values of d, α, and the initial conditions vector *uzero* to generate the orbit shown in Figure 5.3.2 (which I'll explain in a moment). But first, a comment about the command *norm* in the nested function *xyderiv*. **two-body.m** operates just as do the MATLAB differential equation codes you have already seen in Chapter 4, but *norm* is something new for here. *norm* accepts a vector argument and returns the value equal to the square root of the sum of the squares of the elements of the vector. This is a calculation that is done over and over again in (5.3.6b) and (5.3.6d), and so *norm* is a very convenient command for **twobody.m**. (The vector arguments provided to *norm* are *r1* and *r2*, defined in the first line of *xyderiv*.)

```
twobody.m
function twobody
alpha=1;d=2.1;fps=4*pi*pi;
options=odeset('AbsTol',1e-8,'RelTol',1e-5);
tspan=[0 1.8];
uzero=[1;0;0;2*pi];
[t,u]=ode45(@orbit,tspan,uzero,options);
plot(u(:,1),u(:,3),'-k')
title('FIGURE 5.3.2 - Aliens destroy the Earth!','FontSize',16)
xlabel('unit of distance = 1 AU','FontSize',16)
ylabel('unit of distance = 1 AU','FontSize',16)
grid on
        function xyderiv=orbit(t,u)
        r1=[u(1),u(3)];r2=[d-u(1),u(3)];
        denom1=norm(r1)^3;denom2=norm(r2)^3;
        xyderiv=[u(2);fps*((-u(1)/denom1)+(alpha*(d-u(1))/
                denom2));...u(4);-fps*u(3)*((1/denom1)
                +(alpha/denom2))];
        end
end
```

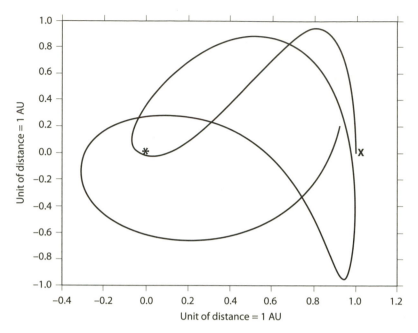

FIGURE 5.3.2. Aliens destroy the Earth!

Now, what is Figure 5.3.2 all about? I am going to have some fun here (I hope you'll agree) by asking you to imagine the following science fiction scenario. Unimaginably awful aliens wish to acquire the Earth for their ever-expanding galactic empire. They want the planet but have no interest in keeping *us* around. They also have no interest in engaging in troublesome warfare; they much prefer to let natural processes do the "get rid of humans" work. So, possessing technology vastly superior to ours, they use a wormhole to transport a "death mass" equal to that of the Sun from their neck of the woods into ours. At the instant that mass suddenly appears in the Solar System (time $t = 0$), the Sun, the Earth, and the "death mass" all lie on the x-axis, with the Earth in between the other two. This so disrupts the Earth's orbit that all life on the planet is exterminated. In particular, at time $t = 0$ the Sun is (as always) 1 AU away from Earth on one side of the planet, and the "death mass" (not shown in Figure 5.3.2) is 1.1 AU on the other side of the Earth. This explains the values of α and d in **two-body.m**. That is, $\alpha = 1$ sets the mass of the death mass equal to the

Sun's mass, and the death mass is $d = 1$ AU $+ 1.1$ AU $= 2.1$ AU from the Sun (take a look back at Figure 5.3.1 for the definition of d).

To understand *uzero* ($= [u_1(0); u_2(0); u_3(0); u_4(0)]$), simply recall that at $t = 0$ we have $u(1) = u_1 = x = 1$ AU, $u(2) = u_2 = \dfrac{dx}{dt} = 0$ because the Earth has no x-axis speed component at that instant, $u(3) = u_3 = y = 0$ because the Earth (marked with an X) is at that instant *on* the x-axis, and $u(4) = u_4 = \dfrac{dy}{dt} = 2\pi$ because at that instant the entire orbital speed of the Earth is perpendicular to the x-axis (and the orbital period is one year to travel an orbital circumference of 2π AU). Figure 5.3.2 then shows the resulting orbit of the Earth for the next 1.8 years (as defined in *tspan*), and we see just how well the evil plan of the aliens would work: the Earth at one point passes so close to the Sun (its fixed location is marked with an asterisk) as to guarantee that the atmosphere and the oceans would boil away! The perturbed orbit of Earth is sufficiently "twisty" that I think it easy to believe that an analytical solution would be pretty complicated.

The assumption implicit in Figure 5.3.2 is, of course, that the Sun and the death mass are sufficiently separated that neither affects the other, at least not for the first 1.8 years after the death mass appears. So, is 2.1 AU a sufficient separation? To avoid even wondering about that, the easiest thing to do would be to simply remove the constraint of both M_1 and M_2 being fixed. We'll do that before this chapter is over. To finish this section, however, let's do a quick-and-dirty partial check of **twobody.m**. If we leave everything in the code unchanged except for setting the death mass to zero, that is, writing *alpha=0*, then of course we know what Earth's orbit should be for the next 1.8 years: just what it was for the *previous* 1.8 years! A circle centered on the Sun with a radius of 1 AU. Figure 5.3.3 shows the orbit calculated by **twobody.m** for $\alpha = 0$, and a circle is just what we get.

5.4 Binary Stars

More realistic than two fixed masses is when the two masses M_1 and M_2 are allowed to move. What Euler did on relaxing the fixed constraints on M_1 and M_2 is to imagine that they are revolving in circular (more generally, elliptic) orbits in a common plane about their center

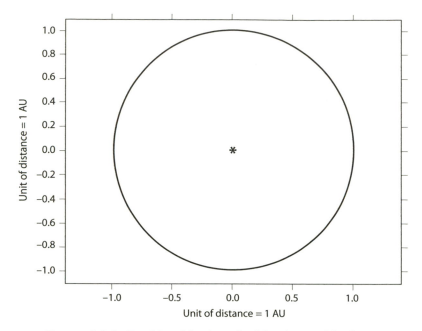

FIGURE 5.3.3. Earth's orbit when the "death mass" is absent

of mass. (Now it is essential we assume that our third mass m is vanishingly small, so it will not have any influence on the circular motions of M_1 and M_2.) Such a situation occurs, for example, when two stars (*binary* stars) orbit each other. This is actually not uncommon, and it is estimated that perhaps half the stars in our galaxy are multiple, gravitationally bound star systems.[17] The Sun's nearest stellar neighbor, for example, Alpha Centauri, about 4.4 light-years distant, is at least a binary system. Alpha Centauri A is very much like the Sun, with a mass just slightly greater than the Sun's, while Alpha Centauri B is just slightly less massive than the Sun. The two stars are in pronounced elliptical orbits, with the distance between them varying from 11 AU to 35 AU over a period of eighty years. A and B may actually be part of a ternary system because a much smaller (0.12 solar masses) third star, Proxima Centauri, is presently associated with them. Since Proxima orbits A and B at a distance of 13, 000 AU, however, it isn't clear if it is gravitationally bound to them or if it will eventually simply wander away into space.

Another example of such coupled motion, in our own Solar System, is that of the Earth and the Moon; the Moon does not simply revolve around the Earth, or vice versa, but rather they both orbit their common center of mass, which, because the two bodies aren't really all that far apart and because the Earth is an extended object much more massive than the Moon, happens to be a point located *inside* the Earth. For two point masses, however, the center of mass will always be between the two masses. Specifically, if at time $t = 0$ M_1 and M_2 are on the x-axis at $x = x_1$ and $x = x_2$, respectively, with $x_2 < x_1$, then their center of mass is at $x = x_c$, defined by

$$x_c (M_1 + M_2) = M_1 x_1 + M_2 x_2,$$

or

$$x_c = \frac{M_1}{M_1 + M_2} x_1 + \frac{M_2}{M_1 + M_2} x_2.$$

If we write

$$\mu = \frac{M_1}{M_1 + M_2}, \quad 0 \le \mu \le 1, \tag{5.4.1}$$

then

$$x_c = \mu x_1 + (1 - \mu) x_2. \tag{5.4.2}$$

As M_1 and M_2 orbit the point $x = x_c$ (at the same angular rate, and so M_1 and M_2 have the same orbital period), the center of mass remains fixed on the x-axis. The orbital radius of M_1 about x_c is $x_1 - x_c$, and the orbital radius of M_2 about x_c is $x_c - x_2$. That is, for M_1 the orbital radius is, using (5.4.2),

$$x_1 - x_c = x_1 - \mu x_1 - (1 - \mu) x_2 = (1 - \mu) x_1 - (1 - \mu) x_2$$
$$= (1 - \mu)(x_1 - x_2),$$

and for M_2 the orbital radius is

$$x_c - x_2 = \mu x_1 + (1 - \mu) x_2 - x_2 = \mu x_1 + x_2[(1 - \mu) - 1]$$
$$= \mu x_1 - \mu x_2 = \mu(x_1 - x_2).$$

As M_1 and M_2 travel on their circular orbits they always remain a fixed distance apart, defined by their original separation on the x-axis at time $t = 0$. Let's write that fixed separation as

$$l = x_1 - x_2, \tag{5.4.3}$$

and so the orbital radii of M_1 and M_2 are, respectively, $(1 - \mu)l$ and μl, measured from the center of mass. Now, with no loss of generality, we'll take the origin of our x,y-coordinate system to be the location of the center of mass, and we have the situation shown in Figure 5.4.1 for time $t = 0$.

The gravitational attraction between M_1 and M_2 provides the force required for the centripetal acceleration of each mass, and so

$$G\frac{M_1 M_2}{l^2} = \frac{M_1 v_1^2}{(1 - \mu)l} \tag{5.4.4}$$

and

$$G\frac{M_1 M_2}{l^2} = \frac{M_2 v_2^2}{\mu l}, \tag{5.4.5}$$

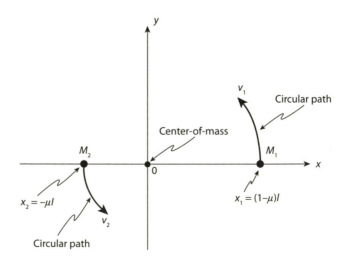

FIGURE 5.4.1. M_1 and M_2 at time $t = 0$

where v_1 and v_2 are the orbital speeds of M_1 and M_2, respectively. Let $M_1 = M$ be one Solar mass and $M_2 = \alpha M$ and so, in our usual system of units (unit of mass is one solar mass, unit of distance is one AU, and unit of time is one year), we have from (5.1.6) and (5.4.4) that

$$\frac{GM_2}{l^2} = \frac{GM\alpha}{l^2} = \frac{4\pi^2\alpha}{l^2} = \frac{v_1^2}{(1-\mu)l},$$

or

$$v_1 = 2\pi\sqrt{\frac{\alpha(1-\mu)}{l}}. \tag{5.4.6}$$

And from (5.4.5) we have

$$\frac{GM_1}{l^2} = \frac{GM}{l^2} = \frac{4\pi^2}{l^2} = \frac{v_2^2}{\mu l},$$

or

$$v_2 = 2\pi\sqrt{\frac{\mu}{l}}. \tag{5.4.7}$$

The orbital period for M_1 is, using (5.4.6),

$$T_1 = \frac{2\pi(1-\mu)l}{v_1} = \frac{2\pi(1-\mu)l}{2\pi\sqrt{\frac{\alpha(1-\mu)}{l}}},$$

or

$$T_1 = l^{3/2}\sqrt{\frac{1-\mu}{\alpha}}. \tag{5.4.8}$$

The orbital period for M_2 is, using (5.4.7),

$$T_2 = \frac{2\pi\mu l}{v_2} = \frac{2\pi\mu l}{2\pi\sqrt{\frac{\mu}{l}}},$$

or

$$T_2 = l^{3/2}\sqrt{\mu}. \tag{5.4.9}$$

Of course, $T_1 = T_2$ (as is easy to show, using (5.4.1) and remembering that $M_1 = M = 1$ and $M_2 = \alpha M$ and so $\alpha = \dfrac{1 - \mu}{\mu}$), and I'll simply write the orbital period of both M_1 and M_2 as

$$T = l^{3/2} \sqrt{\mu}. \tag{5.4.10}$$

Finally, if we write ω as the angular rate (radians/year) of the circular motion for M_1 and M_2 about the origin in Figure 5.4.1, we know $\omega T = 2\pi$ (one complete orbit per period) and so

$$\omega = \frac{2\pi}{l^{3/2} \sqrt{\mu}} \text{ radians/year.} \tag{5.4.11}$$

If at some time $t \geq 0$, then, when M_1 and M_2 have each rotated counter-clockwise through angle θ from their positions at $t = 0$ (as shown in Figure 5.4.1) we have the situation shown in Figure 5.4.2, where

$$\theta = \omega t = \frac{2\pi t}{l^{3/2} \sqrt{\mu}} \tag{5.4.12}$$

and the coordinates of the masses M_1 and M_2 are (x_1, y_1) and (x_2, y_2), respectively, given by

$$x_1(t) = (1 - \mu)l \cos(\theta) = (1 - \mu)l \cos\left(\frac{2\pi t}{l^{3/2} \sqrt{\mu}}\right)$$

$$y_1(t) = (1 - \mu)l \sin(\theta) = (1 - \mu)l \sin\left(\frac{2\pi t}{l^{3/2} \sqrt{\mu}}\right) \tag{5.4.13}$$

and

$$x_2(t) = -\mu l \cos(\theta) = -\mu l \cos\left(\frac{2\pi t}{l^{3/2} \sqrt{\mu}}\right)$$

$$y_2(t) = -\mu l \sin(\theta) = -\mu l \sin\left(\frac{2\pi t}{l^{3/2} \sqrt{\mu}}\right). \tag{5.4.14}$$

The "vanishingly small" mass m, the mass whose motion is the focus of our attention, is shown at its location (x_3, y_3).

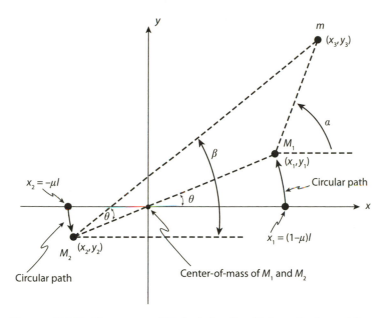

FIGURE 5.4.2. Geometry of Euler's "unfixed" three-body problem

In terms of the notation of Figure 5.4.2 we can write Newton's second law of motion for m as follows, in the same way we did for Euler's *fixed* three-body problem:

$$m\frac{d^2x_3}{dt^2} = -\frac{GM_1m}{(x_3 - x_1)^2 + (y_3 - y_1)^2}\cos(\alpha)$$
$$-\frac{GM_2m}{(x_3 - x_2)^2 + (y_3 - y_2)^2}\cos(\beta)$$

and

$$m\frac{d^2y_3}{dt^2} = -\frac{GM_1m}{(x_3 - x_1)^2 + (y_3 - y_1)^2}\sin(\alpha)$$
$$-\frac{GM_2m}{(x_3 - x_2)^2 + (y_3 - y_2)^2}\sin(\beta),$$

where

$$\cos(\alpha) = \frac{x_3 - x_1}{\sqrt{(x_3 - x_1)^2 + (y_3 - y_1)^2}},$$

$$\sin(\alpha) = \frac{y_3 - y_1}{\sqrt{(x_3 - x_1)^2 + (y_3 - y_1)^2}},$$

$$\cos(\beta) = \frac{x_3 - x_2}{\sqrt{(x_3 - x_2)^2 + (y_3 - y_2)^2}},$$

$$\sin(\beta) = \frac{y_3 - y_2}{\sqrt{(x_3 - x_2)^2 + (y_3 - y_2)^2}}.$$

Or, since $M_1 = M = 1$, $M_2 = \alpha M$, and $G = 4\pi^2$, we have

$$\frac{d^2 x_3}{dt^2} = -4\pi^2 \left[\frac{x_3 - x_1}{\left\{ (x_3 - x_1)^2 + (y_3 - y_1)^2 \right\}^{3/2}} \right.$$
$$\left. + \frac{\alpha(x_3 - x_2)}{\left\{ (x_3 - x_2)^2 + (y_3 - y_2)^2 \right\}^{3/2}} \right]$$

and

$$\frac{d^2 y_3}{dt^2} = -4\pi^2 \left[\frac{y_3 - y_1}{\left\{ (x_3 - x_1)^2 + (y_3 - y_1)^2 \right\}^{3/2}} \right.$$
$$\left. + \frac{\alpha(y_3 - y_2)}{\left\{ (x_3 - x_2)^2 + (y_3 - y_2)^2 \right\}^{3/2}} \right].$$

Or, dropping the subscripts on x_3 and y_3 to simply write x and y, and making the simplifications of writing $a = \mu l$ and $b = (1 - \mu)l$, and using (5.4.12), (5.4.13), and (5.4.14), we have

$$\frac{d^2 x}{dt^2} = -4\pi^2 \left[\frac{x - b\cos(\theta)}{r_1^3} + \frac{\alpha\{x + a\cos(\theta)\}}{r_2^3} \right] \qquad (5.4.15)$$

and

$$\frac{d^2 y}{dt^2} = -4\pi^2 \left[\frac{y - b\sin(\theta)}{r_1^3} + \frac{\alpha\{y + a\sin(\theta)\}}{r_2^3} \right], \qquad (5.4.16)$$

where

$$r_1 = \left[\{x - b\cos(\theta)\}^2 + \{y - b\sin(\theta)\}^2 \right]^{1/2} \qquad (5.4.17)$$

and

$$r_2 = \left[\{x + a\cos(\theta)\}^2 + \{y + a\sin(\theta)\}^2 \right]^{1/2}. \qquad (5.4.18)$$

With these last four equations, along with (5.4.12) for θ, we could now write computer code for the orbit of m, once the values of a and b (that is, μ—which then determines α, and vice versa—and l) are specified. We proceed as usual to set things up for MATLAB by transforming our second-order differential equations into a set of first-order differential equations. So, let's define u_1, u_2, u_3, and u_4 just as we did in the definitions of (5.3.5). That is, I'll write $u_1 = x$, $u_2 = \dfrac{dx}{dt}$, $u_3 = y$, and $u_4 = \dfrac{dy}{dt}$ and so $\dfrac{du_2}{dt} = \dfrac{d^2x}{dt^2}$ and $\dfrac{du_4}{dt} = \dfrac{d^2y}{dt^2}$, and thus the equations we'll feed to MATLAB are:

$$\frac{du_1}{dt} = u_2, \qquad (5.4.19a)$$

$$\frac{du_2}{dt} = -4\pi^2 \left[\frac{u_1 - b\cos(\theta)}{r_1^3} + \frac{\alpha\{u_1 + a\cos(\theta)\}}{r_2^3} \right], \qquad (5.4.19b)$$

$$\frac{du_3}{dt} = u_4, \qquad (5.4.19c)$$

$$\frac{du_4}{dt} = -4\pi^2 \left[\frac{u_3 - b\sin(\theta)}{r_1^3} + \frac{\alpha\{u_3 + a\sin(\theta)\}}{r_2^3} \right], \qquad (5.4.19d)$$

where

$$r_1 = \left[\{u_1 - b\cos(\theta)\}^2 + \{u_3 - b\sin(\theta)\}^2 \right]^{1/2} \qquad (5.4.19e)$$

and

$$r_2 = \left[\{u_1 + a\cos(\theta)\}^2 + \{u_3 + a\sin(\theta)\}^2 \right]^{1/2}. \qquad (5.4.19f)$$

The code **fbinary.m**—the "binary' part of the name should be obvious, and I'll explain why the letter "f" in the next section—implements (5.4.19).

---

fbinary.m

```
function fbinary
alpha=1;l=2.1;fps=4*pi*pi;
mu=1/(1+alpha);a=mu*l;b=(1-mu)*l;w=2*pi/(sqrt(mu)*(l^(3/2)));
options=odeset('AbsTol',1e-8,'RelTol',1e-5);
tspan=[0 2.2];
uzero=[-0.05;0;0;2*pi];
[t,u]=ode45(@orbit,tspan,uzero,options);
plot(u(:,1),u(:,3),'-k')
title('FIGURE 5.4.3 - Aliens destroy the Earth?','FontSize',16)
xlabel('unit of distance = 1 AU','FontSize',16)
ylabel('unit of distance = 1 AU','FontSize',16)
legend('orbit duration = 2.2 years')
        function xyderiv=orbit(t,u)
        theta=w*t;c4=cos(theta);c5=sin(theta);
        r1=[u(1)-b*c4,u(3)-b*c5];denom1=norm(r1)^3;
        r2=[u(1)+a*c4,u(3)+a*c5];denom2=norm(r2)^3;
        c1=1/denom1;c2=alpha/denom2;c3=c1+c2;
        xyderiv=[u(2);-fps*(c3*u(1)-b*c4*c1+a*c4*c2);u(4);...
              -fps*(c3*u(3)-b*c5*c1+a*c5*c2)];
        end
   end
```

---

fbinary.m creates Figure 5.4.3, which is Figure 5.3.2 again, except that now the Sun and the death mass are more realistically modeled as the two *moving* components of a binary star. As in Figure 5.3.2, in the new figure the locations of the Sun (the death mass is not shown) and the Earth *at time* $t = 0$ are marked with an asterisk and an X, respectively. Since **fbinary.m** has the two masses orbiting their center of mass, which is always the origin, then with the death mass exiting the Solar System mouth of the aliens' wormhole (at $t = 0$) a distance of 2.1 AU from the Sun, with the Earth in between the Sun and the death mass and with the Earth 1 AU from the Sun, you can now see where the opening lines of code come from: *alpha=1* because the death mass is

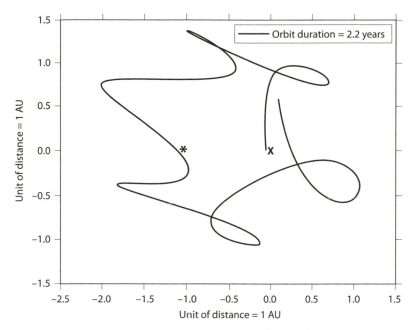

FIGURE 5.4.3. Aliens destroy the Earth?

one Solar mass, and $l = 2.1$ because the death mass appears 2.1 AU from the Sun. Also, after mu is computed (to be 0.5) we see that our two binary masses are initially on the x-axis at $x = -\mu l = -1.05$ (the Sun) and $x = (1 - \mu)l = 1.05$ (the death mass), and so the Earth and the death mass are traveling in the *same* circular orbit, always positioned at the opposite ends of a diameter of that orbit. This means that the Earth, which is 1 AU from the Sun at $t = 0$, is (at $t = 0$) on the x-axis at $x = -0.05$, and that explains the value of the first element of *uzero*.

Now, as interesting as the orbit in Figure 5.4.3 is, it does have one very significant failing. It fails to tell us if the Earth *really is* destroyed via a close encounter with the Sun (or with the death mass—even if it isn't a stellar fusion furnace, getting too close to a solar-sized mass would produce huge, destructive tidal forces in the crust of the Earth),[18] and that's why the figure caption ends with a question mark. The perturbed orbit of the Earth comes quite close to where the Sun is *at time t = 0*, but that doesn't mean anything in itself. Destruction results only when the Earth and the Sun (or the death mass) arrive in the same "small" neighborhood of space *at the same time*. The orbit in Figure 5.4.3 doesn't tell us if that does (or doesn't) occur, and even if the figure showed the common circular orbit of the Sun and the death mass, we still wouldn't have that timing information. In the next section I'll show you a beautiful solution to this failing—and it's due to Euler, too. What a guy—he's *everywhere* in mathematical physics!

5.5 Euler's Problem in Rotating Coordinates

In the previous section the equations for m's motion, under the influence of the total gravitational field of the M_1, M_2 binary, are for a *fixed* coordinate system, what astronomers call the *sidereal* ("relative to the background of the remote stars") system, and what physicists call an *inertial* system. There is nothing wrong with working with those inertial equations but, for historical reasons (and because it is modern practice), I'm now going to show you how to convert to a *rotating* coordinate system (called *synodic*—"relative to the Sun" — by astronomers). This is, in fact, what Euler himself did two and a half centuries ago, in 1772. From a technical point of view, too, you'll see that changing coordinate systems this way will allow us to get rid of the trigonometric time functions on the right-hand sides of our equations (remember, $\theta = \theta(t)$). With modern computers, and software like MATLAB, such time-dependent terms are of no concern, but in Euler's day they *were* burdensome to handle. The price we'll pay for that simplification is that of adding new, first-derivative terms to the left-hand sides of (5.4.15) and (5.4.16), which in fact *always* happens when one goes from an inertial coordinate system to a noninertial (here, a rotating) coordinate system. Physicists associate these additional derivatives with the appearance of a fictitious force, the *Coriolis force*, named after the French physicist Gaspard-Gustave de Coriolis (1792–1843), who wrote of it in 1835. There will also be a new term in each differential equation representing a centrifugal force, but for us all these additional terms will simply be the automatic results of straightforward mathematical manipulations.

In our rotating system the x-axis is always the line passing through M_1 and M_2, and so, of course, this axis rotates counterclockwise at the angular rate ω given by (5.4.11). To distinguish this rotating system from the inertial system (x,y), I'll use the notation (x^*,y^*). Although rotating, the (x^*,y^*) system is still a rectangular system and the y^*-axis is perpendicular to the x^*-axis. The origin for each system is the same fixed point in space, the center of mass of M_1 and M_2. In Figure 5.5.1 I've drawn both sets of axes, with x^*,y^* rotated *counterclockwise* through angle θ with respect to x,y. The masses M_1 and M_2 are, of course, both on the x^*-axis, each the same *fixed* distance from the origin as in Figure 5.4.2, while the mass m is at coordinates (x,y) in the inertial system and at (x^*,y^*) in the rotating system.

Now, just as complex numbers were of great help, earlier in this chapter in explaining the retrograde motion of Mars, they will be central to transforming from sidereal to synodic coordinates.[19] The position vector of M_1 in the inertial system is $(i = \sqrt{-1})$

$$\mathbf{z}_1 = be^{i\theta} \tag{5.5.1}$$

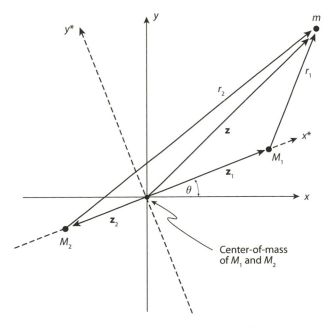

FIGURE 5.5.1. $M_1, M_2,$ and m in the inertial and rotating systems

and the position vector of M_2 in the inertial system is

$$\mathbf{z}_2 = ae^{i(\pi+\theta)} = ae^{i\pi}e^{i\theta} = -ae^{i\theta} \tag{5.5.2}$$

where, as before, $a = \mu l$ and $b = (1 - \mu)l$. The position vector of m in the inertial system is

$$\mathbf{z} = x + iy. \tag{5.5.3}$$

And finally, I've written $\mathbf{r}_1$ and $\mathbf{r}_2$ as the vectors *from* M_1 and M_2, respectively, *to* m. By vector algebra, then, we have

$$\mathbf{z}_1 + \mathbf{r}_1 = \mathbf{z}$$

and

$$\mathbf{z}_2 + \mathbf{r}_2 = \mathbf{z},$$

so

$$\mathbf{r}_1 = \mathbf{z} - \mathbf{z}_1$$

and

$$\mathbf{r}_2 = \mathbf{z} - \mathbf{z}_2.$$

Thus, the *distances* between M_1 and M_2, and m, respectively, are

$$r_1 = |\mathbf{r}_1| = |\mathbf{z} - \mathbf{z}_1| \tag{5.5.4}$$

and

$$r_2 = |\mathbf{r}_2| = |\mathbf{z} - \mathbf{z}_2|. \tag{5.5.5}$$

When we go to the rotating system the position vector of m in the inertial system, $\mathbf{z}$, appears to have rotated *clockwise* through angle θ and so

$$\mathbf{z}^* = \mathbf{z}e^{-i\theta} \tag{5.5.6}$$

is the position vector of m in the rotating system. Equivalently,

$$\mathbf{z} = \mathbf{z}^* e^{i\theta}. \tag{5.5.7}$$

Notice that in the rotating system we see the position vectors of M_1 and M_2 become, because of (5.5.1), (5.5.2), and (5.5.6), the *constants*

$$\mathbf{z}_1^* = \mathbf{z}_1 e^{-i\theta} = b e^{i\theta} e^{-i\theta} = b$$

and

$$\mathbf{z}_2^* = \mathbf{z}_2 e^{-i\theta} = -a e^{i\theta} e^{-i\theta} = -a.$$

That is, in the rotating coordinate system, M_1 and M_2 *don't move*, and this is precisely why we've introduced the rotating system in the first place! *Now* as we watch the Earth's perturbed orbit after the death mass emerges from the aliens' wormhole, we'll always know where the Sun and the death mass are, relative to the Earth.

Using (5.5.1) and (5.5.7) in (5.5.4), we have

$$r_1 = |\mathbf{z}^* e^{i\theta} - b e^{i\theta}| = |(\mathbf{z}^* - b)e^{i\theta}| = |\mathbf{z}^* - b||e^{i\theta}|,$$

or, as $|e^{i\theta}| = 1$,

$$r_1 = |\mathbf{z}^* - b| = |x^* + iy^* - b|,$$

and so

$$r_1 = \sqrt{(x^* - b)^2 + y^{*2}}. \tag{5.5.8}$$

Also,

$$r_2 = |\mathbf{z}^* e^{i\theta} + a e^{i\theta}|,$$

and so, in the same way,

$$r_2 = \sqrt{(x^* + a)^2 + y^{*2}}. \tag{5.5.9}$$

Let's now form a complex quantity using the left-hand sides of (5.4.15) and (5.4.16) as the real and imaginary parts, respectively. (This approach will provide us with a nice example of the power of Euler's famous identity—as I said before, Euler is *everywhere!*) That is, let's write (remembering (5.5.3) and (5.5.7))

$$
\begin{aligned}
\frac{d^2x}{dt^2} + i\frac{d^2y}{dt^2} &= \frac{d^2(x+iy)}{dt^2} = \frac{d^2\mathbf{z}}{dt^2} = \frac{d^2(\mathbf{z}^*e^{i\theta})}{dt^2} = \frac{d}{dt}\left\{ \frac{d}{dt}(\mathbf{z}^*e^{i\theta}) \right\} \\
&= \frac{d}{dt}\left\{ \frac{d\mathbf{z}^*}{dt}e^{i\theta} + \mathbf{z}^*i\frac{d\theta}{dt}e^{i\theta} \right\} \\
&= \frac{d^2\mathbf{z}^*}{dt^2}e^{i\theta} + \frac{d\mathbf{z}^*}{dt}i\frac{d\theta}{dt}e^{i\theta} + \frac{d\mathbf{z}^*}{dt}i\frac{d\theta}{dt}e^{i\theta} \\
&\quad + \mathbf{z}^*i\frac{d^2\theta}{dt^2}e^{i\theta} + \mathbf{z}^*i\frac{d\theta}{dt}i\frac{d\theta}{dt}e^{i\theta} \\
&= e^{i\theta}\left[\frac{d^2\mathbf{z}^*}{dt^2} + i2\frac{d\mathbf{z}^*}{dt}\cdot\frac{d\theta}{dt} + i\mathbf{z}^*\frac{d^2\theta}{dt^2} - \mathbf{z}^*\left(\frac{d\theta}{dt}\right)^2 \right],
\end{aligned}
$$

or, since from (5.4.12) we have

$$
\frac{d\theta}{dt} = \frac{2\pi}{l^{3/2}\sqrt{\mu}} \text{ and so } \frac{d^2\theta}{dt^2} = 0, \tag{5.5.10}
$$

then our complex quantity from the left-hand side of (5.4.15) and (5.4.16) becomes

$$
e^{i\theta}\left[\left\{ \frac{d^2\mathbf{z}^*}{dt^2} - \mathbf{z}^*\left(\frac{d\theta}{dt}\right)^2 \right\} + i2\frac{d\mathbf{z}^*}{dt}\cdot\frac{d\theta}{dt} \right]. \tag{5.5.11}
$$

To see what the expression in (5.5.11) equals, let's now repeat the above process using the *right*-hand sides of (5.4.15) and (5.4.16). That is, using the right-hand sides of those equations as the real and imaginary parts of a complex quantity, respectively, we have

$$- 4\pi^2 \left[\frac{\{x - b\cos(\theta)\} + i\{y - b\sin(\theta)\}}{r_1^3} \right.$$

$$\left. + \frac{\alpha\{x + a\cos(\theta)\} + i\alpha\{y + a\sin(\theta)\}}{r_2^3} \right]$$

$$= -4\pi^2 \left[\frac{x + iy - b\{\cos(\theta) + i\sin(\theta)\}}{r_1^3} \right.$$

$$\left. + \frac{\alpha(x + iy) + \alpha a\{\cos(\theta) + i\sin(\theta)\}}{r_2^3} \right]$$

$$= -4\pi^2 \left[\frac{\mathbf{z} - be^{i\theta}}{r_1^3} + \frac{\alpha\mathbf{z} + \alpha a e^{i\theta}}{r_2^3} \right]$$

$$= -4\pi^2 \left[\frac{\mathbf{z}^* e^{i\theta} - be^{i\theta}}{r_1^3} + \alpha \frac{\mathbf{z}^* e^{i\theta} + ae^{i\theta}}{r_2^3} \right],$$

and so this complex quantity becomes

$$-4\pi^2 \left[\frac{\mathbf{z}^* - b}{r_1^3} + \alpha \frac{\mathbf{z}^* + a}{r_2^3} \right] e^{i\theta}. \tag{5.5.12}$$

Since our two complex quantities, (5.5.11) and (5.5.12), are equal, canceling the common $e^{i\theta}$ factor gives us

$$\frac{d^2\mathbf{z}^*}{dt^2} - \mathbf{z}^* \left(\frac{d\theta}{dt} \right)^2 + i2 \frac{d\mathbf{z}^*}{dt} \cdot \frac{d\theta}{dt} = -4\pi^2 \left[\frac{\mathbf{z}^* - b}{r_1^3} + \alpha \frac{\mathbf{z}^* + a}{r_2^3} \right].$$

$$\tag{5.5.13}$$

And since $\mathbf{z}^* = x^* + iy^*$, equating the real and imaginary parts of (5.5.13) says

$$\frac{d^2x^*}{dt^2} - x^* \left(\frac{d\theta}{dt} \right)^2 - 2\frac{dy^*}{dt} \cdot \frac{d\theta}{dt} = -4\pi^2 \left[\frac{x^* - b}{r_1^3} + \alpha \frac{x^* + a}{r_2^3} \right]$$

and

$$\frac{d^2y^*}{dt^2} - y^*\left(\frac{d\theta}{dt}\right)^2 + 2\frac{dx^*}{dt} \cdot \frac{d\theta}{dt} = -4\pi^2\left[\frac{y^*}{r_1^3} + \alpha\frac{y^*}{r_2^3}\right],$$

or, replacing $\frac{d\theta}{dt}$ with ω (take a look again at (5.4.12)) and rearranging terms, we at last arrive at the synodic equivalent of the inertial sidereal equations of (5.4.15) and (5.4.16), that is, at the equations of motion for m in the rotating (*noninertial*) coodinate system:

$$\frac{d^2x^*}{dt^2} = 2\omega\frac{dy^*}{dt} + \omega^2 x^* - 4\pi^2\left[\frac{x^* - b}{r_1^3} + \alpha\frac{x^* + a}{r_2^3}\right] \tag{5.5.14}$$

and

$$\frac{d^2y^*}{dt^2} = -2\omega\frac{dx^*}{dt} + \omega^2 y^* - 4\pi^2\left[\frac{y^*}{r_1^3} + \alpha\frac{y^*}{r_2^3}\right]. \tag{5.5.15}$$

In these last two equations r_1 and r_2 are given by (5.5.8) and (5.5.9), and the constants ω, a, b, and α are all as previously defined. That is, $\omega = \frac{2\pi}{l^{3/2}\sqrt{\mu}}$, $a = \mu l$, $b = (1 - \mu)l$, and $\mu = \frac{1}{1 + \alpha}$, where $M_2 = \alpha M_1$ and it is understood that $M_1 = 1$ solar mass (and that the units of distance and time are the AU and the year). Note carefully that as promised, the trigonometric time functions in the right-hand sides of (5.4.15) and (5.4.16) are now *not* present in (5.5.14) and (5.5.15), but instead there are now new *first*-derivative (Coriolis) terms on the left-hand sides and new ω^2 terms (centrifugal) on the right-hand sides.

Okay, now we can write code![20] Since it is understood that we are working in the rotating system, I will for convenience drop the asterisk notation that distinguishes x^* from x, and so on, and simply use x and y. But don't confuse this with my earlier use of the nonasterisk nota-tion to denote the inertial system. *We are now in the rotating system.* We proceed as usual to set things up for MATLAB by transforming our second-order differential equations into a set of first-order differential equations. So, let's define u_1, u_2, u_3, and u_4 just as we did in the defini-tions of (5.3.5). That is, I'll write $u_1 = x$, $u_2 = \frac{dx}{dt}$, $u_3 = y$, and $u_4 = \frac{dy}{dt}$

and so $\dfrac{du_2}{dt} = \dfrac{d^2x}{dt^2}$ and $\dfrac{du_4}{dt} = \dfrac{d^2y}{dt^2}$, and thus the equations we'll feed to MATLAB are (with just a bit of rearranging):

$$\frac{du_1}{dt} = u_2, \tag{5.5.16a}$$

$$\frac{du_2}{dt} = 2\omega u_4 + \left[\omega^2 - 4\pi^2\left\{\frac{1}{r_1^3} + \frac{\alpha}{r_2^3}\right\}\right]u_1 + 4\pi^2\left[b\frac{1}{r_1^3} - a\frac{\alpha}{r_2^3}\right],$$
$$\tag{5.5.16b}$$

$$\frac{du_3}{dt} = u_4, \tag{5.5.16c}$$

$$\frac{du_4}{dt} = -2\omega u_2 + \left[\omega^2 - 4\pi^2\left\{\frac{1}{r_1^3} + \frac{\alpha}{r_2^3}\right\}\right]u_3, \tag{5.5.16d}$$

where

$$r_1 = \sqrt{(u_1 - b)^2 + u_3^2} \tag{5.5.16e}$$

and

$$r_2 = \sqrt{(u_1 + a^2 + u_3^2}. \tag{5.5.16f}$$

The code **rbinary.m** implements the equations of (5.5.16), where the "r" in the name distinguishes the *rotating* coordinate system from the *fixed* coordinate system incorporated in the previous section's code **fbinary.m**. **rbinary.m** creates Figure 5.5.2, which shows that the aliens clearly *do* destroy the Earth. That's because the Earth detaches from the Sun (the asterisk) and goes into a precessing orbit around the death mass—at $(1.05, 0)$—with numerous close approaches. Indeed, the Earth probably wouldn't survive the first such close approach, with tidal forces almost surely shredding the Earth. If you compare Figure 5.5.2 with the inertial system orbit of Figure 5.4.3, you will see that it is simply impossible to "see" the precessing, close-approach behavior in the earlier figure. Switching to rotating coordinates has, therefore, been of great aid in improving our understanding of the workings of the death mass.

---

rbinary.m

```
function rbinary
alpha=1;l=2.1;fps=4*pi*pi;
mu=1/(1+alpha);a=mu*l;b=(1-mu)*l;w=2*pi/(sqrt(mu)*(l^(3/2)));
c1=2*w;c2=w*w;
options=odeset('AbsTol',1e-8,'RelTol',1e-5);
tspan=[0 2.2];
uzero=[-0.05;0;0;2*pi];
[t,u]=ode45(@orbit,tspan,uzero,options);
plot(u(:,1),u(:,3),'-k')
title('FIGURE 5.5.2 - Aliens destroy the Earth!','FontSize',16)
xlabel('unit of distance = 1 AU','FontSize',16)
ylabel('unit of distance = 1 AU','FontSize',16)
legend('orbit duration = 2.2 years')
      function xyderiv=orbit(t,u)
      r1=[u(1)-b,u(3)];denom1=norm(r1)^3;
      r2=[u(1)+a,u(3)];denom2=norm(r2)^3;
      c3=1/denom1;c4=alpha/denom2;c5=c2-fps*(c3+c4);c6=b*
        c3-a*c4;
      xyderiv=[u(2);c1*u(4)+c5*u(1)+fps*c6;u(4);-c1*u(2)+c5*u(3)];
      end
end
```

In addition to the physical insight the conversion to a rotating coordinate system gives to the restricted three-body problem there is also a theoretical payoff as well, one that Euler missed (and he didn't miss much). In 1836, more than half a century after Euler's death, Carl Jacobi found that the synodic differential equations can be formally integrated. Here's how. Multiplying through (5.5.14) by $\dfrac{dx}{dt}$ and (5.5.15) by $\dfrac{dy}{dt}$ gives

$$\frac{d^2x}{dt^2} \cdot \frac{dx}{dt} = 2\omega\frac{dy}{dt} \cdot \frac{dx}{dt} + \left\{\omega^2x - 4\pi^2\left[\frac{x-b}{r_1^3} + \alpha\frac{x+a}{r_2^3}\right]\right\}\frac{dx}{dt}$$

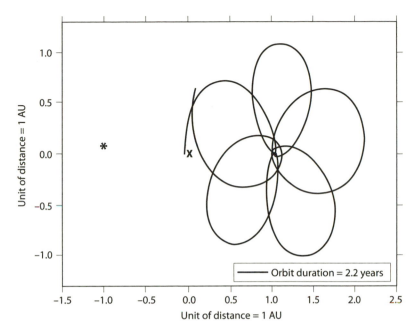

FIGURE 5.5.2. Aliens destroy the Earth!

$$(5.5.17)$$

and

$$\frac{d^2y}{dt^2} \cdot \frac{dy}{dt} = -2\omega \frac{dx}{dt} \cdot \frac{dy}{dt} + \left\{ \omega^2 y - 4\pi^2 \left[\frac{y}{r_1^3} + \alpha \frac{y}{r_2^3} \right] \right\} \frac{dy}{dt}. \quad (5.5.18)$$

Now, imagine that we have some function $U(x,y) = U(t)$, since x and y are themselves functions of t. Then the *total* differential of U is

$$dU = \frac{\partial U}{\partial x} dx + \frac{\partial U}{\partial y} dy,$$

and so it immediately follows that

$$\frac{dU}{dt} = \frac{\partial U}{\partial x} \cdot \frac{dx}{dt} + \frac{\partial U}{\partial y} \cdot \frac{dy}{dt}. \quad (5.5.19)$$

Imagine further that $U(x,y)$ is such that

$$\frac{\partial U}{\partial x} = \omega^2 x - 4\pi^2 \left[\frac{x-b}{r_1^3} + \alpha \frac{x+a}{r_2^3} \right] \quad (5.5.20)$$

and

$$\frac{\partial U}{\partial y} = \omega^2 y - 4\pi^2 \left[\frac{y}{r_1^3} + \alpha \frac{y}{r_2^3} \right]. \tag{5.5.21}$$

Adding (5.5.17) and (5.5.18) and using (5.5.20) and (5.5.21), we then have

$$\frac{d^2x}{dt^2} \cdot \frac{dx}{dt} + \frac{d^2y}{dt^2} \cdot \frac{dy}{dt} = \frac{\partial U}{\partial x} \cdot \frac{dx}{dt} + \frac{\partial U}{\partial y} \cdot \frac{dy}{dt}, \tag{5.5.22}$$

or, since the right-hand side of (5.5.22) is the left-hand side of (5.5.19),

$$\frac{d^2x}{dt^2} \cdot \frac{dx}{dt} + \frac{d^2y}{dt^2} \cdot \frac{dy}{dt} = \frac{dU}{dt},$$

or

$$\frac{d}{dt} \left\{ \frac{1}{2} \left(\frac{dx}{dt} \right)^2 \right\} + \frac{d}{dt} \left\{ \frac{1}{2} \left(\frac{dy}{dt} \right)^2 \right\} = \frac{dU}{dt}$$

which integrates by inspection to give (with $\frac{1}{2}C$ the constant of indefinite integration)

$$\frac{1}{2}C + \frac{1}{2} \left(\frac{dx}{dt} \right)^2 + \frac{1}{2} \left(\frac{dy}{dt} \right)^2 = U,$$

or, at last,

$$2U - \left\{ \left(\frac{dx}{dt} \right)^2 + \left(\frac{dy}{dt} \right)^2 \right\} = C, \tag{5.5.23}$$

where C is called *Jacobi's constant*.

If we know that at some particular time $t = T$, our vanishingly small mass m is at point (x, y) and is moving with velocity $\left(\frac{dx}{dt}, \frac{dy}{dt} \right)$, then its entire future motion for $t \geq T$ is such that the left-hand side of (5.5.23) is a constant, that is, the value of the Jacobi constant *puts constraints on m's motion*. To see this, suppose we impose what is called the *zero velocity condition* on m, that is, $\frac{dx}{dt} = \frac{dy}{dt} = 0$. We then have $2U = C$ at every point on the locus of points along which this condition holds (the locus is called a *zero-velocity curve*), and so this locus separates the x, y-plane into regions where either $2U > C$ or $2U < C$. If, in fact, m *has* a trajectory that one suspects might pass through a region where $2U < C$, then (5.5.23) requires

$$\left(\frac{dx}{dt}\right)^2 + \left(\frac{dy}{dt}\right)^2 < 0,$$

which is impossible, since the sum of two real quantities squared can't be negative. This means that any region of the x, y-plane in which $2U < C$ is a region *forbidden* to m. It can be shown—see Challenge Problem 5.6—that $2U$ has a minimum value of $3(\omega l)^2$, and so, as long as the Jacobian constant $C \leq 3(\omega l)^2$, then there will be no forbidden regions for m in the x, y-plane.

The crucial question left to be answered is, *is there a function* $U(x,y)$ such that (5.5.20) and (5.5.21) are in fact true? So far, all we've done is *assume* that U exists. I'll let you grind through the (routine) algebra, but if we define U as

$$U(x, y) = \frac{2\pi^2}{l^3} \left[r_1^2 + \alpha r_2^2\right]$$

$$+ 4\pi^2 \left[\frac{1}{r_1} + \frac{\alpha}{r_2}\right] \quad (\geq 0 \text{ for all } x \text{ and } y), \quad (5.5.24)$$

and remember that

$$\omega = \frac{2\pi}{l^{3/2}\sqrt{\mu}}, \quad \alpha = \frac{1 - \mu}{\mu}, \quad r_1 = \sqrt{(x - b)^2 + y^2}, \quad \text{and}$$

$$r_2 = \sqrt{(x - a)^2 + y^2},$$

then (5.5.20) and (5.5.21) follow.

5.6 Poincaré and the King Oscar II Competition

By now you should be convinced that the three-body *restricted* problem *in a plane* is an extremely complicated business, despite all the simplifications we've made. The general three-body problem, with no restrictions, in space, is even more difficult. It is true that the American astronomer George William Hill (1838–1914) used the restricted analyses of the previous sections and (much) more to effectively study (1877–1878) the Sun-Earth-Moon system (with the Moon playing the role of m), but by Hill's day theoreticians were becoming increasingly frustrated in their attempts to solve the three-body problem in general. This frustration finally resulted in the following announcement appearing in a number of scientific journals in mid-1885, in various languages, including French and German (below I've reproduced in part the English version, printed in the July 30, 1885, issue of the British journal *Nature*, pp. 302–303):

THE HIGHER MATHEMATICS

PROF. G. MITTAG-LEFFLER, principal editor of the *Acta Mathematica*, forwards us the following communication, which will shortly appear in that journal:—

His Majesty Oscar II., wishing to give a fresh proof of his interest in the advancement of mathematical science, an interest already manifested by his graciously encouraging the publication of the journal *Acta Mathematica*, which is placed under his august protection, has resolved to award a prize, on January 21, 1889, the sixtieth anniversary of his birthday, to an important discovery in the field of higher mathematical analysis.[21] This prize will consist of a gold medal ... bearing his Majesty's image and having a value of a thousand francs, together with a sum of two thousand five hundred crowns.[22]

His Majesty has been pleased to entrust the task of carrying out his intentions to a commission of three members, Mr. Carl Weierstrass in Berlin, Mr. Charles Hermite in Paris, and the chief editor of this journal, Mr. Gösta Mittag-Leffler in Stockholm. The commissioners having presented a report of their work to his Majesty, he has graciously signified his approval of the final propositions of theirs.

Having taken into consideration the questions which from different points of view equally engage the attention of analysts, and the solution of which would be of the greatest interest for the progress of science, the commission respectfully proposes to his Majesty to award the prize to the best memoir on one of the following subjects:—

1. A system being given of a number whatever of particles attracting one another mutually according to Newton's law, it is proposed, on the assumption that there never takes place an impact of two particles, to expand the coordinates of each particle in a series proceeding according to some known functions of time and converging uniformly for any space of time. It seems that this problem, the solution of which will considerably enlarge our knowledge with regard to the system of the universe, might be solved by means of the analytical resources at our present disposition....[23]

In case none of the memoirs tendered for competition on any of the subjects proposed above should be deemed worthy of the prize, this may be adjudged to a memoir sent in for competition that contains a complete solution of an important question of the theory of functions other than those proposed by the Commission.

The memoirs offered for competition should be furnished with an epigraph and, besides, with the author's name and place of residence in a sealed cover, and directed to the chief editor of the *Acta Mathematica* before June 1, 1888.

———◦———

Not quite four years later the following little announcement appeared, a single, easy-to-miss paragraph tucked away in the upper-left-hand corner of p. 396 of the February 21, 1889, issue of *Nature*:

———◦———

In the presence of a distinguished company of men of science, the King of Sweden recently opened the sealed papers containing the names of the two successful competitors for the mathematical prizes offered by him five years ago. The successful competitors were found to be Prof. H. Poincaré, of the Faculté des Sciences, Paris, who receives £160, and M. Paul Appert [this is a misprint: it should have read M. Paul Appell], Professor in the same Faculty, who receives a gold medal valued at £40. The papers, with reports by Profs. Weierstrass and Hermite, will be published in the *Acta Mathematica*. Twelve papers were sent in for the competition.[24]

—◇—

What happened during the years between those two announcements was nothing less than a revolution in celestial mechanics.

Notice that the first proposed problem of the competition is actually the n-body problem (we've only been discussing the special case of $n = 3$), and so apparently the Commission's hope was, "Maybe we'll get lucky and somebody will actually solve the *general* problem and we'll get the three-body solution as a special case!" In fact, only five of the twelve entries attempted the first problem—and none actually solved it, not even the French mathematician Henri Poincaré (1854–1912), who received top honors, and who had specifically limited himself to the restricted, planar three-body case. The competition was a very human affair, full of intrigue (there is strong evidence the fix was in to give Poincaré a decided advantage[25]—see note 21 again) and jealousy. And finally, it ended with that worst of all things that can happen to a mathematician: after being declared the winner, and as he polished his masterpiece for publication in *Acta Mathematica*, Poincaré discovered he had made a serious error.

Poincaré's original competition submission had claimed as a result that the *stability* of the orbit of m in the restricted three-body problem could be determined, and very possibly this result could be extended to include the general n-body case. This was a most important result, getting directly to the question of the long-term gravitational stability (or not) of the Solar System. That is, as multiple massive planets swing about the Sun, with all those bodies constantly pulling at each other with always time-varying forces, would their rhythmic, regular rotation eventually self-destruct? Would one or more of the planets eventually plunge into the Sun or, equally unpleasant, escape entirely from the Sun and shoot off into deep space? Poincaré's claim seemed to hold out the promise that such questions might actually have answers. And then, on November 30, 1889, Mittag-Leffler got a telegram from Poincaré informing him of the horrifying discovery of a mistake! "Hold the presses," was the message from the distraught mathematician, "until I can straighten things out!" Eventually he did, too,[26] and his "straightening out of things" led him to an even more stupendous discovery than was his original (erroneous) result—chaos!

It is commonly believed today that chaos is a relatively new concept, discovered in the form of the predictability of the weather only in the short term. That is, if one tries to predict the weather farther out than a few days (say, a week) into the future, one fails because the solution to the meteorological equations is simply far too sensitive to the initial conditions (which are *not* known with zero error). In fact, however, Poincaré discovered the same disconcerting behavior in Newton's gravitational equations for $n \geq 3$ massive bodies a full decade before the end of the nineteenth century (see note 10 of Chapter 4 again).

The famous "mathematical intelligence" that Laplace claimed (in his 1819 *Essai philosophique sur les probabilities*), if it "at a given instant knew all the forces acting in nature and the positions of every object in the universe—if endowed with a brain sufficiently vast to make all the necessary calculations—could describe with a single formula the motions of the astronomical bodies and those of the lightest atoms," was expelled from Poincaré's worldview. Laplace's next sentence—"To such an intelligence, nothing would be uncertain; the future, like the past, would be an open book"—was denied by Poincaré. And that beggars the additional issue of *what about* that "vast brain" itself—how could it "know" everything about every particle *in itself* (after all, it's part of the universe, too!)? Wouldn't just *thinking* about itself *change* itself?

Poincaré's appreciation for sensitivity to initial conditions in physical systems can be found in his 1897 book *Science and Method*, where he writes, "A very small cause, which escapes us, determines a considerable effect, which we cannot ignore, and we then say that this effect is due to chance." He then uses the weather as an example of this, and then, more interestingly (for us, here), there comes a celestial example: he writes,

> Let us pass to another example, the distribution of the minor planets on the Zodiac. Their initial longitudes may have had some definite order, but their mean motions [angular velocities]

were different and they have been revolving for so long that we say that practically they are distributed *by chance* throughout the Zodiac. Very small initial differences in their distances from the sun ... have resulted in enormous differences in the actual longitudes [remember Kepler's law relating orbit period to distance from the Sun]. A difference of a thousandth part of a second in the mean daily motion will have the effect of a second in three years, a degree in ten thousand years, a whole circumference in three or four million years, and what is that beside the time that has elapsed since the minor planets [formed]?

What, then, is the present status of the future of the Solar System? Are we doomed, or not? To quote one recent commentary on this,

The apparent regularity of the motion of the Earth and Pluto, and indeed the fact that the Solar System has survived for 4.5 billion years, implies that [chaos is not in the immediate future but] ... an error as small as 10^{-10} in the initial conditions will lead to a 100% discrepancy in longitude in 100 [million years]. ... The apparent stability of the current planetary system on billion-year timescales may simply be a manifestation of the fact that the Solar System is in the chaotic sense a dramatically young system.[27]

In other words, maybe the Solar System is stable, but maybe it isn't.

The prize commission's demand for a "uniformly convergent series" of the coordinates of all the gravitating bodies as a function of time is revealing. It was already suspected in 1885 that, unlike the two-body problem, it was impossible to simply integrate the differential equations of the n-body problem to get a solution for $n \geq 3$. Rather, the solution would have to take the form of an *infinite series*. This can be understood as follows. In three-dimensional space the motion of each of the n bodies is described by three second-order differential equations (Newton's second law for each of the three dimensions). Thus, there $(n)(3)(2) = 6n$ arbitrary constants; that is, the general n-body problem is said to have dimension $6n$. So, the general three-body problem has dimension 18.

The key to integrating such a high-dimensional system is to use *constraints* connecting the variables of the problem. For example, the fact that in Newtonian gravity energy and angular momentum are conserved means that no matter how the positions and velocities of all the masses vary with time, the total energy (a one-dimensional scalar) and the total angular momentum (a three-dimensional vector) remain constant. Each of these constraint conditions, usually called *first integrals*, is an algebraic function of the positions and velocities of the bodies which allows one variable (in theory, at least) to be expressed as a function of the remaining variables that appear in that first integral. So, each first integral results (in theory) in a reduction, by one, of the dimensionality of the problem. *If* one has enough independent first integrals, *then* the dimensionality of the problem can eventually (in theory) be reduced to 1, and that means the solution sought can be found by doing a final integration. To reduce the three-body problem of dimension 18 to dimension 1 would, therefore, require seventeen independent first integrals.

Alas, only ten first integrals of the n-body problem are known—three for the motion components of the center of mass (which is either stationary or moving at a constant velocity because there are no external forces acting on the bodies, only their own gravitational forces), three for the components of the conserved linear momentum vector, three for the components of the conserved angular momentum vector, and one more for the total energy—and 1887 the German mathematician Ernst Bruns (1848–1919) showed (with a few missteps that have since been cleaned up) that there *can be no additional first integrals*! So, integrating the differential equations for $n \geq 3$ was ruled out as a means for obtaining a solution, but that still left open the possibility for an infinite series expansion. Poincaré's prize memoir has no such solution, but in 1912 the Swedish-born Finnish mathematician Karl Sundman (1873–1949) *did* find such a solution for the three-body problem. Sundman's series is in powers of $t^{1/3}$, and as such, it converges *very slowly*.[28] It has been estimated that, to calculate to the accuracy normally required in astronomical work, something like $10^{8,000,000}$ terms of the series would have to be used![29] Still, if Sundman had submitted his series to the Oscar Prize Commission (of course, by the competition deadline he was still a teenager!), he almost surely would have been declared the winner, not Poincaré.[30]

So, we can't integrate the equations of motion for $n \geq 3$, and Sundman's series solution is, although mathematically perfectly correct, worthless for practical orbit calculations. What do we do? How can the evolution of actual n-body orbits over time be calculated from a given initial state of the masses? The answer, if one doesn't demand an orbit calculation that extends arbitrarily far into the future, is to use numerical methods. For us, today, that means to use an electronic computer and its modern software.

5.7 Computers and the Pythagorean Three-Body Problem

In 1893, not quite three years after Poincaré's *Acta Mathematica* paper appeared, the German mathematician Ernst Meissel (1826–1895) had a historically important conversation with the Dane Carl Burrau (1867–1947). Burrau, an astronomer, was in Meissel's hometown of Kiel on a trip, and Meissel (who was one of Jacobi's last students at the University of Berlin) mentioned a particular planar three-body problem to his visitor. The geometry of the problem is simple: In some system of units (I'll explain the system of units that we'll use in just a moment), three point masses, $m_1 = 3$, $m_2 = 4$, and $m_3 = 5$, are at time $t = 0$ *at rest* at the vertices of a right triangle with sides 3, 4, and 5. Corresponding masses and sides are opposite each other, as shown in Figure 5.7.1, with the right triangle positioned so that the center of mass for the three masses is at the origin—is that obvious to you from the figure, by inspection?—where it remains for all time (if it isn't moving at $t = 0$, then it *never* moves). The problem is to determine the orbits of each of the masses for $t \geq 0$.[31] Because of the geometry of the problem, it has become known as the *Pythagorean three-body problem*. Meissel's interest in this problem was sparked by his suspicion (for whatever reason) that its solution would have a periodic form, representing a stable configuration of gravitating masses that one might actually find somewhere, some day, in space.[32]

Burrau apparently took some time to mull the problem over, as it wasn't until 1913 that he published a partial numerical integration of the equations of motion. Given the technical constraints of the day—no electronic computers!—it is understandable that he wasn't able to make a lot of headway with that approach, but the limited

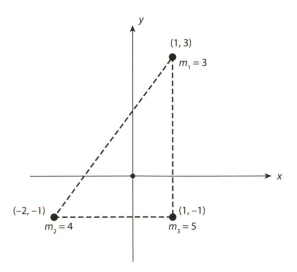

FIGURE 5.7.1. The $t = 0$ geometry of the Pythagorean three-body problem

analysis he was able to perform produced no suggestion of periodicity. The definitive answer would have to wait another half-century (until 1967), however, before the Pythagorean problem was finally successfully handled on a computer.[33] Here's how it was done.

To write the general differential equations of motion for three arbitrary (there are no restrictions now) masses moving in a plane, consider Figure 5.7.2, which shows the position vectors $\boldsymbol{\rho}_1$, $\boldsymbol{\rho}_2$, and $\boldsymbol{\rho}_3$ for masses μ_1, μ_2, and μ_3, respectively. Then, for μ_1, we have Newton's second law of motion (where I am using τ, not t, to represent time—you'll see why soon) in vector form as

$$\mu_1 \frac{d^2 \boldsymbol{\rho}_1}{d\tau^2} = G\mu_1\mu_2 \frac{\boldsymbol{\rho}_2 - \boldsymbol{\rho}_1}{|\boldsymbol{\rho}_2 - \boldsymbol{\rho}_1|^3} + G\mu_1\mu_3 \frac{\boldsymbol{\rho}_3 - \boldsymbol{\rho}_1}{|\boldsymbol{\rho}_3 - \boldsymbol{\rho}_1|^3}.$$

The terms on the right-hand side are as they are because the gravitational forces on μ_1 due to μ_2 and μ_3 are *from μ_1 to μ_2*, and *from μ_1 to μ_3*, respectively. That is, in the directions of the vectors $\boldsymbol{\rho}_2 - \boldsymbol{\rho}_1$ and $\boldsymbol{\rho}_3 - \boldsymbol{\rho}_1$, respectively. Thus,

$$\frac{d^2 \boldsymbol{\rho}_1}{d\tau^2} = G \sum_{j=2}^{3} \mu_j \frac{\boldsymbol{\rho}_j - \boldsymbol{\rho}_1}{|\boldsymbol{\rho}_j - \boldsymbol{\rho}_1|^3}. \tag{5.7.1}$$

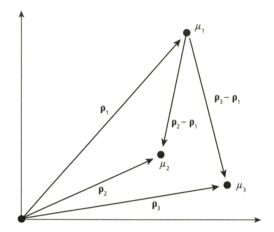

FIGURE 5.7.2. The position vectors of the general three-body planar problem

We can repeat the above procedure for μ_2 and μ_3 and get similar results. If you do that, you'll find that we can write

$$\frac{d^2\boldsymbol{\rho}_i}{d\tau^2} = G \sum_{j=1, j\neq i}^{3} \mu_j \frac{\boldsymbol{\rho}_j - \boldsymbol{\rho}_i}{|\boldsymbol{\rho}_j - \boldsymbol{\rho}_i|^3}, i = 1, 2, 3. \tag{5.7.2}$$

Now, let's scale (5.7.2) in the following way. Define L, M, and T as the units of length, mass, and time, respectively. Then, if we write the scaled versions of the position vectors, of the masses, and of time, as

$$\begin{aligned}
\mathbf{r}_i &= \frac{\boldsymbol{\rho}_i}{L} \\
m_i &= \frac{\mu_i}{M} \\
t &= \frac{\tau}{T}
\end{aligned} \tag{5.7.3}$$

then (5.7.2) becomes

$$\frac{d^2(\mathbf{r}_i L)}{d(tT)^2} = G \sum_{j=1, j\neq i}^{3} m_j M \frac{\mathbf{r}_j L - \mathbf{r}_i L}{|\mathbf{r}_j L - \mathbf{r}_i L|^3},$$

or

$$L\frac{d^2\mathbf{r}_i}{T^2dt^2} = GM \sum_{j=1,j\neq i}^{3} m_j \frac{L(\mathbf{r}_j - \mathbf{r}_i)}{L^3|\mathbf{r}_j - \mathbf{r}_i|^3},$$

or

$$\frac{d^2\mathbf{r}_i}{dt^2} = \frac{GMT^2}{L^3} \sum_{j=1, j\neq i}^{3} m_j \frac{(\mathbf{r}_j - \mathbf{r}_i)}{|\mathbf{r}_j - \mathbf{r}_i|^3}. \tag{5.7.4}$$

Our final scaling step is, with the gravitational constant $G = 6.67 \cdot 10^{-11}\frac{m^3}{kg \cdot s^2}$, to pick *any two* of the three units M, T, and L *any way we wish*, and to let those choices then determine the value of the remaining unit so that

$$\frac{GMT^2}{L^3} = 1, \tag{5.7.5}$$

a value which has an obvious attractiveness for numerical evaluation of (5.7.4). For example, suppose we use the mass of the Sun as the unit mass ($M = 1.99 \cdot 10^{30}$ kg) and one parsec as the unit distance ($L = 3.26$ light-years $= 3.08 \cdot 10^{16}$ m). Then, from (5.7.5) we have the unit of time as

$$T = \sqrt{\frac{L^3}{GM}} = \sqrt{\frac{(3.08 \cdot 10^{16})^3 m^3}{6.67 \cdot 10^{-11}\frac{m^3}{kg \cdot s^2} \cdot 1.99 \cdot 10^{30}\, kg}}$$

$$= \sqrt{22 \cdot 10^{28}\, s^2} = 4.69 \cdot 10^{14}\, s.$$

Since there are $31.536 \cdot 10^6$ seconds in a year, the unit of time is $T = 14.9$ million years. On the other hand, if we use the mass of the Earth as the unit mass ($M = 5.98 \cdot 10^{24}$ kg) and the AU as the unit of distance ($L = 1.49 \cdot 10^{11}$ m), then the unit of time is just a bit more than nine years.

Writing out (5.7.4) in detail, in preparation for writing MATLAB code, we have for the (scaled) mass m_1, with (5.7.5) in force,

$$\frac{d^2x_1}{dt^2} = m_2\frac{x_2 - x_1}{r_{21}^3} + m_3\frac{x_3 - x_1}{r_{31}^3},$$

and

$$\frac{d^2 y_1}{dt^2} = m_2 \frac{y_2 - y_1}{r_{21}^3} + m_3 \frac{y_3 - y_1}{r_{31}^3},$$

where $r_{ji} = \mid \mathbf{r}_j - \mathbf{r}_i \mid = r_{ij}$, and so

$$r_{21} = \sqrt{(x_2 - x_1)^2 + (y_2 - y_1)^2}, \quad r_{31} = \sqrt{(x_3 - x_1)^2 + (y_3 - y_1)^2}.$$

Now, as we've done before, let's make the following definitions:

$$u_1 = x_1, \quad u_2 = \frac{dx_1}{dt}, \quad u_3 = y_1, \quad u_4 = \frac{dy_1}{dt},$$

$$u_5 = x_2, \quad u_6 = \frac{dx_2}{dt}, \quad u_7 = y_2, \quad u_8 = \frac{dy_2}{dt},$$

$$u_9 = x_3, \quad u_{10} = \frac{dx_3}{dt}, \quad u_{11} = y_3, \quad u_{12} = \frac{dy_3}{dt}.$$

Thus,

$$\frac{du_1}{dt} = u_2, \quad \frac{du_2}{dt} = m_2 \frac{u_5 - u_1}{r_{21}^3} + m_3 \frac{u_9 - u_1}{r_{31}^3}, \quad \frac{du_3}{dt} = u_4,$$

$$\frac{du_4}{dt} = m_2 \frac{u_7 - u_3}{r_{21}^3} + m_3 \frac{u_{11} - u_3}{r_{31}^3},$$

where

$$r_{21} = \sqrt{(u_5 - u_1)^2 + (u_7 - u_3)^2}, \quad r_{31} = \sqrt{(u_9 - u_1)^2 + (u_{11} - u_3)^2}.$$

Similar equations follow just as easily from (5.7.4) and the above definitions for the (scaled) masses m_2 and m_3.

The following MATLAB code **burrau.m** implements these equations for the scaled Pythagorean three-body problem, creating Figure 5.7.3. In terms of scaled time (t), Burrau was able to calculate the paths of the three masses for $t = 0$ to $t = 3.35$, and that must have been an exhausting task a hundred years ago. By contrast, **burrau.m** produced Figure 5.7.3, the paths for the three masses for $t = 0$ to $t = 15$, in less than five seconds on my quite ordinary home computer. When you look at the calculated orbits you can't fail to appreciate just how complicated is the planar three-body problem!

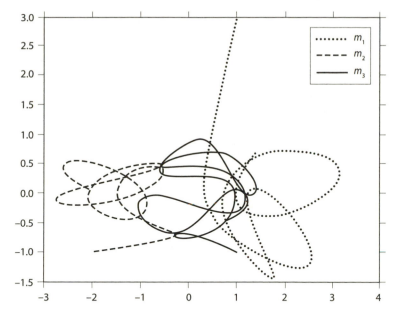

FIGURE 5.7.3. Pythagorean orbits for $0 \le t \le 15$

burrau.m

```
function burrau
global EMAX EMIN
m1=3;m2=4;m3=5;p12=m1*m2;p13=m1*m3;p23=m2*m3;EMAX
   =-769/60;EMIN=-769/60;
options=odeset('AbsTol',1e-14,'RelTol',1e-12);
tspan=[0 15];
uzero=[1;0;3;0;-2;0;-1;0;1;0;-1;0;];
[t,u]=ode45(@orbit,tspan,uzero,options);
EMAX,EMIN
plot(u(:,1),u(:,3),':k')
hold on
plot(u(:,5),u(:,7),'--k')
hold on
```

(Continued)

(Continued)

```
plot(u(:,9),u(:,11),'-k')
title('FIGURE 5.7.3 - Pythagorean orbits for 0  leq t  leq 15',
   'FontSize',16)
legend('m_1','m_2','m_3')
      function xyderiv=orbit(t,u)
       v21=[u(5)-u(1),u(7)-u(3)];r21=norm(v21);denom21=r21^3;
       v31=[u(9)-u(1),u(11)-u(3)];r31=norm(v31);denom31=r31^3;
       v32=[u(5)-u(9),u(7)-u(11)];r32=norm(v32);denom32=r32^3;
       a=(u(5)-u(1))/denom21;b=(u(9)-u(1))/denom31;c=(u(7)-u(3))/
          denom21;
       d=(u(11)-u(3))/denom31;e=(u(9)-u(5))/denom32;f=(u(11)-u(7))/
          denom32;
       KE=(m1*(u(2)^2+u(4)^2)+m2*(u(6)^2+u(8)^2)+m3*(u(10)^2+
          u(12)^2))/2;
       PE=-(p12/r21)-(p13/r31)-(p23/r32);
       E=KE+PE;
       EMAX=max(EMAX,E);EMIN=min(EMIN,E);
       xyderiv=[u(2);m2*a+m3*b;u(4);m2*c+m3*d;u(6);...
       -m1*a+m3*e;u(8);-m1*c+m3*f;u(10);-m1*b-m2*e;...
       u(12);-m1*d-m2*f;];
       end
  end
```

As the code clearly suggests, there is a lot of arithmetic going on in **burrau.m,** and that certainly should make you wonder about how accurate is Figure 5.7.3. As the code goes further and further forward in time, do round-off errors accumulate to the point where they "take control" of the calculations and lead to false orbits that have no connection with the true solution? An obvious way to at least make a partial attempt at answering that concern is, during the calculations, to monitor quantities that we know from theory should remain constant, such as the total energy. This isn't an infallible check, but certainly if the total energy doesn't remain constant as the calculations continually extend the orbits into the future, then you know the resulting solution can't be correct. To understand how **burrau.m** does this, we need to see how energy behaves during scaling.

The starting point is to write, from (5.2.9), the total energy as

$$\sum_{i=1}^{3} \frac{1}{2}\mu_i \left|\frac{d\boldsymbol{\rho}_i}{d\tau}\right|^2 - G\left[\frac{1}{2}\sum_{\substack{i,j=1\\j\neq i}}^{3} \frac{\mu_i\mu_j}{|\boldsymbol{\rho}_i - \boldsymbol{\rho}_j|}\right].$$

From our scaling relations of (5.7.3) we then have the total energy as

$$\frac{ML^2}{T^2}\sum_{i=1}^{3}\frac{1}{2}m_i\left|\frac{d\mathbf{r}_i}{dt}\right|^2 - \frac{1}{2}\cdot\frac{GM^2}{L}\sum_{\substack{i,j=1\\j\neq i}}^{3}\frac{m_im_j}{|\mathbf{r}_i - \mathbf{r}_j|}.$$

From (5.7.5) we have $\dfrac{1}{T^2} = \dfrac{GM}{L^3}$, and so the total energy is

$$\frac{GM^2}{L}\left[\sum_{i=1}^{3}\frac{1}{2}m_i\left|\frac{d\mathbf{r}_i}{dt}\right|^2 - \left\{\frac{m_1m_2}{|\mathbf{r}_1 - \mathbf{r}_2|} + \frac{m_1m_3}{|\mathbf{r}_1 - \mathbf{r}_3|} + \frac{m_2m_3}{|\mathbf{r}_2 - \mathbf{r}_3|}\right\}\right].$$

Since the total energy is constant, and so is $\dfrac{GM^2}{L}$, then we have our scaled statement of the conservation of energy E:

$$E = \left[\sum_{i=1}^{3}\frac{1}{2}m_i\left|\frac{d\mathbf{r}_i}{dt}\right|^2 - \left\{\frac{m_1m_2}{|\mathbf{r}_1 - \mathbf{r}_2|} + \frac{m_1m_3}{|\mathbf{r}_1 - \mathbf{r}_3|} + \frac{m_2m_3}{|\mathbf{r}_2 - \mathbf{r}_3|}\right\}\right]$$

$$= \text{constant}.$$

Since $r_{ji} = |\mathbf{r}_j - \mathbf{r}_i|$ and $\left|\dfrac{d\mathbf{r}_i}{dt}\right|^2 = \left(\dfrac{dx_i}{dt}\right)^2 + \left(\dfrac{dy_i}{dt}\right)^2$, then, in terms of the variables in **burrau.m**, we have the result that the kinetic energy

$$KE = \frac{1}{2}\left\{m_1(u_2^2 + u_4^2) + m_2(u_6^2 + u_8^2) + m_3(u_{10}^2 + u_{12}^2)\right\}$$

and the potential energy

$$PE = -\frac{m_1 m_2}{r_{21}} - \frac{m_1 m_3}{r_{31}} - \frac{m_2 m_3}{r_{32}}$$

must always be such that $E = KE + PE$ remains constant. This constant is easy to calculate for the Pythagorean problem since, initially, $KE = 0$ (the masses start from rest). Thus, from Figure 5.7.1,

$$E = -\frac{(3)(4)}{5} - \frac{(3)(5)}{4} - \frac{(4)(5)}{3} = -\frac{769}{60} = -12.816666 \cdots .$$

If you look at **burrau.m** you can see the code implements the KE, PE, and E calculations as it constructs each new row of the u matrix, and saves both the maximum and the minimum values of E that it has currently encountered (as the variables $EMAX$ and $EMIN$, respectively). These calculations are done in the nested function *xyderivative*, and since they are printed out at the end of program execution in the main function, those variables have earlier been declared to be global (normally, in MATLAB, variables are available only inside the local function in which they are calculated). When **burrau.m** produced Figure 5.7.3 it ended with $EMAX = -12.8069$ and $EMIN = -12.8176$, a variation of less than $\pm 0.1\%$. So, E was not perfectly constant during the calculations, but it was pretty nearly so.

Figure 5.7.3 doesn't show anything to lead one to suspect periodicity, but in fact if we run **burrau.m** farther out in time we see something happen that is far more interesting. Figure 5.7.4 shows the orbits for the interval $0 \leq t \leq 63$ and, after an incredibly wild interplay, the system of three gravitating masses *explodes*! The mass m_1 is ejected in one direction (toward the upper right), while masses m_1 and m_3 depart in the opposite direction as a *binary pair* (the elliptical orbits heading towards the lower left). (The fact that the two departure paths are in the opposite directions is a further indication that the code is working properly, as that is the only possible way to keep the center-of-mass of the expanding system motionless at the origin.) Before this expanding behavior was seen in the 1967 computer solution (see note 33 again) I don't believe it had ever been

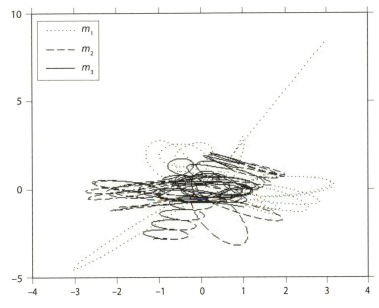

FIGURE 5.7.4. Pythagorean orbits for $0 \leq t \leq 63$

suggested such a thing could happen. It is completely counterintuitive behavior for a system governed totally by forces of attraction.[34] The numerical solution of the Pythagorean problem also provided a possible explanation for a longstanding puzzle: why are binary star systems so common? The Pythagorean three-body problem showed how three individual stars could interact to produce a binary pair through a physical process that, over a time span of hundreds of millions of years, must happen countless times in every galaxy in the universe.

burrau.m worked very hard to create Figure 5.7.4, taking a full forty-nine seconds to complete its calculations (when finished, the u matrix had $184,461$ rows!), during which $EMAX = -12.6008$ and $EMIN = -12.8373$. This is only a variation of $\pm 1.7\%$ about the initial energy, but still, it *is* larger than before. So, we are faced with our old question once again: could inescapable numerical round-off errors have at last caught up with us? Is there some other way to check the computed solution besides monitoring the total energy? Indeed there is, and it is quite elegant. *So* elegant, in fact, that it is considered to be the gold standard for checking computer-generated

n-body solutions. What we'll do is first let **burrau.m** produce a solution for the time interval $0 \le t \le 63$. Then we'll create a new initial state vector *uzero* equal to the *final* state of the solution, that is, equal to the final row in the *u*-matrix. Next, we'll reverse all the velocity components by flipping the signs of the appropriate rows of *uzero*. And finally, we'll solve the problem all over again for the time interval $0 \le t \le 63$, *starting* with the modified *uzero*, that is, starting from the original solution's *final* state vector. The result should be that each mass retraces its path and *ends up back at its starting point, at rest*. I take this idea as physically plausible, but it is not difficult to put together a formal mathematical proof.[35] Since the final state of the masses after this last run of **burrau.m** is the outcome of whatever errors have occurred over twice the desired solution interval, then, if the final state is still pretty close to the original starting state, we can be pretty confident we have a "good" solution for $0 \le t \le 63$.

All of this is quite easy to do: simply insert, immediately before the nested function *xyderivative*, the following additional lines of code into **burrau.m**:

```
[m,n]=size(u)
uzero=[u(m,1);u(m,2);u(m,3);u(m,4);u(m,5);u(m,6);u(m,7);u(m,8);
    u(m,9);... u(m,10);u(m,11);u(m,12)];
clear u;clear t;
uzero(2,1)=-uzero(2,1);uzero(4,1)=-uzero(4,1);uzero(6,1)
    =-uzero(6,1);
uzero(8,1)=-uzero(8,1);uzero(10,1)=-uzero(10,1);uzero(12,1)
    =-uzero(12,1);
[t,u]=ode45(@orbit,tspan,uzero,options);
[m,n]=size(u)
u(m,:)
```

When this modified version of the code was run, it produced the final state vector shown below, in side-by-side comparison with the original starting state vector:

Original State Vector *uzero*	Final State Vector After Velocity Reversal
1	1.0585
0	−0.0264
3	2.9775
0	0.0199
−2	−1.994
0	0.0225
−1	−0.79
0	0.0254
1	0.9601
0	−0.0021
−1	−1.1545
0	−0.0323

We don't have perfect agreement, but the two state vectors are fairly close and I think the calculated solution of Figure 5.7.4 is a reasonably accurate solution.

5.8 Two Very Weird Three-Body Orbits

In this concluding section of the chapter, I want to show you, just for fun, two extremely surprising solutions to the three-body problem, one planar and one in three dimensions. The "explosion" of Burrau's Pythagorean problem was certainly surprising, but the two orbits that follow are *so* surprising that they can be (I think) rightfully called weird!

The first solution, reported in the literature in 2000, is for equal masses.[36] That is, $m_1 = m_2 = m_3 = 1$. The problem starts at $t = 0$ with the initial state vector

```
uzero=[0.97000436;0.93240737/2;-0.24308753;...
0.86473146/2;-0.97000436;0.93240737/2;...
0.24308753;0.86473146/2;0;...
-0.93240737;0;-0.86473146];
```

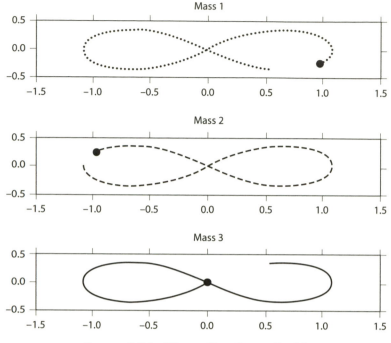

FIGURE 5.8.1. The endless figure-8 orbit

and we'll use *tspan=[0 5.8]*. (The full orbital period, for each mass, is 5.3259····.) With these two changes, the otherwise identical code of **burrau.m** produces Figure 5.8.1, which shows that the three masses chase after one another in an endless figure-8 orbit! The large black dot in each plot marks the $t = 0$ location of each mass.

This orbit has zero angular momentum, and is an example of how in such a system a triple collision (total collapse) is not inevitable (see note 28 again). The gravitational interactions between the three masses is so exquisite that mathematicians often don't use the word **orbit** for the figure-8 but rather borrow from the art of dancing and call it a **choreography**. The probability that such a triple system exists (or has existed, or ever will exist), over the entire history of the entire universe, seems to me to be just about zero. It is just too delicately dependent on all three masses being precisely equal, and on initial conditions for location and velocity being precise to eight decimal places. The point here, though, is would you have ever imagined that Newtonian gravity would even allow such a bizarre possibility?

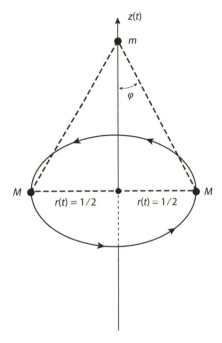

FIGURE 5.8.2. A three-dimensional, three-body problem

For our second weird orbit (and our only three-dimensional three-body problem), consider Figure 5.8.2, which shows two equal masses (M) as a binary pair in a common circular orbit. The center of mass, about which each M orbits, is obviously the center of the orbit. As before, we'll take the center of mass as the origin of our coordinate system. Now, just as we did in Euler's restricted problem, let's suppose we have a third body of *vanishingly small mass* (m) that moves along the line normal to the plane of the binary orbit and that passes through the center of mass (this line will be the z-axis), as shown in the figure. This third body, because of its tiny mass, will not disturb the binary but will itself respond to the gravitational forces of the binary pair. Our problem is to determine how this third body moves.

Without doing any mathematical analysis (yet), there are several qualitative observations we can immediately make about that motion:

1. if m is at the center of mass, with zero speed, then m will stay there;

2. if m is released from from rest at some point $z \neq 0$, then m will oscillate up and down the z-axis;

3. if m is at the center of mass with some positive vertical speed that does not exceed some critical maximum value $= V$, then m will oscillate up and down the z-axis;

4. if m is at the center of mass with some positive speed that does exceed some critical maximum value $= V$ (see Challenge Problem 5.8), then m will move out along the z-axis to plus infinity and never return.

If we define the unit distance to be the diameter of the binary circular orbit (and so the orbital radius is $r(t) = \frac{1}{2}$), the unit mass to be the total mass of the binary (and so $M = \frac{1}{2}$), and take the gravitational constant to be $G = 1$ (a choice that will serve us well), then the orbital period of the binary components is 2π. You should confirm this for yourself but, if you'd rather not, check the notes.[37] You'll notice from Figure 5.8.2 that, because of symmetry, m is always the same distance from each of the binary components (called the *primaries*) no matter where m is on the z-axis or where the primaries are on their orbit, and because of that, the horizontal gravitational forces on m cancel while the vertical forces add (and the sum is always along the z-axis directed toward the center of the binary circular orbit). In the notation of Figure 5.8.2, the total z-axis force on m is (the minus sign is, of course, because the force on m is *downward* when $z > 0$ and *upward* when $z < 0$):

$$-2G \frac{Mm}{z^2 + (\frac{1}{2})^2} \cos(\phi),$$

and so, as

$$\cos(\phi) = \frac{z}{\sqrt{z^2 + (\frac{1}{2})^2}},$$

then from Newton's second law we have

$$m \frac{d^2 z}{dt^2} = -2G \frac{Mmz}{\{z^2 + (\frac{1}{2})^2\}^{3/2}},$$

or, as $G = 1$ and $M = \dfrac{1}{2}$,

$$\frac{d^2z}{dt^2} = -\frac{z}{\left\{z^2 + \frac{1}{4}\right\}^{3/2}}. \tag{5.8.1}$$

This problem can be traced to as far back as 1907, and it is in fact integrable. It doesn't have a "nice" solution—it involves elliptic integrals[38]—but we can still easily generate plots of $z(t)$ using MATLAB. We can numerically solve (5.8.1) with the code **csitnikov.m**—I'll explain the curious name in just a bit—in which we set the initial conditions as usual with *uzero*. For example, Figure 5.8.3 shows $z(t)$ for $0 \le t \le 6\pi$ (that is, for three full orbits of the primaries), for $z(0) = 1.3$ and $\left.\dfrac{dz}{dt}\right|_{t=0} = 0$. As expected, $z(t)$ is a constant-amplitude oscillation but, as a check of the phase-plane portrait for the system will confirm, the oscillations are *not* sinusoidal (also as you'd expect, given the decidedly *nonharmonic* form of (5.8.1)!).

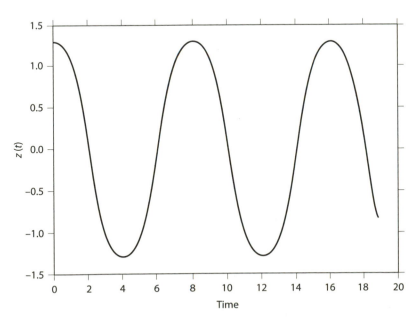

FIGURE 5.8.3. z-axis oscillations in a three-body problem

---

csitnikov.m

```
function csitnikov
options=odeset('AbsTol',1e-10,'RelTol',1e-8);
tspan=[0 6*pi];
uzero=[1.3;0];
[t,u]=ode45(@orbit,tspan,uzero,options);
plot(t,u(:,1),'-k')
xlabel('time','FontSize',16)
ylabel('z(t)','FontSize',16)
title('FIGURE 5.8.3 - z-axis oscillations in a three-body
    problem','FontSize',16)
        function zderiv=orbit(t,u)
        zderiv=[u(2);-u(1)/((u(1)^2+0.25)^1.5)];
        end
end
```

Well, all this is interesting, but certainly not surprising, and absolutely none of it is "weird." What makes the system weird is a seemingly benign twist given in 1960 to the system geometry of Figure 5.8.2 by the Russian mathematician Kirill Sitnikov (born 1926). Instead of imagining the primaries moving on a common *circular* orbit about their center of mass, Sitnikov considered the more general possibility of the primaries orbiting the common center of mass on separate *elliptical* orbits, as shown in Figure 5.8.4. (Now you know why the code **csitnikov.m** has that name; it stands for "circular Sitnikov problem.") The two elliptical orbits collapse into the single circular orbit we considered earlier if the eccentricity ε of the elliptical orbits is zero. The center of mass of the binary system is the common focus for both elliptical orbits, and the z-axis passes, as before, through the center of mass/focus/origin. Each elliptical orbit is "timed" so that the motions of the primaries "mirror" each other; that is, at every instant the distance each primary is from the focus is the same, $r(t)$. In Figure 5.8.4 this distance is given as $r(\theta)$, a function of θ and *not* of time, because that is the functional dependence we derived earlier in (5.2.21). We will have

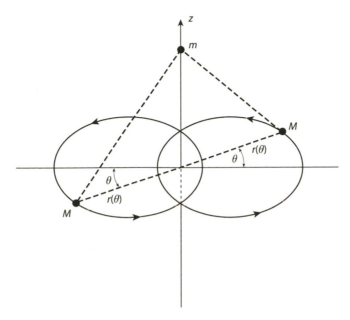

FIGURE 5.8.4. Elliptical Sitnikov problem

to address this issue soon, and you'll be pleased to learn it has a nice solution.

To repeat (5.2.21),

$$r(\theta) = \frac{\alpha}{1 + \varepsilon \cos(\theta - \theta_0)}.$$

If we time the primaries so that the arbitrary constant $\theta_0 = 0$, then we have

$$r(\theta) = \frac{\alpha}{1 + \varepsilon \cos(\theta)}. \qquad (5.8.2)$$

This says that since we'll take $\theta = 0$ when $t = 0$, and since $\varepsilon \geq 0$, we have at $t = 0$ the primaries at their closest approach to the center of mass (the focus/origin).

To determine α, I'll use a general result for elliptical orbits.[39] If we keep the major axis of each orbit as 1 then, when $\varepsilon = 0$, the ellipses collapse into a circle with a diameter of 1, as we had earlier and the closest approach of each primary to the focus is $\frac{1}{2}(1 - \epsilon)$. Thus, from (5.8.2) we have (using $\theta = 0$ at closest approach)

$$\frac{\alpha}{1 + \varepsilon} = \frac{1}{2}(1 - \epsilon),$$

or

$$\alpha = \frac{1}{2}(1 - \epsilon)(1 + \epsilon) = \frac{1}{2}(1 - \epsilon^2).$$

Thus,

$$r(\theta) = \frac{\frac{1}{2}(1 - \epsilon^2)}{1 + \varepsilon \cos(\theta)}, 0 \leq \epsilon < 1. \tag{5.8.3}$$

As stated before, each primary is always the same distance from m on the z-axis, just as in the circular orbit case, and so (5.8.1) is, most generally,

$$\frac{d^2z}{dt^2} = -\frac{z}{\{z^2 + r^2(\theta)\}^{3/2}}. \tag{5.8.4}$$

And finally, we have to address the problem of having $r(\theta)$ and not $r(t)$. As it stands, with both t and θ in (5.8.4) we can't integrate (5.8.4). The solution is to use Kepler's second law of planetary motion (see note 9 again) that we derived earlier as (5.2.15), expressing the conservation of angular momentum, which I'll repeat here:

$$r^2(\theta)\frac{d\theta}{dt} = c, \tag{5.8.5}$$

where the *constant* $c = rv$. r is the instantaneous distance of the primaries from their common focus, and v is the instantaneous orbital speed of each primary (remember, because of the timing of the elliptical orbits r and v are always the same for each primary). What we have, then, is a coupled system of differential equations, the second-order (5.8.4) and the first-order (5.8.5). With (5.8.5) we can numerically solve for $\theta = \theta(t)$ and use it with (5.8.3) and (5.8.4) to find $z(t)$.

So, what's c? If we know r and v at any particular instant of time, then we know c for *all* t, since c is a constant. At the instant of closest approach on an elliptical orbit, for example (see note 39 again), we have

$$r = \frac{1}{2}(1 - \epsilon)$$

and

$$v^2 = \frac{GM}{\frac{1}{2}} \cdot \frac{1 + \epsilon}{1 - \epsilon}.$$

Since $G = 1$ and $M = \frac{1}{2}$ for our problem, then

$$v = \sqrt{\frac{1 + \epsilon}{1 - \epsilon}},$$

and so

$$c = \frac{1}{2}\sqrt{1 - \epsilon^2}.$$

The code **esitnikov.m** (for "elliptical Sitnikov problem") implements all of the above and, for a randomly selected set of parameters—$z(0) = 3$, $\left.\frac{dz}{dt}\right|_{t=0} = 0, \epsilon = 0.15$—plots $z(t)$ over the interval $0 \leq t \leq 100\pi$ (that is, the primaries go through fifty full orbits) in Figure 5.8.5. In the u matrix we have $u_1 = z$, $u_2 = \frac{dz}{dt}$, and $u_3 = \theta$.

```
esitnikov.m
function esitnikov
options=odeset('AbsTol',1e-10,'RelTol',1e-8);
e=0.15;c=sqrt(1-e^2)/2;k=(1-e^2)/2;
tspan=[0 100*pi];
uzero=[3;0;0];
[t,u]=ode45(@orbit,tspan,uzero,options);
plot(t,u(:,1),'-k')
```

(Continued)

(Continued)

```
xlabel('time','FontSize',16)
ylabel('z(t)','FontSize',16)
title('FIGURE 5.8.5 - Elliptical Sitnikov oscillations (/epsilon=0.15)',
    'FontSize',16)
      function zderiv=orbit(t,u)
      r=k/(1+e*cos(u(3)));
      zderiv=[u(2);-u(1)/((u(1)^2+r^2)^1.5);c/r^2];
      end
end
```

No one would call the behavior of $z(t)$ in Figure 5.8.5 "obvious and expected." The increasingly wild swings in the location of m are the result of the non-zero eccentricity of the primary orbits, because now the gravitational force on m depends not only on z, but also on where

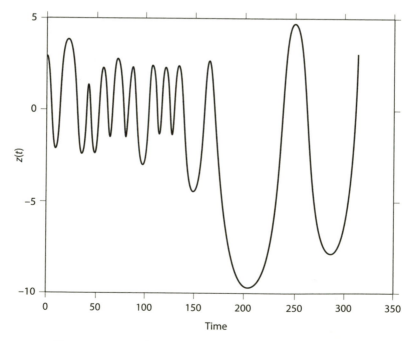

FIGURE 5.8.5. Elliptical Sitnikov oscillations ($\epsilon = 0.15$)

the primaries are in *their* orbits. In the $\epsilon = 0$ case the force on m is always the same for a given z, but in the $\epsilon \neq 0$ case, every time m revisits the same point on the z-axis it almost surely experiences a different force. And that means a tiny change in the initial conditions of how we start m off on its journey will result in a quite different $z(t)$. Figure 5.8.5 illustrates what Poincaré suspected—chaos. Even a restricted three-body problem can be complicated to the point that its behavior is so far beyond analytical calculation as to be simply "weird." As with the figure-8 orbit, the probability something like Sitnikov's system could actually physically exist anywhere in the universe is zero. But it does exist, weird as it is, in theoretical Newtonian gravity.

This chapter opened with three gloomy comments on Newton's headaches over the three-body problem. Those headaches, as I'm sure you now appreciate, have not been cured in the more than three centuries since Newton. So, let me close with this humorous (but still gloomy) quotation from the opening of Laurence Taff's book *Celestial Mechanics* (New York: Wiley-Interscience, 1985):

> Nature is wonderfully perverse. Here [Newton's theory], in the first rational formulation of gravity and mechanics one can solve, completely, the two-body problem. In the next step up (general relativity), one can solve, completely, the one-body problem. No doubt when the super grand unified theory is developed, one will only be able to solve, completely, the zero-body problem.

In an amusing echo of (and one-up on) Taff, the authors of a more recent book note that "We cannot solve the one-body problem [in quantum electrodynamics]. And in quantum chromodynamics we cannot even solve the zero-body problem, the vacuum.[41]

Pretty gloomy stuff, for sure. But, don't forget—we'll always have MATLAB and its wonderful number-crunching *ode45*![42]

5.9 Challenge Problems

CP 5.1. A mass M is fixed at the origin, and a mass m (which is free to move) is at $x = d$ on the positive x-axis at time $t = 0$. If m is at rest at $t = 0$, calculate when ($t = T$) m will collide with M due to Newtonian gravitational attraction.

CP 5.2. Consider the same situation described in the previous problem, but now suppose that the force of gravitational attraction is the general $|F| = GMf(x)$, where $f(x)$ is an *arbitrary* positive function of x. (Newtonian gravity is the particular case $f(x) = \dfrac{1}{x^2}$.) Show that m will always collide with M in a finite time. *Hint:* Do this by computing an upper bound T_m on the actual collision time T, and show that T_m is finite (and so, since $T < T_m$, then T is finite, too). Compare T_m with the actual T you calculated in the previous problem for the specific case of Newtonian gravity. Are they "close"?

CP 5.3. From vector algebra we have the identity $(\mathbf{a} \cdot \mathbf{b})^2 + |\mathbf{a} \times \mathbf{b}|^2 = a^2 b^2$, which you can verify by the straightforward (if tedious) method of expanding the left-hand side. Then, setting $\mathbf{a} = \mathbf{r}$ and $\mathbf{b} = \mathbf{v}$, confirm the claim made in the text, immediately after (5.2.11), that

$$v^2 = \left(\frac{dr}{dt}\right)^2 + \frac{c^2}{r^2}, \text{ where } c = |\mathbf{c}| = |\mathbf{r} \times \mathbf{v}|.$$

CP 5.4. Suppose there is a planetary system somewhere in the universe with a star of exactly the same size as the Sun, with a planet exactly the same size as the Earth, in a circular orbit with a radius of 1 AU. Unlike our Solar System, however, there is also a duplicate Earth directly opposite the Earth-like planet, on the opposite side of the star, in the same circular orbit (radius 1 AU) moving in the same direction as the Earth-like planet. Each planet, therefore, can never be seen from the other. (This amusing gimmick was used in the 1969 British film *Journey to the Far Side of the Sun*.) If we use our usual $GM = 4\pi^2$ units (so that $T = 1$ is the duration of our Earth-year), and T_D is the duration of the orbital period of both our imaginary Earth-like planet and its duplicate, calculate the difference (in seconds) between T and T_D. The masses of the Sun and the Earth are, respectively, $1.99 \cdot 10^{30}$ kg and $5.98 \cdot 10^{24}$ kg.

CP 5.5. Suppose we have three identical point masses (m) at the vertices of an equilateral triangle, each in the same circular orbit of radius R about their center of mass. What is the orbital period (T) of each mass in terms of m and R?

CP 5.6. Prove the claim made near the end of Section 5.5, that $2U$ (U is defined in (5.5.24)) has the *minimum* value of $3(\omega l)^2$.

Here is a suggestion or two on how to proceed for you to think about.

1. Start by rewriting (5.5.24) into the form
   $2U = 4\pi^2 \left[A(r_1) + \frac{1-\mu}{\mu} B(r_2) \right]$, where you should observe that $A(r_1)$ and $B(r_2)$ have the same functional form.

2. Explain why $A(r_1)$ and $B(r_2)$ are both ≥ 0.

3. Establish the truth of the claim that, for $0 \leq \mu \leq 1$, and $A, B \geq 0$, we have $A + \dfrac{1-\mu}{\mu} B \geq \dfrac{\min(A, B)}{\mu}$.

4. At this point, you are only a step or two away from being done.

CP 5.7. Imagine two stars as a binary pair. Then a third star approaches from deep space and gravitationally interacts with the binary. There are several possible scenarios that we can imagine for what happens next. For example, the third star disrupts the binary and all three stars then individually depart into deep space, or the third star forms a new binary with one of the original binary stars while the other original binary star departs, or the third star dances around the binary for a while and then departs, leaving the original binary stars still in a binary relationship. But there is one scenario that is impossible: the formation of a stable triple star system. Use the time reversibility of the Newtonian gravitational equations to explain why this is so. *Hint:* The word *stable* is a big clue, and *no math is required*!

CP 5.8. Calculate the value of V for the circular Sitnikov problem of Figure 5.8.2. V is the critical speed for the tiny mass m that moves along the z-axis; if its speed, when passing through the binary center of mass, exceeds V, then m will never return. That is, V is the *escape speed* for m from the binary along the z-axis.

5.10 Notes and References

1. See, for example, my book *Mrs. Perkins's Electric Quilt and Other Intriguing Stories of Mathematical Physics* (Princeton, N.J.: Princeton University Press, 2009, pp. 136–214).

2. $\mathbf{F} = m\mathbf{a}$ is an important but special case of Newton's second law of motion, which states that force is the rate of change of momentum. That is, $\mathbf{F} = \dfrac{d(m\mathbf{v})}{dt} = m\dfrac{d\mathbf{v}}{dt} + \mathbf{v}\dfrac{dm}{dt}$, which reduces to $\mathbf{F} = m\mathbf{a}$ (where $\mathbf{a} = \dfrac{d\mathbf{v}}{dt}$) if $\dfrac{dm}{dt} = 0$. But this is not true, for example, in the motion of rockets, where m (the mass of the rocket body plus fuel) changes as the fuel is burned and exhausted out the rear. For the motion of astrophysical objects like stars, however, it is a pretty safe assumption unless one is discussing (for example) the evolution of orbits over enormous expanses of time (billions of years?) or is taking into account the exchange of mass between close-orbiting binary stars. $\mathbf{F} = m\mathbf{a}$ is true only in an *inertial* frame of reference, that is, one that is not accelerating with respect to some other frame that is, somehow, known to be "fixed," such as the background of the remotely distant stars. A *rotating* frame of reference, for example, is not inertial, and so to make $\mathbf{F} = m\mathbf{a}$ still work in such a frame it is necessary to include what physicists call *fictitious* or *pseudo*-forces in addition to the real, physical forces that are present. We'll encounter such an accelerated (rotating) frame, and the resulting pseudo-forces, later in this chapter, when we discuss Euler's restricted version of the three-body problem.

3. In all of the three-body discussions I have seen, by physicists and mathematicians alike, the massive bodies are always taken to be point objects. One of Newton's great mathematical accomplishments was to establish that an extended spherically symmetric body of mass m and radius R "looks" like a point object when it gravitationally interacts with another massive body more distant than R from the center of the mass m (see note 1 again).

4. Not all forces are central. When, for example, a charged particle (an electron, say) moves through a magnetic field, there is a force (called the *Lorenz* force, after the Dutch physicist Hendrik Antoon Lorentz [1853–1928], who received a share of the 1902 Nobel Prize) on the electron that is at right angles to both the direction of the field (that is, to the magnetic field vector at the instantaneous location of the electron) and the velocity vector of the electron. We'll not worry here about such a force, even though magnetic fields do exist in space, because a massive electrically charged body is not a likely object in the context of an *astrophysical* three-body problem. General relativity enthusiasts like to talk of the mathematics of an electrically charged massive black hole, but such an object would actually soon neutralize itself as it gobbled up matter (remember, electrical charges of opposite sign attract, and a charged black hole of one polarity would be like a super-vacuum cleaner—gravitationally and electrically—for any charges of the opposite sign that happen by).

5. I've taken this entire example, and the mathematical approach I'm using, from the paper by Donald G. Saari, "A Visit to the Newtonian N-body Problem via Elementary Complex Variables" (*American Mathematical Monthly*, February

1990, pp. 105–119). It also appears in Saari's book *Collisions, Rings, and Other Newtonian N-Body Problems* (New York: American Mathematical Society, 2005).

6. The two-body problem originated in Newton's *Principia*, but it was actually the Swiss mathematician Johann Bernoulli (1667–1748) who first completely solved it (in 1710). There are just a few technical details I am going to expect you to be familiar with here. If **a** and **b** are any two vectors in three-dimensional space, and **i**, **j**, and **k** are the unit vectors along the three perpendicular axes (x, y, and z, respectively) of a rectangular coordinate system in that space, then

(a) the three-dimensional *vector dot product* is

$$\mathbf{a} \cdot \mathbf{b} = (a_x\mathbf{i} + a_y\mathbf{j} + a_z\mathbf{k}) \cdot (b_x\mathbf{i} + b_y\mathbf{j} + b_z\mathbf{k}) = a_xb_x + a_yb_y + a_zb_z;$$

(b) the three-dimensional *vector cross-product* is the 3×3 determinant

$$\mathbf{a} \times \mathbf{b} = \begin{vmatrix} \mathbf{i} & \mathbf{j} & \mathbf{k} \\ a_x & a_y & a_z \\ b_x & b_y & b_z \end{vmatrix}$$

$$= \mathbf{i}\left(a_yb_z - a_zb_y\right) - \mathbf{j}(a_xb_z - a_zb_x) + \mathbf{k}(a_xb_y - a_yb_x)$$

which immediately tells us that $\mathbf{a} \times \mathbf{b} = -\mathbf{b} \times \mathbf{a}$;

(c) the magnitude of **a** is $|\mathbf{a}| = a = \sqrt{a_x^2 + a_y^2 + a_z^2}$, and $|\mathbf{b}| = b = \sqrt{b_x^2 + b_y^2 + b_z^2}$;

(d) the derivative of $\mathbf{a} \cdot \mathbf{a}$ is

$$\frac{d\,(\mathbf{a} \cdot \mathbf{a})}{dt} = \frac{d\mathbf{a}}{dt} \cdot \mathbf{a} + \mathbf{a} \cdot \frac{d\mathbf{a}}{dt} = 2\mathbf{a} \cdot \frac{d\mathbf{a}}{dt} = \frac{d}{dt}(a^2) = 2a\frac{da}{dt}$$

and so

$$\mathbf{a} \cdot \frac{d\mathbf{a}}{dt} = \frac{d\mathbf{a}}{dt} \cdot \mathbf{a} = a\frac{da}{dt};$$

(e) the derivative of $\mathbf{a} \times \mathbf{b}$ is $\dfrac{d}{dt}(\mathbf{a} \times \mathbf{b}) = \mathbf{a} \times \dfrac{d\mathbf{b}}{dt} + \dfrac{d\mathbf{a}}{dt} \times \mathbf{b}.$

7. If the angular momentum $\mathbf{c} = \mathbf{0}$, the mathematical details are just a bit more involved, but the result is the physically obvious one that the motion of m occurs along the fixed, straight line joining m and M, that is, m moves

directly toward M. See, for example, Harry Pollard, *Mathematical Introduction to Celestial Mechanics* (Englewood Cliffs, N.J.: Prentice-Hall, 1966, p. 2).

8. See, for example, Jerry B. Marion and Stephen T. Thornton, *Classical Dynamics of Particles & Systems*, 3rd ed. (New York: Harcourt Brace Jovanovich, 1988, pp. 261–262).

9. Equation (5.2.15) is, in fact, Kepler's second law of planetary motion, the so-called *area law*, that says the position vector **r** sweeps over equal areas in equal time intervals. This physical interpretation isn't of any direct use to us, here, however, and so I won't pursue it. What counts for us is the mathematics of (5.2.15), itself, and we'll use it in a computer code to study an important modern problem in theoretical celestial mechanics in the final section of this chapter.

10. Martin C. Gutzwiller, "Moon-Earth-Sun: The Oldest Three-Body Problem" (*Reviews of Modern Physics*, April 1988, pp. 589–639).

11. Mauri Valtonen and Hannu Karttunen, *The Three-Body Problem* (Cambridge: Cambridge University Press, 2006).

12. Adam Miller Hiltebeitel, "The Problem of Two Fixed Centres and Certain of Its Generalizations" (*American Journal of Mathematics*, 1911, pp. 337–362).

13. A. M. Bork, *FORTRAN for Physics* (Reading, Mass.: Addison-Wesley, 1967); David McCandlish, "Solutions to the Three-Body Problem by Computer" (*American Journal of Physics*, July 1973, pp. 928–929); Albert D. Grauer and N. Bruce Tanis, "Simple Earth-Moon Trajectory Calculations" (*Physics Teacher*, March 1974, pp. 169–172); Arthur W. Luehrmann, "Orbits in the Solar Wind: A Mini-Research Problem" (*American Journal of Physics*, May 1974, pp. 361–371). The interest in the physics education community in numerical orbit calculations seems to have been sparked by one of Feynman's famous Caltech undergraduate lectures; see volume 1 of his *Lectures on Physics* (Reading, Mass.: Addison-Wesley, 1963), specifically the final three sections of Chapter 9.

14. R. M. Eisberg, *Applied Mathematical Physics with Programmable Pocket Calculators* (New York: McGraw-Hill, 1976), and Walter J. Wild, "Euler's Three-Body Problem" (*American Journal of Physics*, April 1980, pp. 297–301). As you might imagine, those handheld calculators from the 1970s were not very fast. Wild reports on one calculation that, to generate an orbital path for two years of simulated time, required over 2.7 *hours* of calculator time!

15. Much of Euler's work in physics is discussed in the book by Dieter Suisky, *Euler as Physicist* (New York: Springer, 2009). Not all of it is discussed, however, as there is not a single word in Suisky's book on the three-body problem.

16. If you look back at the beginning of this chapter you'll see that the universal constant of gravitation is $G = 6.67 \cdot 10^{-11} \dfrac{\text{meters}^3}{\text{kilograms·seconds}^2}$. This is in the mks system of units, where the unit of mass is the kilogram, the unit of time is the second, and the unit of distance is the meter. Those three unit values then determine the above numerical value for G. More generally, however, we can define the values for *any three* of the four parameters (G, unit mass, unit length, unit time) any way we wish and from that determine the fourth. So we are free to pick the unit length to be the AU, the unit time to be the year, *and* the unit mass to be the solar mass ($M = 1$), and *that* determines G (to be $4\pi^2$, which up to now has been the value of the *product GM*—a quantity called the universal gravitational *parameter*). This system of units is quite nice for orbital calculations inside the Solar System. In other applications, however, it is often convenient to make other choices, and we'll do that later in this chapter.

17. Science fiction writers have often used multiple star systems in their stories, and astrophysical research seems to support the idea that such systems could indeed have planets in orbits sufficiently stable to allow the emergence of life. Perhaps the most famous of such tales is Isaac Asimov's 1941 "Nightfall," about a civilization on a planet in a *six*-star system. Unfortunately, the planet has a moon that, every 2,049 "years" (what is a "year" on that planet, in Earth-years, is not stated), just happens to totally eclipse the only star of the six in the sky at that time, thus temporarily plunging the planet into total darkness. Indeed, the planet's astronomers are predicting the next eclipse is about to happen, and that the occurrence of total darkness will drive everybody on the planet insane. Asimov goes to great lengths to have a character explain why that would happen, but I personally found it unconvincing. On the other hand, the professional organization of the Science Fiction Writers of America (of which I myself was once a proud member) once voted "Nightfall" the best short science fiction story of all time, so read the story—it has been anthologized numerous times—and make up your own mind.

18. A vivid description of the tidal forces resulting from the Earth's close encounter (and eventual collision) with a rogue planetary-sized mass arriving from deep space is in the 1933 novel *When Worlds Collide*, by Philip Wylie and Edwin Balmer. The novel was later (1951) made into a well-received film, winning the Academy Award for Special Effects. Gravitational tidal force is thought to be the physical mechanism behind the creation of Saturn's rings, from the debris of a moon or a passing mass that got too close to the huge planet. See also the end of note 28.

19. My use of complex variables to transform the differential equations of motion from the inertial sidereal system to the rotating synodic system was

motivated by their use in the definitive work by Victor Szebehely, *Theory of Orbits: The Restricted Problem of Three Bodies* (New York: Academic Press, 1967, pp. 14–15). That book's presentation is pretty terse (the notation is a bit hard to keep straight, too), however, and so I've filled in a lot of missing steps. Szebehely (1921–1997) was a Hungarian-born American pioneer in orbit calculation by computer.

20. You can find more on the computer solution of (5.5.14) and (5.5.15) in the paper by James Blackburn, M.A.H. Nerenberg, and Y. Beaudoin, "Satellite Motion in the Vicinity of the Triangular Libration Points" (*American Journal of Physics*, November 1977, pp. 1077–1081). The synodic differential equations are not derived in that paper but are simply given, there is no computer code shown (but the authors do state that they used a fourth-order variable step-size Runge-Kutta integration algorithm, which is also the basis for MAT-LAB's *ode45*), and different units are used from the ones I have given here. (The authors define the unit mass to be $M_1 + M_2$, the unit length to be l, and the unit time to be $1/\omega$.) They do, however, treat a practical Solar system problem—a claim I'm a bit reluctant to make for my "aliens destroy the Earth with a wormhole and death mass" example!—and they do agree with my basic thesis for this book when they write, "A student interested in [the restricted three-body problem] must turn to rather advanced books [as an example of such a book they cite Szebehely's text: see my previous note] which for instructional purposes are a kind of 'overkill.' A computer simulation of the restricted three-body system offers an attractive alternative for directly revealing the orbital dynamics." In the more recent paper by Nicholas Johann Harmon, Christine Leidel, and John F. Lindner, "Optimal Exit: Solar Escape as a Restricted Three-Body Problem" (*American Journal of Physics*, September 2003, pp. 871–877), the problem is to study how a "small" mass (a rocket, our *m*) can totally escape to infinity from two mutually rotating masses (the Sun and the Earth). In this paper, the synodic differential equations *are* derived (in a way somewhat different from my presentation here), and reference *is* made to a computer program (but no listing is provided). A website is given, however, and the latest version of that program code, written for the Macintosh operating system in C++, is available for download.

21. In recent years there have been several nice expositions on the King Oscar Competition (particularly in the book by Florin Diacu and Philip Holmes, *Celestial Encounters: The Origins of Chaos and Stability* [Princeton, N.J.: Princeton University Press, 1997, pp. 23–30]), but all are based on the original historical work of June Barrow-Green, "Oscar II's Competition and the Error in Poincaré's Memoir on the Three Body Problem" (*Archive for History of Exact Sciences*, 1994, pp. 107–131). She later expanded her paper into the book-length *Poincaré and the Three Body Problem* (New York: American Mathematical Society,

1997). All of my historical commentary in Section 5.6 is taken from both of Barrow-Green's works. The competition gets a nice fictional treatment in the murder mystery novel—with a title having a double meaning—by Catherine Shaw (the pen name for a PhD mathematician), *The Three Body Problem: A Cambridge Mystery* (London: Allison & Busby, 2004). The king himself makes an appearance. If you like Anne Perry's Victorian London inspector Pitts and William Monk mystery novels, you'll like Shaw's work as well.

22. The value of the prize can be appreciated by noting that the total 1882 salary of Mittag-Leffler, a senior mathematics professor, was $7,000$ crowns.

23. There then followed three more problems; two of them involved questions in differential equations, and the third involved the theory of functions. Only one entrant attempted one of those three problems. All the others who didn't attempt the n-body problem submitted memoirs on questions of their own choosing.

24. There was, in fact, a thirteenth entry, but it arrived after the submission deadline. It wouldn't have mattered anyway, as its author (one Cyrus Legg of Chapham, London), as Barrow-Green (see note 21) put it, "belonged to that indefatigable band of angle trisectors!" Of the twelve competitors, the identity of only two others besides Poincaré and Appell is known today.

25. As per the original competition rules, each submitted memoir was to be unsigned, bearing only an epigraph that would, *only after the selection of the winning submission*, be used to open the matching sealed envelope (also bearing that epigraph), which would at last reveal the author's name and address. To quote Barrow-Green (see note 21), however, "When Poincaré's entry arrived it was clear that his reading of the regulations had been somewhat perfunctory. As required, he had inscribed his memoir with an epigraph [*Nunquam præscriptos transibunt sidera fines*, Nothing exceeds the limits of the stars], but, instead of enclosing a sealed envelope containing his name, he had written and signed (!) a covering letter and had also sent a personal note to Mittag-Leffler." Poincaré wasn't the only one playing games, either: the second-place winner, Appell, also submitted a cover note with his entry, stating it had been written by someone "well-known to him." Well, who would ever say mathematicians have no sense of humor?

26. In 1985 the University of Minnesota mathematician Richard McGehee discovered, in the papers of Mittag-Leffler, the sole surviving copy of the erroneous original *Acta Mathematica* version of Poincaré 's submission, with this note written on it, in Mittag-Leffler's hand: "The whole edition was destroyed." The corrected version finally appeared in a new print run (the cost of which, greater than the value of the prize, was borne entirely by Poincaré)

at the end of 1890. More on Poincaré's error can be found in the paper (which immediately follows Barrow-Green's 1994 paper; see note 21) by K. G. Anderson, "Poincaré's Discovery of Homoclinic Points," pp. 133–147.

27. Imke de Pater and Jack J. Lissauer, *Planetary Sciences* (Cambridge: Cambridge University Press, 2001, p. 30). The astronomer Gregory Laughlin (University of California at Santa Cruz) authored a recent quite interesting essay on the stability (or not) of the Solar System: "Hanging in the Balance" (*Sky & Telescope*, April 2010, pp. 26–33). There he writes, "[E]ven if the present-day positions of the planets could be determined to a precision of less than the width of an atom, it still becomes impossible to make any concrete statement about where the planets will be in 100 million years. We have no way of knowing whether January 1, 100,000,000 AD will occur in winter or summer."

28. Sundman's infinite series solution will always converge if there are no *singularities*, which physically are *collisions* (either binary or triple) among the three bodies. Even when there are no collisions in a three-body problem, it is usually the case that two of the bodies will make repeated close approaches, which, because gravity is inverse square, means the forces on those bodies (and so the accelerations and velocities of those bodies) can become very large. This increase in the "rate at which things happen" is the physical reason behind the enormous number of terms required. Only if the total angular momentum of the three bodies is zero is there even a possibility for a triple collision (what's called a *total collapse*) of the three bodies. You can find a proof of this in Pollard's book (see note 7), pp. 43–44, where it is also shown that *if* total collapse in a zero angular momentum system does occur, then it will not take forever to happen. See also Sérgio B. Volchan, "Some Insights from Total Collapse in the *N*-Body Problem" (*American Journal of Physics*, November 2008, pp. 1034–1039). Total collapse is not inevitable with zero angular momentum, however, as there are three-body solutions for such systems that are forever collision-free (at the end of this chapter I'll show you one). Besides a triple collision, there is the possibility of binary collisions, which Sundman showed how to mathematically handle if they should occur. It has also been shown, however, that the set of initial conditions that result in collisions of *any* sort has measure zero. That is, the probability a set of randomly chosen initial conditions for the positions and velocities of the bodies will actually result in a collision of either type is zero (which does *not* mean impossible, just very unlikely). See Donald Gene Saari, "Improbability of Collisions in Newtonian Gravitational Systems" (*Transactions of the American Mathematical Society*, December 1971, pp. 267–271, with corrections in the July 1973 issue, pp. 351–368). A zero probability for collision is no reason for a sigh of relief. As Professor Laughlin (see the previous note) observes, a near miss could still be the end of the world. In the

specific case of a Mars-like mass missing Earth by a few hundred kilometers, the tidal flexing of the Earth's crust (see note 18 again) would dissipate enormous energy during the encounter, with power levels exceeding what could be many millions of gigawatts! Laughlin's description of the end result is sobering: "The tidal stretching and squeezing during the close approach would heat Earth's interior enough to completely melt the planet's mantle and crust. Earth's oceans would become a crushing atmosphere of steam above a global sea of glowing lava."

29. Mauri Valtonen and Hannu Karttunen, *The Three-Body Problem* (Cambridge: Cambridge University Press, 2006, p. 39).

30. Sundman's result was recognized as a great technical achievement, despite its lack of practical value, and in 1913 the French Academy of Sciences awarded him its *Prix Pontécoulant* for his infinite series solution, with *double* the normal award. Sundman's solution holds, however, only for the $n = 3$ case. Finally, in 1991, a Sundman-like infinite series solution for the general n-body problem was discovered by the Chinese-born mathematician Quidong Wang, now a professor at the University of Arizona.

31. Mathematicians say they have the solution to an n-body problem if they know the position for each mass for *all* time, that is for $-\infty < t < \infty$. For the Pythagorean three-body problem, indeed for any n-body problem *where all the masses are at rest* at $t = 0$, the location of each mass at $t = -t^* < 0$ is the location it has at $t = t^* > 0$ and so, if we have the locations for $0 < t < \infty$ we also have them for $-\infty < t < 0$. This is not hard to show and you can find a pretty proof in Victor Szebehely, "Burrau's Problem of Three Bodies" (*Proceedings of the National Academy of Sciences*, July 1967, pp. 60–65, esp. p. 63).

32. Whole books have been written on these particular solutions. See, for example, Kenneth R. Meyer, *Periodic Solutions of the N-Body Problem* (New York: Springer, 1999). See also note 36.

33. Victor Szebehely and C. Frederick Peters, "Complete Solution of a General Problem of Three Bodies" (*Astronomical Journal*, September 1967, pp. 876–883).

34. If the masses in the Pythagorean problem are huge, and if one uses relativistic mechanics and not Newtonian, then numerical studies have shown that a disintegration of the three body system may *not* occur. With black holes as the three bodies, if the total mass exceeds something like 10^8 solar masses (the unit mass is on the order of 10^7 solar masses) then total collapse into a single black hole is the outcome. See M. J. Valtonen, S. Mikkola, and H. Pietilä, "Burrau's

Three-Body Problem in the Post-Newtonian Approximation" (*Monthly Notices of the Royal Astronomical Society*, 273, 1995, pp. 751–754).

35. See the paper by Szebehely, "Burrau's Problem of Three Bodies."

36. Alain Chenciner and Richard Montgomery, "A Remarkable Periodic Solution of the Three-Body Problem in the Case of Equal Masses" (*Annals of Mathematics*, November 2000, pp. 881–901). See also Charles Simó, "Dynamical Properties of the Figure Eight Solution of the Three-Body Problem," in *Celestial Mechanics* (American Mathematical Society, 2002, pp. 209–228). These two papers are by mathematicians but, as they state, the figure-8 orbit was first discovered in passing during numerical experiments conducted by the physicist Christopher Moore: see his paper "Braids in Classical Dynamics" (*Physical Review Letters*, June 14, 1993, pp. 3675–3679). Moore writes, at the end of his paper, that his studies have allowed him "to directly construct periodic solutions to the *n*-body problem."

37. Call the radius of the binary circular orbit R. Then, equating the gravitational force of attraction between the primaries to the centripetal acceleration force on each mass M in the binary, we have $\dfrac{GM^2}{(2R)^2} = \dfrac{Mv^2}{R}$ where v is the orbital speed of each primary. Thus, $v = \dfrac{1}{2}\sqrt{\dfrac{GM}{R}}$ and so the orbital period is $T = \dfrac{2\pi R}{v} = 4\pi R\sqrt{\dfrac{R}{GM}}$. If we take $M = \dfrac{1}{2}$, $G = 1$, and $2R = 1$ (and so $R = \dfrac{1}{2}$), then we have $T = 2\pi$ as claimed.

38. See, for example, W. D. MacMillan, "An Integrable Case in the Restricted Problem of Three Bodies" (*Astronomical Journal*, nos. 625–626, 1911, pp. 11–13).

39. Victor G. Szebehely, *Adventures in Celestial Mechanics: A First Course in the Theory of Orbits* (Austin: University of Texas Press, 1989; see esp. Chapter 4, pp. 35–43).

40. For theoretical work, what is usually done at this point is to use an analytically tractable *approximation* for $r(t)$ that is reasonably good for *small* ϵ: $r(t) = \dfrac{1}{2}\{1 - \epsilon\cos(t)\} + O(\epsilon^2)$, which means the ignored correction terms beyond the first term are of order ϵ^2 or higher (and so are *very* small). See Jürgen Moser, *Stable and Random Motions in Dynamical Systems: With Special Emphasis on Celestial Mechanics* (Princeton, N.J.: Princeton University Press, 1973, p. 85). With MATLAB, however, and the code **esitnikov.m**, we don't have to make such an approximation, and the code works over the *entire* $0 \leq \epsilon < 1$ interval.

41. Douglas Heggie and Piet Hut, *The Gravitational Million-Body Problem: A Multidisciplinary Approach to Star Cluster Dynamics* (Cambridge: Cambridge University Press, 2003). This book takes a refreshingly enthusiastic position on the use of computers in doing n-body problems and, as the title says, it treats the fascinating case of *large n*.

42. Mathematicians are, generally, not enthusiastic about computer studies of the n-body problem. The University of California at Santa Cruz mathematician Richard Montgomery (see note 36) expresses that position quite clearly in his book review of Shaw's murder mystery novel, *The Three Body Problem* (see note 21), in *Notices of the American Mathematical Society* (October 2006, pp. 1031–1034).

6
ELECTRICAL CIRCUIT ANALYSIS AND COMPUTERS

Question: "What was God doing before the Creation?"
Answer: "He was doing Pure Mathematics and thought it would be a pleasant change to do some Applied."
—John Littlewood

Rutherford [Ernest Rutherford (1871–1937), winner of 1908 Nobel Prize in chemistry], at least one other physicist and myself got into a hopeless muddle over the alleged "penny and feather in a vacuum" experiment [which says both will fall at the same rate]. Was the experiment a bluff? Rutherford said apologetically that he *thought* he had seen it done as a boy. The muddle continued until, after dinner, we were put out of our misery by an engineer.
—Another amusing (for engineers!) tale from pure mathematician Littlewood

6.1 Electronics Captures a Teenage Mind

In this chapter I want to show you a high-intensity number-crunching computer software application that is very different from MATLAB. It is one used every day by professional electrical engineers worldwide, but, like MATLAB (and personal computers and video game consoles, for that matter), it would have been considered a fantasy when I was in high school. (Just to get the terminology straight—a *fantasy* is even more "far out" than is *science fiction*!) In addition, this chapter will also discuss how electronic circuits were first used in the 1940s to solve mathematical problems on what are called *analog* computers. So, in the spirit of Littlewood's two anecdotes, we are going to be really practical in this chapter.

To set the stage for the rest of what I'll do here, let me start by telling you a personal story about the beginnings of my own fascination with electricity. I did that before in one of my earlier books, in the original preface to the hardcover edition of *An Imaginary Tale: The Story of* $\sqrt{-1}$, and I'll start by repeating just the beginning of what I wrote there.

L ong ago, in a year so far in the past (1954) that my life then as a high school freshman now seems like a dream, my father gave me the gift of a subscription to a new magazine called *Popular Electronics*. He did this because he was a scientist, and his oldest son seemed to have talents in science and mathematics that were in danger of being subverted by the evil of science fiction. I had, in fact, given him plenty of reason for such concern. I devoured science fiction in those days, you see, often sitting in the kitchen at eleven at night eating a huge sandwich and reading a novel set on Mars a million years in the future. Dad, of course, would have preferred that I be reading a book on algebra or physics.

Being a clever man, he decided not to simply forbid the science fiction, but rather to outflank the science fiction stories by getting me to read *technical* stories, like the "Carl and Jerry" tales that appeared each month in *Popular Electronics*. Carl and Jerry were two high school electronics whiz kids—geeks or nerds, in today's unattractive terms—who managed each month to get involved in some exciting adventure in which their technical knowledge saved the day. They were a 1950s amalgamation of the Hardy Boys and Tom Swift. My father's plan was to get me to identify with Carl and Jerry, instead of with Robert Heinlein's neurotic time travelers.

Well, Dad's devious plan worked (although I never completely gave up the science fiction), and I got hooked not only on Carl and Jerry[1] but also on the electronic construction projects the magazine featured in each issue. I learned how to read electrical schematics from the magazine, whose editors used the same exploded-view, pictorial wiring diagrams that became so well known to all who ever built an electronic, mail-order kit.[2] I constructed a home workshop in the garage behind the house, and a lot of amazing gadgets were built there—although not all of them worked, or at least not in the way the original designers had intended.

Carl and Jerry had a wonderful electronics workshop in which they cooked up their gadgets each month. My electronics workshop—converted from building model airplanes from balsa-wood kits during my junior high years—was inspired by their fictional one and, although not nearly so grand as Carl and Jerry's, I have many fond memories of the hundreds of hours I spent in it during my high school years of 1954 to 1958. I went from inhaling the fumes of wood glue and airplane dope to those of hot solder and resin. My high school was torn down many years ago to make way for a huge shopping mall, but for decades after, my workshop remained in the garage. I would often go out to it when I visited my parents, just to sit there by myself and, at least for a while, imagine that it was 1954 again and Steven Spielberg was a seven-year-old kid somewhere and Elvis was just getting started and . . .

I built a primitive analog multiplier on that workbench, out of a couple of potentiometers, a current meter, and a battery, and some simple digital games using toggle switches and relays as memory elements. I built an applause meter (a *Popular Electronics* project) that was used one year for judging a high school talent show. Boy, I built a *lot* of weird stuff on that bench. Now both of my parents have passed on, the house has been sold, the garage workshop is long gone. But that sanctuary for a teenage boy who lacked some critical social skills remains, in delicate detail, in my memory.

6.2 My First Project

The very first of those projects of mine was one that appeared in the third issue of *Popular Electronics* (December 1954, a copy of which I still own), titled "The Twinkling Christmas Tree." Figure 6.2.1 shows the first two pages of the article, and the reason the project still has

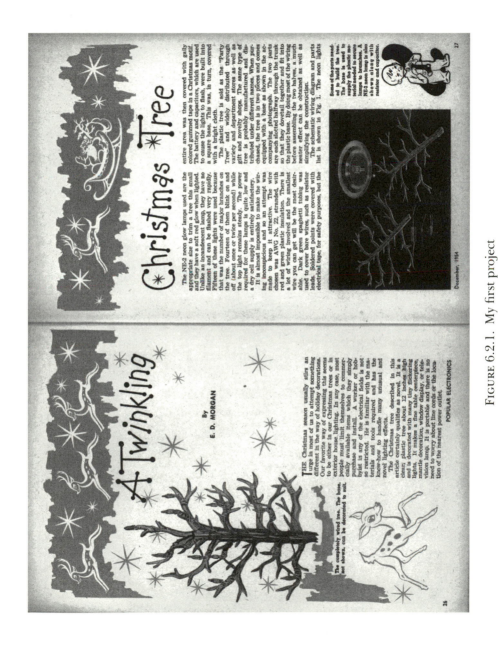

FIGURE 6.2.1. My first project

fascination for me is that it is based on what is perhaps the simplest possible relaxation oscillator that is, an oscillator in which sudden, periodic *switching* occurs. In 1954 I still had years to go before I would understand what its differential equation was all about, but the words *relaxation oscillator* did appear in the *Popular Electronics* article. I didn't know what that meant, either, but I recall that the words sounded neat! The operation of a neon bulb relaxation oscillator circuit, as shown in Figure 6.2.2, was sufficiently simple enough, however, that even as a high school freshman I readily grasped the basic idea of what was happening when the switch S was closed and the bulb began to magically blink on and off.

The component at the far right in Figure 6.2.2 is a bulb filled with neon gas. The bulbs in the *Popular Electronics* project were the venerable NE-2, still available today for about fifteen cents each (if bought in lots of ten or more). The behavior of the bulb is highly nonlinear: here is how I roughly understood things in 1954. Suppose that the bulb is initially off (that is, not conducting current), and then we close S. Then:

1. If the bulb voltage drop v across the bulb terminals is less than some critical voltage that I'll call V_s (and it will be if we assume that C is initially uncharged), then the gas is a nonconductor and the bulb remains an open circuit.

2. The capacitor C begins to be charged by the 90-volt battery through the resistor R, and so the capacitor voltage (that is, v) rises.

3. When the capacitor voltage v reaches V_s the gas ionizes or *strikes* (that is, the electric field inside the bulb becomes large enough that valence electrons in the neon atoms are ripped free from their orbits

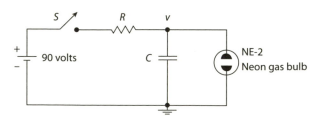

FIGURE 6.2.2. The basic blinking neon gas bulb circuit (**not** the *Popular Electronics* circuit)

and the gas becomes a mix of free negative charges and positive ions). The free electrons thus possess, temporarily, extra energy. The gas is now a good conductor.

4. Once ionized, the now highly conductive gas presents a low-resistance path across the capacitor, and so the capacitor very rapidly dumps its built-up charge through the bulb.

5. With the capacitor discharged, v drops to zero (or close to it), the ionized bulb gas recombines, and the once again orbital electrons give up their extra energy as radiation at a wavelength characteristic of the gas (reddish yellow for neon; it would be blue if the gas was argon instead).

6. The recombined gas is again a nonconductor and the circuit behavior starts over at step (1).

The visual result is a periodically blinking bulb.

Well, despite my crude understanding of the above six steps, the circuit of the *Popular Electronics* twinkling Christmas tree remained a mystery to me in 1954, and would continue to be one for several more years. That's because the circuit in the magazine was *not* the simple one of Figure 6.2.2 but rather was the one shown in Figure 6.2.3. There were fourteen blinking bulbs in the twinkling Christmas tree (each gently "melted" into the tree's plastic branches with the very delicate touch of a soldering iron), which would have required fourteen capacitors if the circuit of Figure 6.2.2 had been used. With the circuit of Figure 6.2.3, however, a single capacitor was used for each *pair* of bulbs, and so only seven capacitors were required. While not understanding how that modified circuit worked, I could still follow

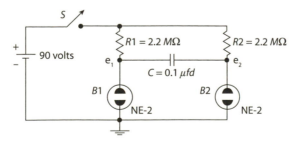

FIGURE 6.2.3. The *Popular Electronics* circuit—how does it work?

the published schematic in the magazine, and so my tree worked—
but why? What happens when switch S is closed in the circuit of Figure
6.2.3?

It wouldn't be until the first few months of 1961 that I learned
the answer to that question. I can pinpoint the time of revelation
because that was when three separate things finally came together
in my mind, as an electrical engineering major in the third year of
the undergraduate program at Stanford. First, the previous year I had
completed my first course in ordinary differential equations (see note
4 in Chapter 3) and so could solve the simple differential equations
for Figure 6.2.3 (even though, at the time, I didn't really know how
to derive them). Second, in my junior year I took the required three-
quarter sequence in electronics (EE 150, 151, and 152) and refined my
understanding of what was happening in a gas discharge tube such as
the NE-2. And finally, in the Winter Quarter of my junior year (January
to spring break of 1961), I took EE 116, another class required of all
electrical engineering majors. That was a neat little two-unit course in
electrical transients.

About 90% of EE 116 revolved around two fundamental principles,
applied in what seemed like ten thousand homework problems (see
Challenge Problem 6.1): neither (1) the voltage drop across a capacitor nor (2) the current in an inductor can change instantly. That's
because if either did change instantly, then an instantaneously infinite
current (capacitor) or voltage drop (inductor) would be required. But
engineers don't believe anything *physical* ever becomes truly infinite in
the "real world."[3] Principle 1 in particular, is crucial to understanding
the circuit of Figure 6.2.3.

The other essential component to understanding the circuit of
Figure 6.2.3 comes from a refinement of what happens *after* a gas discharge tube strikes. It's a seemingly modest twist on step 4 of my 1954
understanding of the NE-2, but it makes all the difference between
understanding and not understanding. Once the tube ionizes the gas
doesn't simply "short" but rather maintains a constant voltage drop
across the tube's electrodes, a constant drop that I'll write as V_E, where
$V_E < V_S$. As long as the external circuitry is capable of providing a voltage drop that is at least equal to V_E, then the drop across the bulb
will be V_E and the external circuitry will "absorb" the excess voltage

in whatever way necessary to be consistent with Kirchhoff's laws. If the external circuitry cannot support a voltage drop of V_E across the bulb, then the bulb will *extinguish* and become a nonconductor again. For the NE-2, the values for V_S and V_E are typically somewhere in the intervals $65 \text{ V} < V_S < 75 \text{ V}$ and $45 \text{ V} < V_E < 55 \text{ V}$. It is virtually certain that two randomly picked bulbs will have different values of V_S and V_E.

It should be carefully understood that the NE-2 is a *glow* discharge, and not an arc. The feature of the arc that made it so attractive to the pre-vacuum tube radio pioneers was the negative resistance region of its operation, where an *increase* in current is associated with a *decrease* in voltage drop. The *constant* voltage drop behavior of an ionized gas glow tube, however, made it useful in the design of circuits called *voltage-regulated power supplies*.[4] The term "glow" is very descriptive of the NE-2, too, as when conducting the emitted light looks just like that, a soft glow rather than the intensely bright, harsh glare that is characteristic of an arc.[5] To push an NE-2 into operating as an arc would almost certainly heat the enclosed neon to the point where the glass bulb would explode.

Now, with all this said, let me reveal the real reason why I want to take you through the *Popular Electronics* circuit. It isn't for nostalgic purposes at all but rather because, as few components as the circuit of Figure 6.2.3 has, it will nevertheless require some fairly subtle arguments for us to eventually figure out just what is going on. Then, if you imagine that you are faced with a circuit that has even more stuff in it than does our little two-neon-bulb one, I don't think I'll have to work very hard at all to convince you that matters might easily be considerably more involved.

So, wouldn't it be nice to have some sort of computer analysis tool that can "figure out" circuits?! There are such tools, too. But that's for the following sections of this chapter, where I'll show you how to "build" the modern version of Figure 6.2.3 *on a computer*. For now, let's analyze the original *Popular Electronics* circuit. To be quite specific in the discussion, I'll take V_S and V_E for both bulbs to be right in the middle of their respective ranges. That is, I'll take $V_E = 50 \text{ V}$ and $V_S = 70 \text{ V}$.

Just before S is closed we'll take C as uncharged, and so initially C has a zero voltage drop across its terminals, and the bulbs $B1$ and $B2$ are both off (that is, nonconducting). Then we close S. The neon bulb node voltages e_1 and e_2 instantly become equal to 90 V, but, because $B1$ and $B2$ are not perfectly identical, one will strike before the other. Suppose it is $B1$. Then the voltage drop across $B1$ is 50 V *and will remain so* as long as $B1$ conducts current; the $B1$ node voltage e_1 is, as electrical engineers put it, *clamped* at 50 V. Since the voltage drop across C cannot change instantly, the $B2$ node voltage $e_2 = 50$ V, too, to keep the instantaneous voltage drop across C equal to zero, but e_2 is *not* clamped and can change. There is now current in both R_1 and R_2, and both currents flow entirely through $B1$, but it is only the current in R_2 that flows through C and so *charges* C, causing e_2 to rise. To be analytical about this point, let's write Kirchhoff's current law at the e_2 node. Remembering that $B2$ is off and so that bulb's current is zero, we have (where I'll write R for both R_1 and R_2, since the two resistors were equal in the *Popular Electronics* circuit)

$$\frac{90 - e_2}{R} = C\frac{d(e_2 - e_1)}{dt} = C\frac{de_2}{dt} - C\frac{de_1}{dt},$$

or, because e_1 is clamped (which means that $\frac{de_1}{dt} = 0$), we have

$$90 - e_2 = RC\frac{de_2}{dt}. \tag{6.2.1}$$

Taking the Laplace transform of (6.2.1)—look back at note 6 in Chapter 4 for a brief discussion of the transform—where $E_2 = E_2(s) = \mathcal{L}\{e_2(t)\}$,

$$\frac{90}{s} - E_2 = RC\left[sE_2 - e_2(0)\right],$$

or, as $e_2(0) = 50$,

$$\frac{90}{s} - E_2 = RCsE_2 - 50RC.$$

Thus,

$$E_2\left(1 + RCs\right) = \frac{90}{s} + 50RC,$$

or

$$E_2 = \frac{90}{s(RCs + 1)} + \frac{50RC}{RCs + 1}. \tag{6.2.2}$$

Making a partial fraction expansion of the first term on the right-hand side of (6.2.2), we have (with A and B as constants to be determined)

$$\frac{90}{s(RCs + 1)} = \frac{A}{s} + \frac{B}{RCs + 1} = \frac{ARCs + A + Bs}{s(RCs + 1)}$$
$$= \frac{A + (ARC + B)s}{s(RCs + 1)}. \tag{6.2.3}$$

Equating the coefficients of equal powers of s in the first and last numerators of (6.2.3), after noticing that in both fractions we have the same denominator, gives $A = 90$ and $ARC + B = 0$. So, $B = -ARC = -90RC$, and therefore, from (6.2.2) and (6.2.3):

$$E_2 = \frac{90}{s} - \frac{90}{RCs + 1} + \frac{50RC}{RCs + 1} = \frac{90}{s} - \frac{40RC}{RCs + 1},$$

or

$$E_2 = \frac{90}{s} - 40\frac{1}{s + \frac{1}{RC}}. \tag{6.2.4}$$

Returning to the time domain,

$$e_2(t) = 90 - 40e^{-t/RC}\,\text{V}, \quad t \geq 0. \tag{6.2.5}$$

So, e_2 does indeed rise (as claimed), from 50 V (at $t = 0$) toward 90 V.

It never gets anywhere near 90 V, however, because when e_2 reaches 70 V, $B2$ strikes and therefore *clamps* e_2 to 50 V. That is, e_2 instantly drops by 20 V (= 70 V−50 V). Thus, e_1 must instantly drop by 20 V, too, because the voltage drop across C cannot change instantly. That is, e_1 goes from 50 V to 30 V, which, being less than $V_E = 50$ V, causes $B1$ to turn off. The situation in the circuit is now as follows: $e_1 = 30$ V, $B1$ is off, e_2 is clamped to 50 V, and $B2$ is on. Without any loss in generality in the calculations that follow, we can now think of this new situation as defining a new $t = 0$.

Just as before in our initial calculations, there is current in both R_1 and R_2 (now both are flowing in $B2$), with the current in R_1 charging C, which causes e_1 to rise. And, as before, to be analytical on this point I'll write Kirchhoff's current law at the e_1 node as follows (remember, $B1$ is off, so that bulb's current is zero):

$$\frac{90 - e_1}{R} = C\frac{d(e_1 - e_2)}{dt} = C\frac{de_1}{dt} - C\frac{de_2}{dt},$$

or, because e_2 is clamped (and so $\frac{de_2}{dt} = 0$), we have

$$90 - e_1 = RC\frac{de_1}{dt}, \tag{6.2.6}$$

or, as $e_1(0) = 30$, then taking the Laplace transform of (6.2.6) gives

$$\frac{90}{s} - E_1 = RC\,[sE_1 - 30] = sRCE_1 - 30RC.$$

Solving for E_1 and doing a partial fraction expansion as before (I'll let you fill in what should now be the easy details) gives

$$E_1 = \frac{90}{s} - 60\frac{1}{s + \frac{1}{RC}}. \tag{6.2.7}$$

Returning to the time domain,

$$e_1(t) = 90 - 60e^{-t/RC}\text{V}, \quad t \geq 0. \tag{6.2.8}$$

What (6.2.8) tells us is that e_1 rises from 30 V (at $t = 0$) toward 90 V. But, when e_1 reaches 70 V, $B1$ strikes, and therefore clamps e_1 to 50 V. Since e_1 has suddenly dropped by 20 V, then so must e_2, and so $e_2 = 30$ V, and thus $B2$ turns off. That is, *the conditions in the circuit are just as they were at the previous $t = 0$ instant, but with the roles of e_1 and e_2, and $B1$ and $B2$, reversed.* What happens from now on is clear: the bulbs $B1$ and $B2$ will alternately turn on and off at a rate determined by how long it takes the voltage $90 - 60e^{-t/RC}$ V, starting at $t = 0$ with a value of 30 V, to rise to 70 V. If we call this time T, then

$$90 - 60e^{-T/RC} = 70,$$

or

$$60e^{-T/RC} = 20,$$

or

$$e^{-T/RC} = \frac{20}{60} = \frac{1}{3},$$

or, at last,

$$T = RC \ln(3). \tag{6.2.9}$$

For the *Popular Electronics* project, $R = 2.2 \cdot 10^6 \, \Omega$ and $C = 0.1 \mu fd = 10^{-7}$ F, and so $T = 0.22 \ln(3) = 0.24$ seconds.

The neon bulbs in each pair switch back and forth like two kids on a teeter-totter, alternating between being on and being off, spending about a quarter of a second in each state. Since the values of each of the resistors and capacitors are almost surely different because of manufacturing variations, the actual values of T will be different for each bulb. In addition, as I mentioned earlier, the values of V_S and V_E for each bulb are virtually certain to be different as well, which will further influence the values of T. For each pair of bulbs in the *Popular Electronics* circuit, $B1$ and $B2$ were positioned on tree branches distant from one another, so that the pair's coupled, out-of-phase blinking wouldn't be obvious. The net, overall result was a truly "randomly" twinkling Christmas tree.

As a final note, the *Popular Electronics* circuit is best known among electrical engineers not as a "twinkling Christmas tree" circuit, but rather as a *free-running astable multivibrator*. If I had known that in 1954, I'm absolutely positive I wouldn't have had the slightest idea of what those words meant—but I would have been extremely impressed![6]

6.3 "Building" Circuits on a Computer

Before a computer software application can "process" a circuit—I'm reluctant to use the word *analyze*, as that seems a reach for a pile of inanimate matter—it must somehow "know" everything about that circuit's structure—that is, what components are connected to what components. There are two fundamentally different ways to accomplish that: with a *wire list* or by *schematic capture*. Both approaches are used today to study circuits having a complexity that far exceeds that of the twinkling Christmas tree. The wire list approach is the more obvious (and the easier to implement) of the two. The name says it all: one simply prepares, according to some set of syntactical rules, a detailed list of what connects to what. In an earlier book, in fact, I took that approach myself in a simple MATLAB code that analyzes any resistor network that contains a single dc voltage source. Lots of people like the wire list approach, but I don't.[7] Looking at a wire list tells you nothing about how the circuit "looks," and that can tell you a lot before you write even the first line of a theoretical analysis.

Far better, in my opinion, is schematic capture. With a schematic capture code you can literally draw the circuit right on your computer screen, using the mouse as your pen. The schematic capture program I use in this chapter is actually a pretty old application called Electronics Workbench (version 5), but that simply means more recent releases just have more jazzy bells and whistles. The simulation physics in EWB, v.5 will give solutions that are just as correct today (2010, as I write) as it did when first released more than fifteen years ago.[8]

EWB, v.5 is easy—indeed *fun*—to use. When opened, EWB produces a mostly blank screen, with a working panel at the top of the screen that resembles the typical toolbar that one sees, for example, in Windows or Word. In the case of EWB, however, that working panel is the equivalent of Carl and Jerry's electronic workshop's supply cabinet! By that I mean there are icons in the working panel that a mouse click can individually open, revealing each icon to be a "parts bin" containing a virtually unlimited supply of a particular kind of component. One bin has resistors, capacitors, inductors, relays, transformers, and switches in it; another has several types of diodes; another has all kinds of different transistors, yet another is full of various differential amplifiers (which I'll tell you about and we'll use in the final sections of this chapter); there is a bin full of connection dots ("blobs of solder") with which to connect the various components together, grounds, and ac and dc voltage and current sources; and there are still more bins with integrated circuits of various kinds, and logic gates and flip-flops for building combinatorial and sequential digital circuits.

You can build just about any circuit you wish with the full array of stuff that the marvelous, never empty parts bins of EWB,v.5 provide. Once a bin is opened, you can use your mouse to drag selected components from the bin into the initially blank screen and then, once all the parts have been selected, wire them together to form the schematic of your circuit. You also can adjust the values for each individual component as you put the schematic together.

Finally, there is one more bin that has some really special items in it: it is the *instrument* bin. In it are a *multimeter* for measuring ac and dc voltages and currents and dc resistances; a *function generator* that can produce sine, square, and triangular waveforms of any frequency and amplitude; a dual-trace, triggered *oscilloscope*; and another wonderful device called a *Bode plotter*, about which I'll tell you more later.

Here's my first example to show you how EWB,v.5 does its thing. Consider the infinite resistor ladder circuit of Figure 6.3.1. The numbers next to each resistor are the individual resistance values, in ohms. The circuit (the reason for the term *ladder* is, I think, obvious) is imagined to go on forever to the right, with each new section doubling the resistance values from those in the section at the immediate left. The question we ask is, what is the input resistance to this ladder, that is, what is the resistance R between the terminals at the far left?

The direct calculation of R, using a brute force application of Kirchhoff's current law at each of the upper nodes, is clearly not a reasonable approach. There is, after all, an *infinity* of such nodes! The

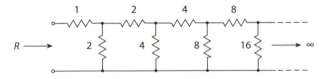

FIGURE 6.3.1. An infinite ladder circuit

clever way to calculate R is to notice that the entire ladder is simply the first (leftmost) section, followed by the rest of the ladder. But the rest of the ladder is also infinite, with each resistance in it double that of the equivalent resistance in the original ladder. (This is the condition not satisfied in the ladders of Chapter 2, and is the reason why this trick won't help on those ladders.) So, if R is input resistance to the original ladder, then $2R$ must be the input resistance to the "rest of the ladder." This observation leads us to the finite but equivalent situation of Figure 6.3.2.[9] From the usual parallel/series formulas for resistors we can immediately write

$$R = 1 + \frac{4R}{2 + 2R},$$

which, with some very minor algebra, gives the quadratic equation

$$R^2 - 2R - 1 = 0,$$

from which we find (keeping only the positive root because we know, *physically*, $R > 0$) that

$$R = 1 + \sqrt{2} = 2.4142 \ \Omega.$$

Now, even with EWB's enormous parts bins we can't really construct an infinite ladder (the computer screen is just too small, for one thing), but we can argue that if we keep a sufficiently large but finite number of ladder sections, we ought to get a pretty good approximation to the infinite case. In Figure 6.3.3 you see EWB's simulation (**ladder.EWB**)—at the top of the computer screen an on/off power switch is displayed that can be flipped on with a mouse click to start the simulation—of the first four sections, with the multimeter attached

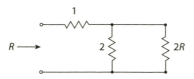

FIGURE 6.3.2. The finite equivalent of the infinite ladder in Figure 6.3.1

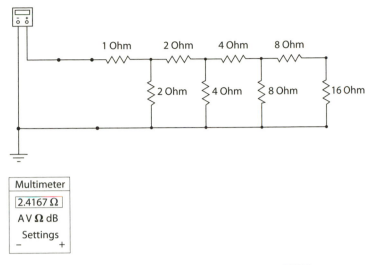

FIGURE 6.3.3. EWB simulation (**ladder.EWB**)

to the input terminals. (The power switch does not appear in the final simulation images.) You should note that the meter display of $2.4167\ \Omega$ is close to but slightly larger than the result we calculated for the infinite ladder. That makes physical sense, too, since the omitted ladder sections are in parallel with the first four sections and so, if present, would tend to *reduce* the measured resistance of the four sections.

Even more interesting to note is that EWB knows nothing of our original problem of an *infinite* ladder; we simply asked it to measure the input resistance of a four-section ladder, and that's just what it did. And what, in fact, *is* the theoretical input resistance of the four-section approximation? The answer is $\dfrac{29}{12} = 2.41666\ldots\Omega$.[10] That is, to four decimal places the input resistance is $2.4167\ \Omega$, which is precisely what EWB says.

My use of a resistor ladder circuit for our first EWB example might seem, at first glance, to be simply because it's easy to theoretically analyze. In fact, such ladders are used in actual practice. A wonderfully ingenious resistor ladder circuit is the heart of, for example, the so-called *R/2R digital-to-analog converter* (DAC) circuit you'll find in a Challenge Problem that I'll refer you to a little later in this chapter.

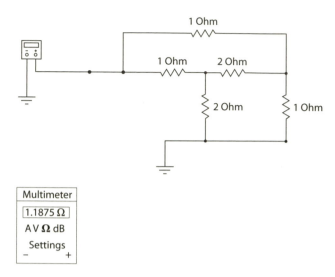

FIGURE 6.3.4. EWB simulation (**bridgedtee.EWB**)

Such DAC circuits let a digital computer "talk" to the outside analog world.

As a second, quick EWB example, now of a finite resistor circuit that (as far as I know) doesn't yield to a trick analysis but rather *requires* application of Kirchhoff to calculate its input resistance R because there are no purely series or parallel connections present, take a look at Figure 6.3.4. There you'll see what electrical engineers call a *bridged tee*. See if you can show that the multimeter in **bridgedtee.EWB** is indeed displaying the correct answer of 1.1875 Ω (see Challenge Problem 6.2).

6.4 Frequency Response by Computer Analysis

The resistance of a resistor is frequency independent, and so the behavior of all-resistor circuits is "the same" at dc ($\omega = 0$) as it is for any $\omega > 0$. So, upping the ante, what can EWB do with circuits that *do* contain frequency-dependent components, such as capacitors and inductors? As far as EWB is concerned, they are just as easy to handle as are all-resistor circuits. I'll show you three frequency-dependent circuits here, which we'll first analyze theoretically, and then we'll see what EWB says about them. The theoretical analyses are not trivial, and yet in mere seconds EWB will painlessly provide us with computer solutions that will be in excellent agreement with theory.

As my first example, consider the circuit of Figure 6.4.1, where an ac voltage $\mathbf{V}_{in}$ of frequency ω is applied to the left (input) terminals and an ac voltage $\mathbf{V}_{out}$ of frequency ω appears at the right (output) terminals. The two capacitors are equal, and the two inductors are also equal. Our problem here is to determine the so-called *frequency response* of the circuit, that is, the ratio of the output voltage to the input voltage as a function of ω. This is also called the circuit's *transfer function*.

Before I start writing equations, let me summarize what I will assume on your understanding of elementary ac circuit analysis. First, since the circuit of Figure 6.4.1 consists only of linear capacitors and inductors, the frequency of all voltages and currents is everywhere the same (ω). Second, the impedance of a capacitor C is $\mathbf{Z}_C = \dfrac{1}{i\omega C}$, where $i = \sqrt{-1}$, and the impedance of an inductor L is $\mathbf{Z}_L = i\omega L$. The "i" is present in both impedances because the instantaneous ac voltage drop across, and current in, for both components are *always* 90° out of phase. And finally, the (imaginary) ac impedance of a C or an L is analogous to the resistance of a resistor (a resistor has a real impedance because the voltage drop across it, and the current in it, are always *in* phase), and all impedances obey the familiar parallel/series formulas because Kirchhoff's laws apply to all circuits. That is, if we agree to treat $\mathbf{V}_{out}$ and $\mathbf{V}_{in}$ as complex quantities (with "i" present in the impedances, the complex nature of $\mathbf{V}_{out}$ and $\mathbf{V}_{in}$ shouldn't be a surprise), then we can write Kirchhoff's equations just as if we were dealing with nothing but resistors. In other words, we can mathematically treat capacitors and inductors as simply resistors with imaginary resistances!

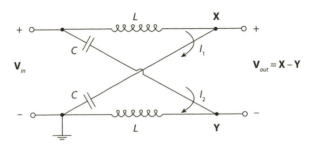

FIGURE 6.4.1. A frequency-dependent circuit

The transfer function of the circuit is written as

$$\mathbf{H}(i\omega) = \frac{\mathbf{V}_{out}}{\mathbf{V}_{in}},$$

which will be, of course, complex-valued because $\mathbf{V}_{out}$ and $\mathbf{V}_{in}$ are (by now you should have noticed that I am writing complex-valued quantities in bold). Since $\mathbf{H}(i\omega)$ is a complex quantity, then in the complex plane it is represented by a vector (different for every value of ω) of length $|\mathbf{H}(i\omega)|$ that makes angle $\phi(\omega)$ with the real axis. $|\mathbf{H}(i\omega)|$ is called the *amplitude response* of the circuit, and $\phi(\omega)$ is called the *phase response* of the circuit. Both $|\mathbf{H}(i\omega)|$ and $\phi(\omega)$ are, of course, real-valued functions of ω. More specifically, if we write the complex $\mathbf{H}(i\omega)$ with explicit real and imaginary parts as

$$\mathbf{H}(i\omega) = \mathrm{Re}\,\{\mathbf{H}(i\omega)\} + i\mathrm{Im}\,\{\mathbf{H}(i\omega)\},$$

then both $\mathrm{Re}\,\{\mathbf{H}(i\omega)\}$ and $\mathrm{Im}\,\{\mathbf{H}(i\omega)\}$ are real-valued functions of ω, and

$$|\mathbf{H}(i\omega)| = \sqrt{\mathrm{Re}^2\,\{\mathbf{H}(i\omega)\} + \mathrm{Im}^2\,\{\mathbf{H}(i\omega)\}},$$

$$\phi(\omega) = \tan^{-1}\left[\frac{\mathrm{Im}\,\{\mathbf{H}(i\omega)\}}{\mathrm{Re}\,\{\mathbf{H}(i\omega)\}}\right].$$

Let's see how all of this applies to the circuit of Figure 6.4.1.

By the symmetry of the circuit, the two currents $\mathbf{I}_1$ and $\mathbf{I}_2$ are equal, and so I'll write both as just $\mathbf{I}$. We can then express the output voltage as follows. To start,

$$\mathbf{V}_{out} = \mathbf{X} - \mathbf{Y}, \tag{6.4.1}$$

where $\mathbf{X}$ and $\mathbf{Y}$ are the voltages at the upper and lower output terminals, respectively. For convenience, I've shown the lower input terminal as grounded, which simply means that terminal serves as the zero voltage reference point for all other voltages in the circuit. In particular,

$$\mathbf{X} = \mathbf{I}\mathbf{Z}_C = \frac{\mathbf{I}}{i\omega C} = -i\frac{\mathbf{I}}{\omega C} \tag{6.4.2}$$

and

$$\mathbf{Y} = \mathbf{I}\mathbf{Z}_L = i\omega L\mathbf{I}. \tag{6.4.3}$$

Thus, from (6.4.1),

$$\mathbf{V}_{out} = -i\frac{\mathbf{I}}{\omega C} - i\omega L\mathbf{I} = -i\mathbf{I}\left(\frac{1}{\omega C} + \omega L\right),$$

where

$$\mathbf{I} = \frac{\mathbf{V}_{in}}{\mathbf{Z}_L + \mathbf{Z}_C} = \frac{\mathbf{V}_{in}}{i\omega L + \frac{1}{i\omega C}} = \frac{\mathbf{V}_{in}}{i\left(\omega L - \frac{1}{\omega C}\right)},$$

and so

$$\mathbf{V}_{out} = -i\frac{\mathbf{V}_{in}}{i\left(\omega L - \frac{1}{\omega C}\right)}\left(\frac{1}{\omega C} + \omega L\right) = \mathbf{V}_{in}\frac{\frac{1}{\omega C} + \omega L}{\frac{1}{\omega C} - \omega L}.$$

Thus,

$$\mathbf{H}(i\omega) = \frac{\mathbf{V}_{out}}{\mathbf{V}_{in}} = \frac{1 + \omega^2 LC}{1 - \omega^2 LC},$$

or, if we define $\omega_0 = \frac{1}{\sqrt{LC}}$, we have

$$\mathbf{H}(i\omega) = \frac{1 + \left(\frac{\omega}{\omega_0}\right)^2}{1 - \left(\frac{\omega}{\omega_0}\right)^2}, \quad \omega_0 = \frac{1}{\sqrt{LC}}. \tag{6.4.4}$$

We see from (6.4.4) that the transfer function $\mathbf{H}(i\omega)$ is always real (for this particular circuit). For $\omega < \omega_0$ $\mathbf{H}(i\omega) > 0$ and for $\omega > \omega_0$ $\mathbf{H}(i\omega) < 0$. Thus, the phase function is given by

$$\phi(\omega) = \begin{cases} 0, & \omega < \omega_0 \\ \\ \pi, & \omega > \omega_0. \end{cases} \tag{6.4.5}$$

At $\omega = \omega_0$ the amplitude response $|\mathbf{H}(i\omega)|$ blows up, that is, $\lim_{\omega \to \omega_0} |\mathbf{H}(i\omega)| = \infty$ (physicists and electrical engineers call these amplitude blow-ups *resonances*), while $\phi(\omega_0)$ is undefined (that is, $\phi(\omega)$ is discontinuous at $\omega = \omega_0$).

Figures 6.4.2 and 6.4.3 show the EWB simulation of Figure 6.4.1 (which, because of the circuit's appearance, is often called a *lattice*), for the particular values of $L = 2$ mH $= 2 \cdot 10^{-3}$ H and $C = 0.05\mu fd = 5 \cdot 10^{-8}$ F. This sets the value of ω_0 at $100,000$ radians per second, and so $f_0 = 15.915$ kHz. That's the resonant frequency at which the amplitude response should blow up, and at which the phase response should switch between a $0°$ and a $180°$ shift from input to output. The instrument attached to the input *and* the output terminals in the schematic is called a *Bode plotter*,[11] and it displays either the amplitude response (Figure 6.4.2) or the phase response (Figure 6.4.3) over a specified interval of frequency; the choice of display is made by clicking on one of the button icons at the top of the plotter's front panel. The input terminals are also connected to the function generator, which provides the sinusoidal input signal which is automatically varied in frequency

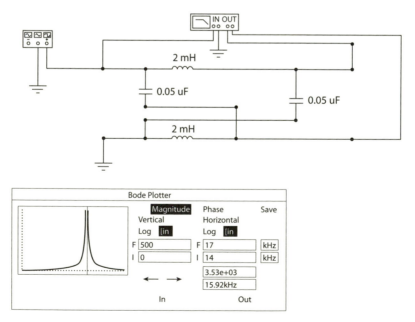

FIGURE 6.4.2. EWB amplitude simulation (**lattice.EWB**)

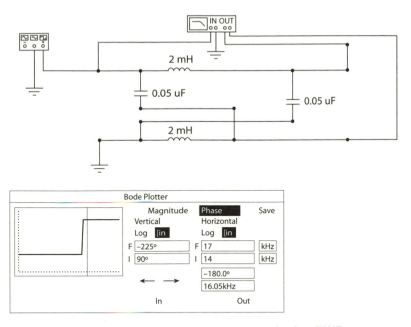

FIGURE 6.4.3. EWB phase simulation (**lattice.EWB**)

between the two limits set on the Bode plotter (in the simulations you can see that I set those frequency limits at 14 kHz and 17 kHz, which puts f_0 right in the middle of the displayed amplitude response curve).

Once EWB has created either the amplitude or the phase response on the Bode Plotter screen, it also displays a vertical line. That line can be positioned anywhere across the screen you wish, either by dragging it with the mouse or by clicking the left or right arrow buttons on the plotter's front panel. At every moment, either the frequency or the phase (as appropriate) for where the line intersects the Bode plot is displayed on the panel. In Figure 6.4.2 I placed the vertical line as close as I could to where the amplitude response blows up and the displayed frequency is 15.92 kHz, very close to the theoretical resonance frequency. And in Figure 6.4.3 I placed the vertical line just *above* f_0 (at 16.05 kHz), and the displayed phase shift is, indeed, π radians. (If the vertical line is moved to just *below* f_0 the displayed shift becomes zero.) The phase discontinuity at 15.92 kHz is clearly visible—as a nearly vertical transition—in Figure 6.4.3.

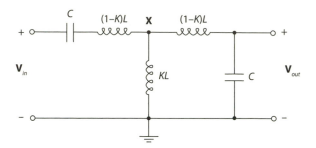

FIGURE 6.4.4. A doubly resonant circuit $(0 \le k \le 1)$

Matters are just a bit more complicated in the circuit of Figure 6.4.4, both in its analysis and in the theoretical results. As we'll show, this circuit has *two* resonant frequencies at which its amplitude response blows up, and at both of the resonant frequencies the phase response is discontinuous. EWB will handle it all with ease. But first, the theory.

Writing Kirchhoff's current law at the node labeled with $\mathbf{X}$ as its voltage, we have

$$\frac{\mathbf{V}_{in} - \mathbf{X}}{\frac{1}{i\omega C} + i\omega(1-k)L} = \frac{\mathbf{X}}{i\omega kL} + \frac{\mathbf{X} - \mathbf{V}_{out}}{i\omega(1-k)L}. \tag{6.4.6}$$

Since the current into the output node (labeled $\mathbf{V}_{out}$) is equal to the current out of that node, we also have

$$\frac{\mathbf{X} - \mathbf{V}_{out}}{i\omega(1-k)L} = \frac{\mathbf{V}_{out}}{\frac{1}{i\omega C}} = i\omega C \, \mathbf{V}_{out}. \tag{6.4.7}$$

From (6.4.7) we therefore have

$$\mathbf{X} - \mathbf{V}_{out} = i\omega C \, \mathbf{V}_{out} i\omega(1-k)L = -\omega^2(1-k)LC \, \mathbf{V}_{out},$$

or

$$\mathbf{X} = \mathbf{V}_{out}\left[1 - \omega^2(1-k)LC\right]. \tag{6.4.8}$$

Returning to (6.4.6),

$$\frac{i\omega C\,(\mathbf{V}_{in} - \mathbf{X})}{1 - \omega^2(1-k)LC} = \frac{\mathbf{X}}{i\omega kL} + \frac{\mathbf{X} - \mathbf{V}_{out}}{i\omega(1-k)L}$$

$$= \mathbf{X}\left[\frac{1}{i\omega kL} + \frac{1}{i\omega(1-k)L}\right] + i\frac{\mathbf{V}_{out}}{\omega(1-k)L}.$$

$$(6.4.9)$$

Concentrating on just the first term on the right in (6.4.9),

$$\mathbf{X}\left[\frac{1}{i\omega kL} + \frac{1}{i\omega(1-k)L}\right] = \mathbf{X}\left[\frac{i\omega(1-k)L + i\omega kL}{-\omega^2 k(1-k)L^2}\right]$$

$$= \mathbf{X}\frac{i\omega L}{-\omega^2 k(1-k)L^2}$$

$$= -\mathbf{X}\frac{i}{\omega k(1-k)L},$$

and so (6.4.9) becomes, when $\mathbf{X}$ is replaced by its expression in (6.4.8),

$$\frac{i\omega C\,\mathbf{V}_{in}}{1 - \omega^2(1-k)LC} - i\omega C\,\mathbf{V}_{out} = -i\,\mathbf{V}_{out}\frac{1 - \omega^2(1-k)LC}{\omega k(1-k)L}$$

$$+ i\,\mathbf{V}_{out}\frac{1}{\omega(1-k)L},$$

or

$$\frac{i\omega C\mathbf{V}_{in}}{1 - \omega^2(1-k)LC} = i\mathbf{V}_{out}\left[\omega C - \frac{1 - \omega^2(1-k)LC}{\omega k(1-k)L} + \frac{1}{\omega(1-k)L}\right],$$

or

$$\frac{\omega C\mathbf{V}_{in}}{1 - \omega^2(1-k)LC} = \mathbf{V}_{out}\left[\frac{\omega^2 k(1-k)LC - 1 + \omega^2(1-k)LC + k}{\omega k(1-k)L}\right],$$

or, after a minor bit of algebra in the numerator on the right,

$$\frac{\omega C\,\mathbf{V}_{in}}{1 - \omega^2(1-k)LC} = \mathbf{V}_{out}\frac{\omega^2 LC(1+k) - 1}{\omega k L}.$$

Thus, the transfer function of the circuit in Figure 6.4.4 is

$$\mathbf{H}(i\omega) = \frac{\mathbf{V}_{out}}{\mathbf{V}_{in}} = \frac{\omega^2 kLC}{\left[1 - \omega^2(1-k)LC\right]\left[\omega^2 LC(1+k) - 1\right]}.$$

$$(6.4.10)$$

From (6.4.10) we see that the transfer function is always real, and that it blows up at two distinct frequencies, that is, when $1 - \omega^2(1-k)LC = 0$ and when $\omega^2 LC(1+k) - 1 = 0$. These two resonant frequencies are therefore given by

$$\omega_{01} = \frac{1}{\sqrt{1-k}\sqrt{LC}} = \frac{\omega_0}{\sqrt{1-k}}$$

and

$$\omega_{02} = \frac{1}{\sqrt{1+k}\sqrt{LC}} = \frac{\omega_0}{\sqrt{1+k}},$$

where

$$\omega_0 = \frac{1}{\sqrt{LC}}.$$

Obviously, $\omega_{02} < \omega_0 < \omega_{01}$. If, for example, $L = 1\,\mathrm{mH} = 10^{-3}\,\mathrm{H}$ and $C = 0.1\mu fd = 10^{-7}$ F, then $\omega_0 = 100,000$ rads/s, and so $f_0 = 15.915$ kHz. If, for example, $k = 0.7$, then

$$f_{01} = \frac{15,915}{\sqrt{1-0.7}} = 29,057\,\mathrm{Hz}$$

and

$$f_{02} = \frac{15,915}{\sqrt{1+0.7}} = 12,206\,\mathrm{Hz}.$$

Going back to (6.4.10) and making the substitutions

$$LC = \frac{1}{\omega_0^2}, \quad 1 - k = \left(\frac{\omega_0}{\omega_{01}}\right)^2, \quad 1 + k = \left(\frac{\omega_0}{\omega_{02}}\right)^2,$$

we have

$$\mathbf{H}(i\omega) = \frac{\omega^2 k \dfrac{1}{\omega_0^2}}{\left[1 - \omega^2 \left(\dfrac{\omega_0}{\omega_{01}}\right)^2 \dfrac{1}{\omega_0^2}\right]\left[\omega^2 \dfrac{1}{\omega_0^2}\left(\dfrac{\omega_0}{\omega_{02}}\right)^2 - 1\right]},$$

or

$$\mathbf{H}(i\omega) = -k\left(\frac{\omega}{\omega_0}\right)^2 \frac{1}{\left[1 - \left(\dfrac{\omega}{\omega_{01}}\right)^2\right]\left[1 - \left(\dfrac{\omega}{\omega_{02}}\right)^2\right]}.$$

This last expression makes it clear that

$$\mathbf{H}(i\omega) \begin{array}{ll} < 0 & \text{for } \omega < \omega_{02} \\ > 0 & \text{for } \omega_{02} < \omega_0 < \omega_{01}, \\ < 0 & \text{for } \omega > \omega_{01} \end{array}$$

and so

$$\phi(\omega) = \begin{array}{ll} \pi & \text{for } \omega < \omega_{02} \\ 0 & \text{for } \omega_{02} < \omega_0 < \omega_{01} \\ \pi & \text{for } \omega > \omega_{01}. \end{array}$$

Figures 6.4.5 and 6.4.6 show EWB simulations of the circuit in Figure 6.4.4, with the Bode plotter set to operate between 7 kHz and 33 kHz so as to span both resonances. The figures show that EWB *is* displaying a double resonance and that the code has located the resonant frequencies at 12.2 kHz and 29.1 kHz, very near the calculated theoretical values.

For the final example of this section on the power of EWB, consider the circuit of Figure 6.4.7, made of three identical capacitor-resistor sections in cascade. This circuit, besides being computationally interesting (as I'm about to show you), has historical value as well, and I'll elaborate on that later in the chapter. For now, our problem is, if we apply a sinusoidal voltage of some given (but arbitrary) amplitude

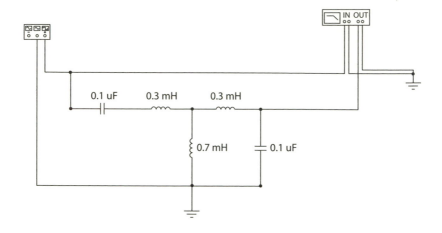

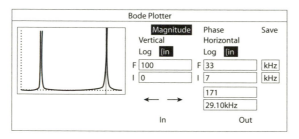

FIGURE 6.4.5. EWB amplitude simulation 1 (**doublepeak.ewb**), $k = 0.7$

to the input terminals, a smaller amplitude, phase shifted sinusoidal voltage will appear at the output terminals. Both the reduction in amplitude and the phase shift will be functions of the frequency. In particular, at what frequency will the phase shift be 180° and, at that frequency, what is the amplitude reduction factor? As you'll see, EWB will be able to handle this problem easily.

As usual, I'll use Kirchhoff to first answer our questions with theory, but rather than using the node-current law, just to be different here I'll instead use his loop-voltage law (conservation of energy): the sum of all the individual voltage drops around any closed loop in any circuit is identically zero. Using the clever artifice (an idea due to the great Maxwell, himself) of first writing the so-called *loop currents* $\mathbf{I}_1$, $\mathbf{I}_2$, and $\mathbf{I}_3$ (which are in bold because they are complex-valued), then traveling around each loop in the clockwise sense (as shown by the three loop

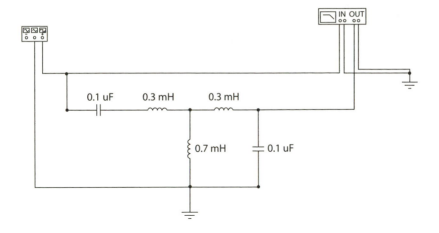

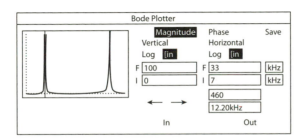

FIGURE 6.4.6. EWB amplitude simulation 2 (**doublepeak.ewb**), $k = 0.7$

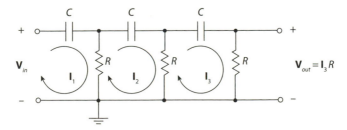

FIGURE 6.4.7. At what frequency is the phase-shift $180°$?

arrows) and writing a voltage *rise* as a negative voltage drop, gives us

$$-\mathbf{V}_{in} + \frac{1}{i\omega C}\mathbf{I}_1 + R(\mathbf{I}_1 - \mathbf{I}_2) = 0, \tag{6.4.11}$$

$$(\mathbf{I}_2 - \mathbf{I}_1)R + \frac{1}{i\omega C}\mathbf{I}_2 + R(\mathbf{I}_2 - \mathbf{I}_1) = 0, \tag{6.4.12}$$

and

$$(\mathbf{I}_3 - \mathbf{I}_2)R + \frac{1}{i\omega C}\mathbf{I}_3 + \mathbf{I}_3 R = 0. \tag{6.4.13}$$

Rearranging these simultaneous algebraic equations, with the idea of getting them conditioned for solution by Cramer's determinant rule (just as we did back in Chapter 4), results in

$$
\begin{aligned}
\left(R + \frac{1}{i\omega C}\right)\mathbf{I}_1 &- & R\,\mathbf{I}_2 &+ & 0\,\mathbf{I}_3 &= & \mathbf{V}_{in} \\
R\,\mathbf{I}_1 &- & \left(2R + \frac{1}{i\omega C}\right)\mathbf{I}_2 &+ & R\,\mathbf{I}_3 &= & 0 \\
0\,\mathbf{I}_1 &+ & R\,\mathbf{I}_2 &- & \left(2R + \frac{1}{i\omega C}\right)\mathbf{I}_3 &= & 0
\end{aligned}
$$

The system determinant is

$$\mathbf{D} = \begin{vmatrix} R + \frac{1}{i\omega C} & -R & 0 \\ R & -\left(2R + \frac{1}{i\omega C}\right) & R \\ 0 & R & -\left(2R + \frac{1}{i\omega C}\right) \end{vmatrix},$$

which I'll let you expand to confirm that

$$\mathbf{D} = R^3 - \frac{5R}{(\omega C)^2} - i\left[\frac{6R^2}{\omega C} - \frac{1}{(\omega C)^3}\right]. \tag{6.4.14}$$

The importance of (6.4.14) is that Cramer's rule then tells us

$$\mathbf{I}_3 = \frac{\begin{vmatrix} R + \frac{1}{i\omega C} & -R & \mathbf{V}_{in} \\ R & -\left(2R + \frac{1}{i\omega C}\right) & 0 \\ 0 & R & 0 \end{vmatrix}}{\mathbf{D}},$$

which I'll let you confirm says that (you should actually see it by inspection)

$$\mathbf{I}_3 = \frac{\mathbf{V}_{in}R^2}{\mathbf{D}}. \tag{6.4.15}$$

Since $\mathbf{V}_{out} = \mathbf{I}_3 R$, then (6.4.15) tells us that

$$\mathbf{V}_{out} = \frac{\mathbf{V}_{in}R^3}{\mathbf{D}}, \tag{6.4.16}$$

and so, combining (6.4.14) and (6.4.16), we have the transfer function of our three-section RC circuit as

$$\mathbf{H}(i\omega) = \frac{\mathbf{V}_{out}}{\mathbf{V}_{in}} = \frac{R^3}{R^3 - \dfrac{5R}{(\omega C)^2} - i\left[\dfrac{6R^2}{\omega C} - \dfrac{1}{(\omega C)^3}\right]}. \qquad (6.4.17)$$

Now, here's the crucial observation. *If the imaginary part of the denominator vanishes, that is, if*

$$\frac{6R^2}{\omega C} - \frac{1}{(\omega C)^3} = 0,$$

which is equivalent to saying *if*

$$\omega = \frac{1}{RC\sqrt{6}} = \omega_0,$$

then $\mathbf{H}(i\omega_0) < 0$, which means the phase shift is the required $180°$. Further, at the frequency $\omega = \omega_0$ the transfer function is

$$\mathbf{H}(i\omega_0) = \frac{R^3}{R^3 - \dfrac{5R}{\dfrac{1}{6R^2}}} = \frac{R^3}{R^3 - 30R^3} = -\frac{1}{29}.$$

That is, *if* a sinusoidal voltage is applied to the input of the circuit of Figure 6.4.7 at a frequency of ω_0, *then* the output voltage will be a sinusoid $180°$ out of phase with the input and reduced in amplitude by a factor of 29. That's what theory says. What does EWB say?

Figure 6.4.8 shows an EWB computer simulation (**RCphase-shifter.EWB**) of the circuit of Figure 6.4.7, with the values of R and C selected to give a $180°$ phase shift frequency of

$$f_0 = \frac{1}{2\pi RC\sqrt{6}}$$
$$= \frac{6.5 \cdot 10^{-2}}{RC} \text{Hz} = 10 \text{ kHz } (R = 100 \, k\Omega \text{ and } C = 65 \, \mu\mu fd$$
$$= 65 \text{ picofarad} = 65 \cdot 10^{-12} \text{ F}).$$

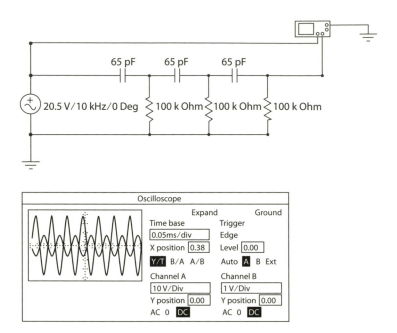

FIGURE 6.4.8. EWB computer simulation (**RCphaseshifter.EWB**)

The input signal is provided by an ac signal source set to 10 kHz, with an amplitude of 58 volts, peak to peak.[12] The dual-trace oscilloscope shows both the input voltage (on channel A, the larger signal) and the output voltage (on channel B, the smaller signal). The two traces clearly show the 180° phase difference, and the gain settings on channel A and channel B show the amplitude reduction factor of 29 (as closely as one can read the screen).

As a final remark on this circuit, while the *total* phase shift is 180°, the three individual sections do not contribute equally. That is, the shift through each of the three RC sections is not 60°. For example, if you look at just the final RC section of Figure 6.4.7, its transfer function is

$$\frac{R}{R + \frac{1}{i\omega C}} = \frac{i\omega RC}{1 + i\omega RC} = \frac{i\omega RC[1 - i\omega RC\}}{1 + (\omega RC)^2} = \frac{(\omega RC)^2 + i\omega RC}{1 + (\omega RC)^2},$$

and so the phase shift through the third section is

$$\phi_3(\omega) = \tan^{-1}\left\{\frac{\omega RC}{(\omega RC)^2}\right\} = \tan^{-1}\left\{\frac{1}{\omega RC}\right\}.$$

At $\omega = \omega_0 = \dfrac{1}{RC\sqrt{6}}$, we therefore have

$$\phi_3(\omega_0) = \tan^{-1}\left\{\sqrt{6}\right\} = 67.792°.$$

The other two individual sections each have yet different phase shifts (see Challenge Problem 6.3). The reason for this, of course, is that each section is in a different situation. The third section has nothing connected to its output. The second (middle) section has the third section connected to its output. And the output of the first section is, as electrical engineers put it, "loaded down" by both the second and the third sections.

6.5 Differential Amplifiers and Electronic Circuit Magic

Until the 1960s electronic circuit designers worked exclusively with resistors, capacitors, inductors, transformers, diodes, vacuum tubes, and the early transistors. All of those components were physically big. When I worked as a logic designer in the early 1960s in the Southern California aerospace industry, for example, it was routine to find just four individual circuits (logic gates or flip-flops) on a plug-in card the size of a man's hand. With the development of integrated circuit fabrication technology, however, it became possible to photoetch microelectronic circuits onto tiny silicon wafers. Such circuits are today produced at densities so high that each of the popular video game consoles contains hundreds of millions of transistors. There is, though, one glaring omission to this transition from discrete to integrated. Inductors. It is very difficult to make even a traditional inductor (wire wound around an iron core) that behaves like a mathematical inductor ($v = L\dfrac{di}{dt}$), and to fabricate an integrated circuit inductor is exponentially more difficult.

It is easy, however, to fabricate integrated circuit resistors and capacitors that are nearly mathematically perfect, as well as transistors and diodes, too. With just those components it then proves to be possible to make amplifiers with extremely large voltage gains (in the hundreds of thousands).[13] The big theoretical question then was, were there any restrictions on the nature of the circuits that could be constructed with just those components? More to the point, did not having inductors limit designers? The answer isn't a priori obvious, I think, but it is *no*. With just Rs, Cs, and high gain amplifiers, you can build anything, including circuits that seem to require inductors! In the rest of this section I'll show you by example how it is done, and when we finish I think you'll agree with my use of the words "circuit magic."

You know how Rs and Cs work; here's what's going on with a high gain amplifier. Our symbol for such a device is shown in Figure 6.5.1, as a triangle pointing from its two input voltages (v_1 and v_2) to its output voltage v_{out}, where all three voltages are measured with respect to a common reference (shown as the ground symbol). There are also, of course, as in any real piece of electronic hardware, connections for a power supply, but those are usually not shown explicitly but rather are simply assumed to be there. The R_{in} in Figure 6.5.1 is the resistance between the two input terminals of the diff-amp, and the current i is the input current to the diff-amp. The positive voltage gain of the amplifier is denoted by A and, *by definition*,

$$v_{out} = A(v_2 - v_1), \quad A > 0. \tag{6.5.1}$$

The device defined by (6.5.1) is called a *differential amplifier* (diff-amp), as v_{out} is the gain times the *difference* between the voltage on the positive input terminal and the voltage on the negative input terminal. Now, this may seem just too simple to be of any use.[14] Not so, and here are some classic examples to show you why that first impression

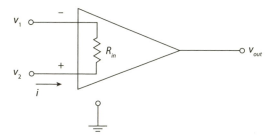

FIGURE 6.5.1. The differential amplifier

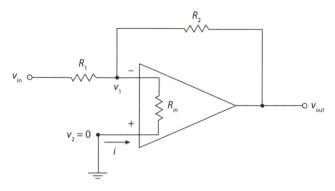

FIGURE 6.5.2. A simple (but *most* useful) diff-amp circuit

is wrong. In Figure 6.5.2 I've drawn my first example, where $v_2 = 0$ because the positive input terminal is connected directly to ground. Using Kirchhoff's current law at the negative input terminal of the diff-amp, we have

$$\frac{v_{in} - v_1}{R_1} = \frac{v_1 - v_{out}}{R_2} + \frac{v_1}{R_{in}}. \qquad (6.5.2)$$

Also, from (6.5.1) with $v_2 = 0$,

$$v_{out} = -Av_1. \qquad (6.5.3)$$

From (6.5.3) we have

$$v_1 = -\frac{v_{out}}{A}$$

and if we insert this into (6.5.2) we get

$$\frac{v_{in} + \dfrac{v_{out}}{A}}{R_1} = \frac{-\dfrac{v_{out}}{A} - v_{out}}{R_2} - \frac{\dfrac{v_{out}}{A}}{R_{in}},$$

which, with some simple algebra, becomes

$$v_{in} = -\frac{v_{out}}{A}\left[1 + \frac{R_1}{R_{in}} + \frac{R_1}{R_2}(1 + A)\right]. \qquad (6.5.4)$$

At this point we've made no assumptions. But in an actual diff-amp A is quite large, and so I'll now assume both that $A \to \infty$ and that v_{out} remains finite. This last assumption is experimentally observed to be the case whenever there is a connection between the output and the *negative* input terminal (as there is in Figure 6.5.2). That is, v_{out} will always be finite in a diff-amp circuit with *just* negative feedback. That's good, too, because if v_{out} didn't remain finite, we wouldn't have much of a useful circuit! (If positive feedback is also present the circuit may not be stable. An example of this is given at the end of the chapter.) Making the assumption of $A \to \infty$ is the *ideal* diff-amp assumption, and with it (6.5.4) becomes

$$v_{in} = -\frac{R_1}{R_2} v_{out}$$

or, if we define the so-called *feedback gain* of the entire circuit (diff-amp plus external resistors) as

$$A_f = \frac{v_{out}}{v_{in}}$$

then

$$A_f = -\frac{R_2}{R_1}. \tag{6.5.5}$$

For a real diff-amp, of course, A isn't actually infinite, but it can be quite large; for example, $A > 200,000$ is not at all unusual. The EWB ideal diff-amp model uses $A = 10^6$.

The result in (6.5.5) may look simple, but there are two amazing implications because of it. First, once A becomes "large," the effective gain of the circuit (A_f) becomes *independent* of A and is determined only by the external resistors R_1 and R_2. And second, once A becomes "large," the specific value of R_{in}, even if "small," becomes irrelevant! To understand this second claim, look back at (6.5.3) and observe that

$$\lim_{A \to \infty} v_1 = -\lim_{A \to \infty} \frac{v_{out}}{A} = 0,$$

because, remember, v_{out} is always finite. That is, as $A \to \infty$ the voltage on the negative input terminal of the diff-amp in Figure 6.5.2

approaches ground potential. That terminal is not really ground, of course, as it is separated from true ground by R_{in}. For that reason the negative input terminal is said to be a *virtual* ground. More generally, if the positive input terminal of the diff-amp did not connect directly to ground, then we couldn't automatically argue that v_2 is zero, but it would still be true that $\lim_{A \to \infty} v_1 \to v_2$. This is a direct result (because of the very definition of a diff-amp as given by (6.5.1)) of requiring that v_{out} remain finite as $A \to \infty$. The amazing conclusion from this, then, is that since $v_2 - v_1$ is the voltage drop across the diff-amp's input resistance R_{in}, then it must be true that the input current i to the diff-amp goes to *zero* as $A \to \infty$, *independent* of the value of R_{in}!

When external circuitry is connected around a diff-amp, the *entire* circuit that results is called an *operational amplifier* (op-amp). That's because the transfer function of the entire circuit is generally one of some mathematical operation (summation, integration, differentiation, sign changing, absolute value, and so on). The first use of such circuits (originally built in the 1940s using vacuum tubes) was that of solving differential equations on analog computers (not digital computers). For example, suppose you want to solve the following coupled pair of first-order *nonlinear* differential equations

$$\begin{aligned} \frac{dx}{dt} &= ax - bxy \\ \frac{dy}{dt} &= -cy + dxy \end{aligned}, \tag{6.5.6}$$

where a, b, c, and d are all positive constants. I haven't simply pulled (6.5.6) out of the air; these equations are the *Lotka-Volterra predator-prey* equations, famous to all who work in the field of ecology.[15]

Imagine that $x(t)$ and $y(t)$ are the sizes of two populations of animals, one of which (y) eats the other (x). That is, x is the *prey* of the *predator y*. Classic examples for x are rabbits and sheep, while y would be foxes and coyotes, respectively. The constants a and c are the growth and death rates, respectively, for x and y, respectively. If b and d were zero, then x would increase exponentially without limit, while y would decrease exponentially toward zero. In other words, the woods and hills would soon swarm with rabbits and sheep, while the foxes and the coyotes would starve to death. But of course, in the real world b and d

are not zero, and the product term xy in the predator-prey equations is a measure of the likelihood of an *encounter* between predator and prey. (*Encounter* is a euphemism for "a predator eats a prey.") Such an encounter tends to reduce the rate at which the prey population increases (hence the minus sign in $-bxy$), as well as to decrease the rate at which the predator population dies (hence the plus sign in $+dxy$). The specific values of b and d influence the significance of an encounter between prey and predator, respectively.

There are no known theoretical solutions to the Lotka-Volterra equations, but we can study the behaviors of $x(t)$ and $y(t)$ with EWB as follows. Our method will be to mimic the analog computer approach that would have been used in pre-MATLAB days. That is, what we'll do in the rest of this section is literally (almost) build an analog computer circuit that EWB will then simulate on a digital computer. (That opens an obvious philosophical question that I'll leave for better minds than mine to ponder.) We start by assuming that there are two wires in front of us, one with the voltage $\dfrac{dx}{dt}$ on it and the other with the voltage $\dfrac{dy}{dt}$. (Just where these two wonderful wires come from will be revealed momentarily!) Then, *if* we have two integration circuits available we could integrate the voltages on these two wires to produce the solutions we are after, $x(t)$ and $y(t)$. As you've probably begun to suspect, we can indeed make an integrator out of a diff-amp and some external circuitry, and I'll soon show you how. For now, however, let's just blissfully continue on our "supposing" ways and imagine that we next (somehow) multiply the x and y together to get xy, multiply the x by a to get ax and the y by $-c$ to get $-cy$, multiply the xy by $-b$ and also by d to get both $-bxy$ and dxy, and then add the ax to the $-bxy$ and the $-cy$ to dxy. If we do all that, we'll end up with two wires, one with the voltage $ax - bxy$ on it and the other with the voltage $-cy + dxy$ on it. Now you know where the two wonderful wires that we initially assumed existed actually come from.

They are the two wires we just made! That's right, just connect those two wires back to the inputs of the integrators and then watch the solutions $x(t)$ and $y(t)$ appear at the integrator outputs. Talk about pulling yourself up by your bootstraps—the solution voltages $x(t)$ and $y(t)$ seem to appear out of nowhere with the sole purpose of generating themselves! This invariably strikes students, when they first see the line of argument I've just made, as borderline supernatural. For now, just take my word for it, but by the end of this section you'll see it happen with your own eyes. But right now we have some obvious questions to answer. Specifically, how do we make integrators, multipliers, and summers?

In Figure 6.5.3 I've replaced the feedback resistor R_2 in Figure 6.5.2 with the capacitor C. With the ideal diff-amp assumption that $A \to \infty$ we have zero current into the diff-amp and so, applying Kirchhoff's current law at the negative input terminal (which is virtual ground), we have (writing just R for R_1)

$$\frac{v_{in} - 0}{R} = \left(C\frac{d}{dt}0 - v_{out} \right) = -C\frac{dv_{out}}{dt}.$$

Thus,

$$\frac{dv_{out}}{dt} = -\frac{1}{RC}v_{in},$$

or

$$v_{out}(t) = -\frac{1}{RC} \int_0^t v_{in}(u)du + v_{out}(0). \tag{6.5.7}$$

That is, the diff-amp circuit of Figure 6.5.3 is an operational amplifier that performs integration with the *negative* gain of $-\frac{1}{RC}$. I'll come back, in just a bit, to discuss what to do about "$v_{out}(0)$," the initial condition term in (6.5.7). What if we don't want negative gain when we integrate, but rather positive gain instead? One easy answer is to simply take the output of the circuit in Figure 6.5.3 and use it as the input to the circuit of Figure 6.5.2 with $R_1 = R_2$, which (6.5.5) tells us has gain -1 (and so its purpose is just to change algebraic sign). See Challenge Problem 6.4 for another way to make a positive gain integrator.

To build a summer, we'll use the circuit of Figure 6.5.4, which is simply the circuit of Figure 6.5.2 with multiple input resistors connected to the negative terminal virtual ground (now also called the *summing junction*) of the diff-amp, one resistor for each term in the sum. Writing R_f for the diff-amp feedback resistor and $R_1, R_2, \ldots, R_n$ for the

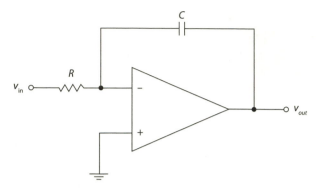

FIGURE 6.5.3. The diff-amp integrator

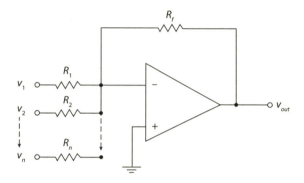

FIGURE 6.5.4. The diff-amp summer

input resistors in an n-term sum, then Kirchhoff's current law at the diff-amp's negative input terminal is

$$\frac{v_1 - 0}{R_1} + \frac{v_2 - 0}{R_2} + \cdots + \frac{v_n - 0}{R_n} = \frac{0 - v_{out}}{R_f},$$

or

$$v_{out} = -\left[\frac{R_f}{R_1}v_1 + \frac{R_f}{R_2}v_2 + \cdots + \frac{R_f}{R_n}v_n\right]. \tag{6.5.8}$$

In Challenge Problem 6.5 the diff-amp summer is used to make the $R/2R$ ladder digital-to-analog converter that I mentioned back in Section 6.3.

Finally, to build a multiplier is just a bit more difficult. But not by much. One way to do it, which was actually used in the earliest days of analog computers, is to assume we have a device that squares its single input. Then, to make a multiplier (which of course has two inputs, which I'll call x and y), we use the mathematical identity

$$xy = \frac{1}{4}\left[(x+y)^2 - (x-y)^2\right]. \qquad (6.5.9)$$

With the circuitry we've already discussed we can perform all the operations in (6.5.9) except, of course, the squaring. That crucial task can be accomplished by using appropriately connected voltage sources and diodes to construct a network whose output is a close approximation to the square of its input (I am not going to pursue the details of these diode circuits here, but they are not at all difficult to build in integrated circuitry). The resulting circuit is called a *quarter-square multiplier*. For EWB, of course, simulating a multiplier is easy; all the code does is *actually multiply* x and y (remember, EWB is running on a digital computer, where multiplication is readily available). For how multipliers are constructed today, in real hardware, see any good modern text on integrated circuits.

Okay, back to the Lotka-Volterra differential equations of (6.5.6). Let's agree to now forget about rabbits, foxes, sheep, and coyotes and to simply take the equations as they stand. I'm taking this position because I, for one, have no interest in looking into the birth rates of rabbits or the starvation rates of wild foxes to use to figure out what the coefficients a, b, c, and d should be (and the same personal reluctance to search for this data holds for sheep and coyotes). From now on, a, b, c, and d are just numbers, with no particular physical significance. Once we have constructed our analog computer circuit to solve the equations of (6.5.6) you'll see how easy it is to select any particular values you wish for a, b, c, and d. Figure 6.5.5 shows the complete analog computer circuit as I drew it on my computer screen, using EWB's schematic capture feature. The handwritten entries on the schematic are my notes as to what voltage is present on each wire. In the circuit I have arbitrarily used the values $a = c = 1$ and $b = d = 100$.

Notice, in particular, that I have taken the inputs to the two integrators at the far left to be $-\dfrac{dx}{dt}$ and $-\dfrac{dy}{dt}$, rather than $\dfrac{dx}{dt}$ and $\dfrac{dy}{dt}$, and that's because of the *negative* gains (-1, since the RC product is $10^6 \cdot 10^{-6} = 1$) of the integrators. That way the integrator outputs are the solutions $x(t)$ and $y(t)$, which are then fed directly to the dual-trace oscilloscope ($y(t)$ is on channel A, and $x(t)$ is on channel B). EWB allows the user to specify the gain of the multiplier, too, and I have used a gain of $\dfrac{1}{100} = 0.01$. Finally, since the Lotka-Volterra equations are each *first* order, we need to specify the initial values for both $x(t)$ and $y(t)$. EWB defaults to using zero for all the initial voltages in a circuit at the start of a simulation, but does allow the option of user-defined initial voltage values for each wire in a circuit. The oscilloscope traces in Figure 6.5.5 are for the initial conditions $x(0) = 200$ volts and $y(0) = 50$ volts.[16] The solutions displayed clearly show *non*-sinusoidal

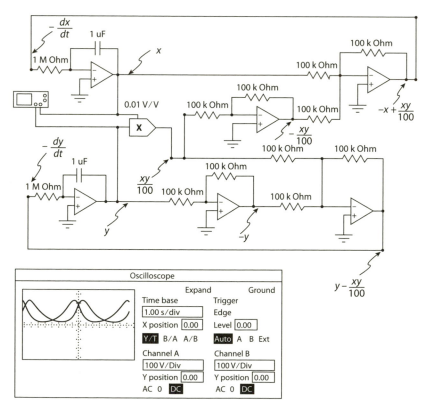

FIGURE 6.5.5. EWB "solves" the Lotka-Volterra coupled differential equations

oscillations for both $x(t)$ and $y(t)$, with $y(t)$ reaching its maximum values later in time than does $x(t)$; as electrical engineers put it, $y(t)$ *lags* $x(t)$ or, alternatively, $x(t)$ *leads* $y(t)$. And that is how people once "solved" complicated differential equations with electronic *analog* computers, in the years before MATLAB![17]

Let me end this section with some practical comments about analog computer circuits. We solved the Lotka-Volterra equations by assuming we had available the highest-order derivative, for both $x(t)$ and $y(t)$, and then integrating. Why couldn't we have instead gone in the opposite direction, that is, why not assume we have $x(t)$ and $y(t)$ available and then *differentiate* to form all the required derivatives? It is, in fact, easy to make a diff-amp differentiator; just swap the R and the C in the integrator of Figure 6.5.3 and you should be able to quickly show that

$$v_{out} = -RC\frac{dv_{in}}{dt}.$$

From a purely mathematical point of view, this is perfectly fine. With some care, an EWB analog computer simulation with differentiators would work. But in actual practice, with real hardware, the use of differentiators is avoided. Why?

In any real electronic circuit there is unavoidable noise that is *always* present. The characteristic signature of noise is that of brief, sudden, unpredictable changes in voltage levels. Its origins are numerous, including the very atomic structure of matter, as well as unrelated signals radiated into the circuit (sometimes called "*&%$$#@ interference" by frustrated engineers!). Besides differentiating the voltages representing the problem variables you are interested in, the noise will also be differentiated, perhaps producing even larger changes (more noise!) that have nothing to do with the mathematical behavior of the actual equations. For example, if the voltage level into a differentiator randomly fluctuates by just 10^{-2} V in 10 microseconds $= 10^{-5}$ seconds, then the *rate of change* of that fluctuation introduces an extraneous component of $1,000$ V/s at the output of the differentiator! Integrators, on the other hand, tend to smooth away noise fluctuations.

Also, from a practical point of view, incorporating initial conditions in a real hardware circuit using integrators is easy and obvious: just place an initial charge on the feedback capacitor of an integrator to establish the initial output voltage of the integrator. It is true that in an EWB simulation on a computer one can set the initial output of a differentiator just as easily as with an integrator, but it is not at all obvious how to set a non-zero initial condition of a differentiator in real hardware.

6.6 More Circuit Magic: The Inductor Problem

In this section I'll show you how diff-amps can handle the "inductor problem" I mentioned in the last section—that is, how inductor-free circuits can be built with integrated circuitry. But first, as fun, warmup exercises, here are two more new diff-amp circuits with what I think are surprising behaviors (similar to Challenge Problems 6.4 and 6.5, both of which you should try your hand at by the end of this section). My first example is the easy to analyze (but most useful) circuit in Figure 6.6.1. I've written the voltage at the negative input terminal of the ideal diff-amp as x, but because of the negative feedback, we know that $x = v_{in}$. Notice that here we have an example of a circuit in which the negative input terminal is *not* virtual ground, because the positive input terminal is not connected to true ground. Applying Kirchhoff's current law at the negative input terminal, and remembering that the input current to an ideal (infinite gain) diff-amp with negative feedback is zero, we have

$$\frac{0 - v_{in}}{R_1} = \frac{v_{in} - v_{out}}{R_2},$$

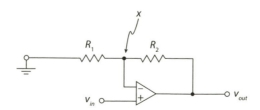

FIGURE 6.6.1. A *positive*-gain diff-amp circuit

or, with a little algebra, the overall voltage gain of the circuit is

$$A_f = \frac{v_{out}}{v_{in}} = 1 + \frac{R_2}{R_1}. \tag{6.6.1}$$

Notice that if $R_1 = \infty$ (that is, simply remove that resistor from the circuit) and/or $R_2 = 0$ (that is, replace that resistor with a short, directly connecting the diff-amp output back to its negative input terminal), then $A_f = 1$. That is, $v_{out} = v_{in}$ and the resulting circuit, one with no external hardware at all (except for the feedback wire!), is called a *voltage follower*. It is often used as a buffer, to isolate one part of a circuit from another to avoid "loading" problems.

In Figure 6.6.2 you'll find a circuit that behaves like a *negative capacitor,* something you'll never find in an electronics store. I've placed the entire circuit inside a dashed outline, which you are to imagine is a sealed black box. All you have available to you are the two terminals at the left, with a voltage drop of v across them (which you've applied) and the resulting input current i. I say this circuit is a negative capacitor because the relationship between v and i is

$$i = -C_{eq}\frac{dv}{dt}, \quad C_{eq} > 0, \tag{6.6.2}$$

where C_{eq} is determined by R_1, R_2, and C (a real capacitor). This is easy to show, as follows.

With the diff-amp output voltage denoted by x, and remembering that an ideal diff-amp with negative feedback has the same voltage on

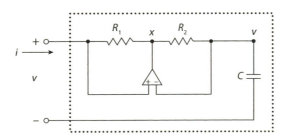

FIGURE 6.6.2. A *negative* capacitor!

both of its input terminals, we can write

$$i = \frac{v - x}{R_1} \quad \text{and} \quad \frac{x - v}{R_2} = C\frac{dv}{dt}.$$

That is,

$$iR_1 = v - x \quad \text{and} \quad x - v = R_2C\frac{dv}{dt} = -(v - x),$$

$$R_2C\frac{dv}{dt} = -iR_1,$$

and so

$$i = -\frac{R_2C}{R_1} \cdot \frac{dv}{dt},$$

which is (6.6.2) with

$$C_{eq} = \frac{R_2C}{R_1}.$$

The amazing circuit of Figure 6.6.2 is actually a bit more amazing than this calculation shows. To see that, try replacing the C with an arbitrary complex impedance $\mathbf{Z}$ and then computing the input impedance $\mathbf{Z}_{in}$ "seen" looking into the input terminals to the black box, where

$$\mathbf{Z}_{in} = \frac{\mathbf{V}}{\mathbf{I}}$$

and where $\mathbf{V}$ and $\mathbf{I}$ are now *sinusoidal ac* quantities (take a look back at Section 6.4) and are not arbitrary functions of time (as in the analysis for the negative capacitor). You should now be able to show that

$$\mathbf{Z}_{in} = -\frac{R_1}{R_2}\mathbf{Z}.$$

Electrical engineers call this circuit a *negative impedance converter* (NIC).

Now we can take a look at how to build a mathematically (almost) perfect inductor in integrated circuitry. Take a look at Figure 6.6.3,

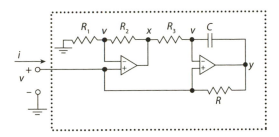

FIGURE 6.6.3. A simulated *grounded* inductor

which consists of four resistors, one capacitor, and two ideal diff-amps. Again, I've drawn the entire circuit inside a dashed outline, with the idea that only the two input terminals at the left are available to you. Notice *carefully* that one of those terminals is grounded. What I'll now show you is that this circuit simulates what electrical engineers call a *grounded inductor*, that is, an inductor with one end that *must be* connected to ground. This is in contrast to the more general case of a *floating inductor*, which means *each* end can freely be connected to *any point* in a circuit (perhaps ground, but that's your choice). At the conclusion of this section I'll show you how the simulation of a floating inductor can often be accomplished.

When I say the circuit of Figure 6.6.3 simulates an inductor, I mean that the relationship between the voltage v and the current i is

$$v = L_{eg}\frac{di}{dt},\qquad(6.6.3)$$

where L_{eg} is determined by R_1, R_2, R_3, R, and C. In Figure 6.6.3 I've labeled the three node voltages in the upper-bus as, from left to right, v, x, v. The first and last of these (the two v's) follow from the rule that the input terminal voltages on an ideal diff-amp with negative feedback are equal. The output voltage of the rightmost diff-amp is y. The defining equations for our circuit, with this notation, are

$$i = \frac{v-y}{R},\quad \frac{0-v}{R_1} = \frac{v-x}{R_2},\quad \frac{x-v}{R_3} = C\frac{d(v-y)}{dt}$$

or

$$y = v - iR,\quad -vR_2 = vR_1 - R_1x,\quad x - v = R_3C\frac{d(v-y)}{dt}.$$

So,

$$x = v\frac{R_1 + R_2}{R_1} = v + v\frac{R_2}{R_1}.$$

Thus,

$$v + v\frac{R_2}{R_1} - v = R_3 C\frac{d(v - v + iR)}{dt},$$

or

$$v\frac{R_2}{R_1} = R_3 RC\frac{di}{dt},$$

and so, finally,

$$v = \frac{R_1 R_3 RC}{R_2} \cdot \frac{di}{dt},$$

which is (6.6.3) with

$$L_{eg} = \frac{R_1 R_3 RC}{R_2}.$$

For example, if all the resistors are equal to $10,000$ ohms $(= 10^4\ \Omega)$ and C is 0.1 microfarad $\left(= 10^{-7}F\right)$, then

$$L_{eg} = \frac{\left(10^4\right)^3 10^{-7}}{10^4} = 10\,\text{H}.$$

This is a fairly big inductor, available in real hardware only as a large, heavy, and almost certainly mathematically imperfect component, while the simulation circuit of Figure 6.6.3 (built using just two ideal diff-amps, four resistors, and a capacitor) is virtually perfect.

So, now we can build a *grounded* inductor in integrated circuitry with just diff-amps, resistors, and capacitors. But what if we have a circuit that, in real hardware, uses a *floating* inductor? How do we build that circuit without an inductor? The key is the circuit of Figure 6.6.4, which

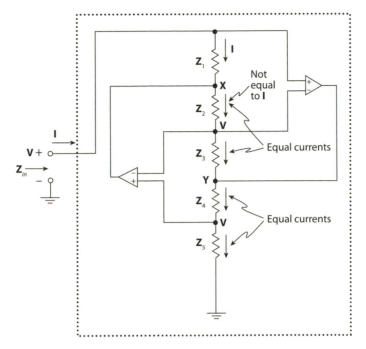

FIGURE 6.6.4. The GIC

electrical engineers call a *generalized impedance converter* (GIC). When you look at the GIC you might be surprised to see that one of its input terminals at the far left is still grounded. How, you may now be asking yourself, is that going to make a floating inductor? The short answer is that there is still one more trick I haven't yet shown you. Stay with me for just a bit more, and the clouds will clear! First, though, let's see just what the GIC actually does.

Our analysis will be for the case of ac quantities, that is, I'll be assuming that the input voltage **V** and the resulting current **I** are sinusoidal with frequency ω. The diff-amp output voltages are **X** and **Y**, and the applied input voltage **V** appears as shown in the vertical stack of five arbitrary impedances because of the rule for diff-amps that have negative feedback: there is zero voltage drop across their input terminals. The input impedance Z_{in}, "seen looking into" the two input terminals at the far left, can be found by writing the following equations, remembering that an ideal diff-amp with negative feedback has zero input current:

$$I = \frac{V - X}{Z_1},\tag{6.6.4}$$

$$\frac{X - V}{Z_2} = \frac{V - Y}{Z_3},\tag{6.6.5}$$

and

$$\frac{Y - V}{Z_4} = \frac{V}{Z_5}.\tag{6.6.6}$$

These are three equations in four variables (V, I, X, and Y) which means that we can calculate the *ratio* $Z_{in} = V/I$. Note *carefully* that we do not use Kirchhoff's current law at any node to which a diff-amp *output* is connected. That's not because the law doesn't apply at such a node—the current law is *always* true, everywhere in any electric circuit at every instant of time—but rather because we simply don't know what the diff-amp output currents are.

From (6.6.6) we have

$$Y - V = \frac{Z_4}{Z_5}V,$$

or

$$Y = V\left(1 + \frac{Z_4}{Z_5}\right).\tag{6.6.7}$$

From (6.6.5) we have

$$X - V = \frac{Z_2}{Z_3}V - \frac{Z_2}{Z_3}Y,$$

or, using (6.6.7),

$$X - V = \frac{Z_2}{Z_3}V - \frac{Z_2}{Z_3}V\left(1 + \frac{Z_4}{Z_5}\right),$$

and so

$$X = V\left[1 + \frac{Z_2}{Z_3} - \frac{Z_2}{Z_3} - \frac{Z_2 Z_4}{Z_3 Z_5}\right],$$

or

$$\mathbf{X} = \mathbf{V} - \mathbf{V}\frac{\mathbf{Z}_2\mathbf{Z}_4}{\mathbf{Z}_3\mathbf{Z}_5}. \qquad (6.6.8)$$

Putting (6.6.8) into (6.6.4) gives

$$\mathbf{I} = \frac{\mathbf{V} - \mathbf{V} + \mathbf{V}\frac{\mathbf{Z}_2\mathbf{Z}_4}{\mathbf{Z}_3\mathbf{Z}_5}}{\mathbf{Z}_1} = \mathbf{V}\frac{\mathbf{Z}_2\mathbf{Z}_4}{\mathbf{Z}_1\mathbf{Z}_3\mathbf{Z}_5},$$

and so, at last, we have the input impedance of the GIC as

$$\mathbf{Z}_{in} = \frac{\mathbf{V}}{\mathbf{I}} = \frac{\mathbf{Z}_1\mathbf{Z}_3\mathbf{Z}_5}{\mathbf{Z}_2\mathbf{Z}_4}. \qquad (6.6.9)$$

Notice that if either $\mathbf{Z}_2$ or $\mathbf{Z}_4$ (but not both) is a capacitor, and all the other $\mathbf{Z}$s are resistors, then the GIC behaves like an inductor (that is, $\mathbf{Z}_{in}$ behaves as $i\omega$ times "something real"). A *grounded* inductor, however, and so, as matters stand, we still haven't addressed the *floating* inductor issue. Therefore, we are now ready for that trick I mentioned earlier that will solve the problem, and to set it up, let's consider a specific circuit (so we can really see how the numbers work) with a floating inductor that we wish to put into inductor-free form. As we move through the analysis it will appear that we've traded the floating inductor puzzle for something even more perplexing—and then, suddenly and at the last possible moment, the GIC will save the day.

Consider, then, the circuit of Figure 6.6.5, consisting of two resistors, a capacitor, and a floating inductor. The circuit has two sections: a series connection of the resistor R_a and the inductor L (which I've lumped together as $\mathbf{Z}_a$), and a parallel connection of the resistor R_b and the capacitor C (which I've lumped together as $\mathbf{Z}_b$). I've used EWB's schematic capture feature to draw the circuit with specific component values ($R_a = 10^6 \ \Omega$, $L = 1$ H, $R_b = 8 \cdot 10^6 \ \Omega$, and $C = 10^{-12}$ F), and then used the code to display on the Bode plotter—see Figure 6.6.6—the magnitude of the circuit's transfer function over the frequency interval 10 kHz $\leq f \leq$ 300 kHz. From that transfer function you can see why electrical engineers call this circuit a *low-pass filter*; sinusoids of lower frequency are transmitted from input to output with

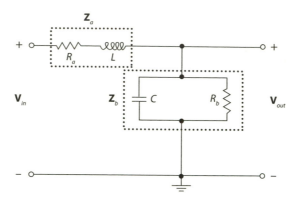

FIGURE 6.6.5. A floating inductor circuit

less attenuation than are sinusoids of higher frequency. There is one particular frequency at which the transfer function magnitude has its maximum value, and I've dragged the Bode plotter cursor (the vertical line) over the display screen to show that maximum magnitude is 0.989 at 111.5 kHz (see Challenge Problem 6.6). We can use this "experimental" result as a partial check on the equivalence of the behavior of the inductor-free version of the circuit we are about to derive.

The transfer function of the circuit in Figure 6.6.5 is

$$\mathbf{H} = \frac{\mathbf{V}_{out}}{\mathbf{V}_{in}} = \frac{\mathbf{Z}_b}{\mathbf{Z}_a + \mathbf{Z}_b},$$

where

$$\mathbf{Z}_a = R_a + i\omega L, \quad \mathbf{Z}_b = \frac{R_b \frac{1}{i\omega C}}{R_b + \frac{1}{i\omega C}}.$$

I'll now define $s = i\omega$, and so the transfer function is

$$\mathbf{H} = \frac{\dfrac{R_b \frac{1}{sC}}{R_b + \frac{1}{sC}}}{R_a + sL + \dfrac{R_b \frac{1}{sC}}{R_b + \frac{1}{sC}}}. \tag{6.6.10}$$

The next step is the long promised trick, and to appreciate it, just keep one idea in mind: if we build a circuit that is inductor-free that

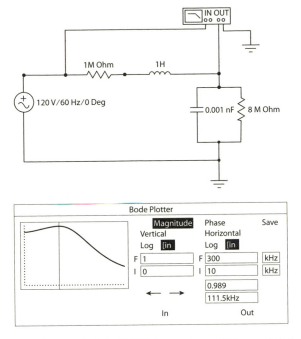

FIGURE 6.6.6. EWB simulation of Figure 6.6.5

is supposed to be *equivalent* to the one in Figure 6.6.5, then, while it will *look* different, it must still have exactly the same transfer function. Now, if we divide the numerator and the denominator of (6.6.10) by the same thing, we clearly change nothing; let's divide by s. Thus,

$$\mathbf{H} = \cfrac{\cfrac{R_b \frac{1}{s^2 C}}{R_b + \frac{1}{sC}}}{\cfrac{R_a}{s} + L + \cfrac{R_b \frac{1}{s^2 C}}{R_b + \frac{1}{sC}}}. \qquad (6.6.11)$$

Next, observe that

$$\frac{R_b \frac{1}{s^2 C}}{R_b + \frac{1}{sC}} = \frac{\frac{R_b}{s} \cdot \frac{1}{s^2 C}}{\frac{R_b}{s} + \frac{1}{s^2 C}},$$

where to get the right-hand side I've simply divided the numerator and the denominator of the left-hand side by s and so, again, nothing has changed. With this, (6.6.11) becomes

$$\mathbf{H} = \cfrac{\cfrac{\frac{R_b}{s} \cdot \frac{1}{s^2 C}}{\frac{R_b}{s} + \frac{1}{s^2 C}}}{\cfrac{R_a}{s} + L + \cfrac{\frac{R_b}{s} \cdot \frac{1}{s^2 C}}{\frac{R_b}{s} + \frac{1}{s^2 C}}}. \tag{6.6.12}$$

Look carefully at what (6.6.12) is saying. Since an impedance with no s in it is a resistor, and an impedance with $\frac{1}{s}$ in it is a capacitor, then $\frac{R_a}{s} + L$ is the impedance of a capacitor of value $\frac{1}{R_a}$ $(= \frac{1}{10^6} = 10^{-6}$ F$)$ *in series* with a resistor of value L $(= 1\Omega)$. And $\frac{\frac{R_b}{s} \cdot \frac{1}{s^2 C}}{\frac{R_b}{s} + \frac{1}{s^2 C}}$ is the impedance of a capacitor of value $\frac{1}{R_b}$ $(= \frac{1}{8 \cdot 10^6} = 0.125 \cdot 10^{-6}$F $= 125$ nF$)$ *in parallel* with some component that has an impedance with $\frac{1}{s^2}$ in it, a component I have to admit that we seem not to have encountered before! Which is too bad, because you'll notice that the floating inductor has been transformed into a floating resistor (which is no problem to build in real hardware, including integrated circuitry), while the grounded capacitor C has been transformed into a grounded "whatever an $\frac{1}{s^2 C}$ component is."

But not all is lost. We actually have, in fact, just recently run into this strange component. Since $s = i\omega$, the $\frac{1}{s^2 C}$ component has the impedance $-\frac{1}{\omega^2 C}$. That's a *real* quantity and so the component is a resistor (of sorts). It's a negative resistor, of course, and its value varies with frequency. Electrical engineers call it a *frequency-dependent negative resistor* (FDNR), and its grounded form is easily constructed from a GIC! All you have to do, if you look back at (6.6.9), is pick any two of $\mathbf{Z}_1, \mathbf{Z}_3$, or $\mathbf{Z}_5$ in Figure 6.6.4 as capacitors and all the other $\mathbf{Z}$s as resistors. Suppose, to be specific, we pick $\mathbf{Z}_3$ and $\mathbf{Z}_5$ to be equal capacitors (C^*) and the other $\mathbf{Z}$s to be equal resistors (R^*). Then, $\mathbf{Z}_{in}$ for the GIC is

$$\mathbf{Z}_{in} = \frac{R^* \left(\frac{1}{i\omega C^*}\right) \left(\frac{1}{i\omega C^*}\right)}{(R^*)^2} = -\frac{1}{\omega^2 R^* (C^*)^2}.$$

Since $\dfrac{1}{s^2 C}$ in (6.6.12) is, for the circuit of Figure 6.6.6, equal to

$$\frac{1}{(i\omega)^2 C} = \frac{1}{-\omega^2 10^{-12}},$$

then

$$R^* \left(C^*\right)^2 = 10^{-12}.$$

If we arbitrarily pick $R^* = 10,000$ ohms ($10^4\ \Omega$), then $(C^*)^2 = 10^{-16}$, and so $C^* = 10^{-8}$ F (10 nF). Figure 6.6.7 shows an EWB simulation of the inductor-free circuit using a GIC. As you can see by comparing the

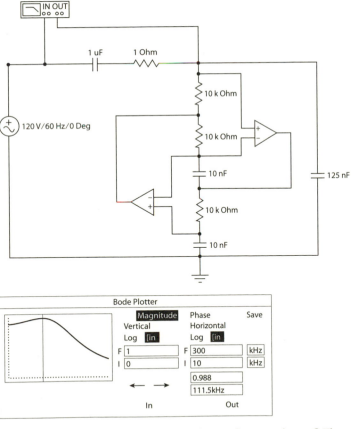

FIGURE 6.6.7. EWB simulation of the inductor-free version of Figure 6.6.6
(**gicfloatinginductor.ewb**)

Bode plots of the two EWB simulations, the transfer function magnitudes are an almost exact match, with just the slightest deviation in the maximum value of the transfer function magnitude (0.988 versus 0.989).

The purpose of this chapter is not to turn you into an electronic circuit design engineer. Indeed, our inductor-free design for a low-pass filter, while it "works" in an EWB simulation, does have some glaring practical difficulties. To name just one, any circuit design that calls for a resistor of just (and exactly) 1Ω should raise an eyebrow (or maybe both)! For another, the one-microfarad capacitor *is* pretty large for construction in integrated circuitry. But these are problems that electrical engineers have long since discovered how to avoid, through the use of circuits even more astonishing than is the GIC (the *current conveyor* and the *differential voltage current-controlled source*, to name just two, neither of which I am going to pursue). Rather, the point here was simply to demonstrate the impressive capabilities of a number-crunching scientific computer code, EWB, one very different from a more traditional number-crunching code like MATLAB.

6.7 Closing the Loop: Sinusoidal and Relaxation Oscillator by Computer

In this final section of the chapter I want to "close the loop" on two of the circuits that we discussed earlier. (The words "closing the loop" have a double meaning, as you'll see.) First, look back at Figure 6.4.8, the EWB simulation of the three-section *RC* phase-shifter. As shown earlier, a sinusoidal signal of a particular frequency "goes in" the input terminals and "comes out" the output terminals reduced in amplitude by a factor of 29 and shifted in phase by 180°. What would happen if we then took that attenuated output, multiplied it back up by 29, shifted it another 180° to arrive at an exact replica of what went into the input terminals, and then used that reconstituted signal *as the input*? Well, with a little imagination, you might well find it easy to believe that the signal would just go round and round, from "input-to-output-to-input-to..." That is, we would have a sinusoidal oscillator at the frequency for which the three-section phase-shifter has a 180° phase shift. This is the

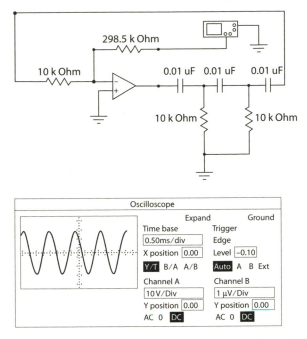

FIGURE 6.7.1. EWB phase-shift oscillator simulation (**phaseshiftosc.ewb**)

first interpretation of what "closing the loop" means. In Figure 6.7.1 you'll find an EWB simulation of this, where the required gain and second 180° phase shift are simultaneously achieved with the *negative* gain diff-amp circuit of Figure 6.5.2 (take a look back at (6.5.5), too). With $C = 10^{-8}$ F and $R = 10^4$ Ω, the theoretical oscillation frequency is

$$\frac{1}{2\pi\sqrt{6}RC} = \frac{10,000}{2\pi\sqrt{6}} \text{ Hz} = 650 \text{ Hz}.$$

The oscilloscope display shows (as close as I can read it) $2\frac{1}{2}$ cycles of oscillation over eight divisions of time (that is, 4 milliseconds at the oscilloscope time-base sweep-rate of 0.5 milliseconds per division). The period of one cycle of oscillation, then, is 1.6 milliseconds, which is a measured oscillation frequency of 625 Hz, a less than 4% deviation from theory.

There are several comments I need to immediately make about Figure 6.7.1. First, notice that the gain is not -29 but rather -29.85

(−298.5/10). That's because I simply could not get the circuit to oscillate for any gain magnitude less than 29.85, a result I attribute not to the failure of theory but to the details of the EWB code itself that remain unknown to me.[18] Second, notice how the final 10 KΩ resistor in the third section of the RC phase-shift network serves a double purpose: it is also the input resistor to the negative gain amplifier (remember, the negative input terminal to which the 10 KΩ resistor connects is virtual ground). And third, a real hardware version of this oscillator circuit would immediately break into oscillation at the instant power is applied, but the EWB simulation does not! That's because in the simulation, by default, all the circuit voltages start at zero. And so, as EWB evolves its circuit simulation equations through time, all the voltages *remain* at zero. To get the simulation going I had to tell EWB to start with a non-zero diff-amp output voltage (I used 0.1 V). In real hardware, that mathematical trick is automatically performed by the unavoidable presence of noise fluctuations. Fourier theory tells us that such fluctuations have energy at all frequencies, and so we are assured that there will be a non-zero voltage at the 180° phase-shift frequency of the RC network. All other frequencies will also 'go round-and-round' the loop, too, but they will experience phase shifts other than 180°, *destructively* interfere, and so die out.

For the second meaning of "closing the loop," take a look at Figure 6.7.2. There you'll see a circuit that is the diff-amp version of the *Popular Electronics* twinkling Christmas tree, neon-bulb astable multivibrator that I told you about at the beginning of this chapter. When we first turn power on to this circuit, let's assume that the capacitor C is uncharged. That is, the voltage x at the negative input terminal of the diff-amp is zero. Notice that x cannot change instantly, as x is the voltage drop across a capacitor whose other end is connected directly to true ground, and *that's* a voltage that is definitely *not* going to change, *ever*! We also know that, in any actual circuit, because of noise fluctuations, the voltage y at the positive input terminal will be non-zero. For sake of argument, let's take $y > 0$. (I'll let you run through the following argument for the $y < 0$ case, and you'll see the conclusions are unchanged.) Thus, $y \neq x$ and so here's a case where the diff-amp input terminal voltages are *not* equal (even with negative feedback present because of R_1) because we also have a lot of positive feedback as well.

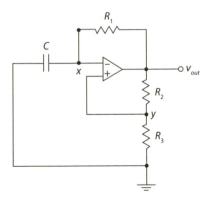

FIGURE 6.7.2. A *positive* feedback diff-amp astable multivibrator

With zero voltage on the negative input terminal and a positive voltage on the positive input terminal, the diff-amp output voltage v_{out} will almost instantly go to its maximum possible positive value.[19] This maximum value is determined by the diff-amp's power supply voltages which, by default in EWB, are ±20 V. (You'll recall that in the Lotka-Volterra differential equation example I changed the default power supply voltages, but now they are back to ±20 V.) So, v_{out} snaps to +20 V, and so (because of both Ohm's law and the fact that our ideal diff-amp has zero input current)[20] we have

$$y = \frac{R_3}{R_2 + R_3} v_{out} = \frac{R_3}{R_2 + R_3} V, \qquad (6.7.1)$$

where I'm writing V for +20 V.

So, here's the state of the circuit the instant after we apply power: $v_{out} = V$ (+ 20 V), the voltage on the negative input diff-amp terminal is 0, and the voltage on the positive input diff-amp terminal is $y > 0$, as given by (6.7.1). This creates a current from the diff-amp output back through the feedback resistor R_1, a current that flows totally into C and so charges C. That is, the voltage x begins to rise, exponentially, toward V, starting from zero. So,

$$x(t) = V - Ve^{-t/R_1 C}, \qquad t \geq 0.$$

But x never gets to V. The instant it *just* exceeds the y of (6.7.1) there will be a larger voltage on the diff-amp's negative input terminal than there is on the positive input terminal, and so the diff-amp output will snap almost instantly to its most negative possible value of $-V$. (Electrical engineers call the "snapping" of a diff-amp output from one extreme voltage to the other *railing*.)

So, here's the state of the circuit now: $v_{out} = -V$ (-20 V), the voltage on the negative input diff-amp terminal is $x = \dfrac{R_3}{R_2 + R_3}V > 0$, and the voltage on the positive input diff-amp terminal is $-\dfrac{R_3}{R_2 + R_3}V < 0$. That is, the capacitor voltage x is *greater* than the diff-amp output voltage, so the capacitor now begins to discharge through the feedback resistor R_1. This results in the voltage x beginning to fall, exponentially, toward $-V$, starting from $\dfrac{R_3}{R_2 + R_3}V$. Calling the instant at which this decrease begins our new $t = 0$ instant, we have

$$x(t) = V\left[1 + \frac{R_3}{R_2 + R_3}\right]e^{-t/R_1 C} - V, \quad t \geq 0.$$

But x never gets to $-V$. The instant it *just* goes more negative than the voltage on the positive input terminal the diff-amp will again rail to an output of $+V$. Let's call the time it takes to do that T. Then,

$$V\left[1 + \frac{R_3}{R_2 + R_3}\right]e^{-T/R_1 C} - V = -\frac{R_3}{R_2 + R_3}V,$$

or

$$\left[1 + \frac{R_3}{R_2 + R_3}\right]e^{-T/R_1 C} = 1 - \frac{R_3}{R_2 + R_3} = \frac{R_2}{R_2 + R_3},$$

or

$$e^{-T/R_1 C} = \frac{\frac{R_2}{R_2 + R_3}}{1 + \frac{R_3}{R_2 + R_3}} = \frac{R_2}{R_2 + 2R_3}.$$

So,

$$e^{T/R_1 C} = \frac{R_2 + 2R_3}{R_2} = 1 + 2\frac{R_3}{R_2},$$

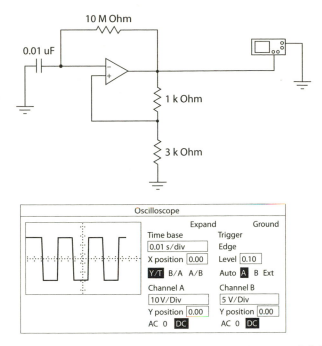

FIGURE 6.7.3. EWB simulation of square-wave oscillator (**astable.ewb**)

or

$$T = R_1 C \ln\left(1 + 2\frac{R_3}{R_2}\right).$$ (6.7.2)

Now you can see that the circuit has fallen into a periodic pattern, railing from $+V$ to $-V$ to $+V$, etc., with the time interval between consecutive railings being T as given in (6.7.2). The output voltage of the diff-amp is a *square wave* with amplitude $\pm V$ and period $2T$. Figure 6.7.3 shows an EWB simulation of the astable multivibrator, with the values $C = 10^{-8} F$, $R_1 = 10^6 \Omega$, $R_2 = 10^3 \Omega$, and $R_3 = 3 \cdot 10^3 \Omega$. Our theoretical analysis gives a period of $2 \cdot 10^{-2} \ln(1 + 6) = 0.039$ seconds, and that's pretty nearly what the computer code says, too. As close as I can read the oscilloscope display, there are slightly less than four divisions per period, which, with a sweep rate of 0.01 seconds/division, gives a period of just less than 0.04 seconds.

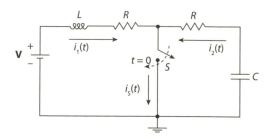

FIGURE 6.8.1. What is $i_S(t), t \geq 0$?

6.8 Challenge Problems

CP 6.1. Here's a problem typical of the ten thousand homework problems I solved in EE 116. In the circuit of Figure 6.8.1 the switch
S has been open (as shown) for a very long time. Then, at time
$t = 0$, S is closed. Calculate the current $i_S(t)$ in the switch for $t \geq 0$.
Before you write even a single line of mathematics, explain why
the switch current the instant after S is closed ($t = 0+$), and the
switch current after a long time has passed, are always equal. That
is, explain why $i_S(t = 0+) = i_S(t = \infty)$. Further, show from your
mathematical analysis that if R, L, and C satisfy a certain condition, then $i_S(t = 0+) = i_S(t = \infty) = i_S(t)$ for *all* $t \geq 0$; that is, the
switch current will be a *constant*. And finally, in general $i_S(t)$ will
vary with t, and so there will be some $t = T$ for which $i_S(T)$ is the
extreme current. For the particular values of $V = 100\,\text{V}, R = 1,000\,\Omega$,
$C = 0.01\mu fd = 10^{-8}$ F, and $L = 100$ mH $= 0.1$ H, calculate the values of both T and $i_S(T)$, and thus determine if $i_S(T)$ is the *maximum*
current or if it is the *minimum* current. Note, carefully, that once S is
closed the voltage drop across it is *always* zero, no matter what $i_S(t)$
may be. (But that does *not* mean the current in the switch is zero
because, remember, the switch resistance is also zero and so Ohm's
law is indeterminate.)

CP 6.2. Show that EWB's answer for the input resistance to the circuit of Figure 6.3.4, $1.1875\,\Omega$, is *exactly* correct. *Hint*: Apply a dc
voltage V at the input terminals, write I as the input current (and
so $R_{in} = \frac{V}{I}$), and then write Kirchhoff's current law at the obvious

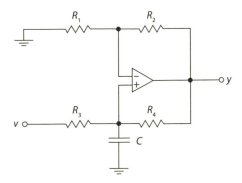

FIGURE 6.8.2. A positive gain integrator

node. You should then have three equations in four unknowns, which is sufficient to solve for the value of the ratio $\frac{V}{I}$.

CP 6.3. Compute, for the three-section RC phase-shift circuit of Figure 6.4.7, the individual phase shifts $\phi_1(\omega_0)$ and $\phi_2(\omega_0)$ caused by the first and second sections, respectively, at the 180° phase-shift frequency $\omega_0 = \frac{1}{RC\sqrt{6}}$.

CP 6.4. Show, assuming zero differential input voltage to the diff-amp, that the circuit of Figure 6.8.2 is a positive gain integrator if $R_1R_4 = R_3R_2$.

CP 6.5. Figure 6.8.3 shows what is called an *R/2R ladder DAC*. Each of the switches represents one digital bit; when a switch is thrown to the right that bit is a 0, and when it is thrown to the left that bit is a 1. S_0 is the LSB (least significant bit) and S_3 is the MSB (most significant bit). Explain why the diff-amp output, v_{out}, is equal to the binary fraction $\frac{1}{2}S_3 + \frac{1}{4}S_2 + \frac{1}{8}S_1 + \frac{1}{16}S_0$ of the negative of the reference voltage V_{ref}. (In an actual computer the switches would be realized as transistor circuits that can be switched between the 0 and the 1 positions at electronic speeds.) *Hint*: Notice that, independent of the individual switch positions, the bottom ends of all the 2R resistors are *always* connected to ground (either true ground in the 0 position or the diff-amp's virtual ground in the

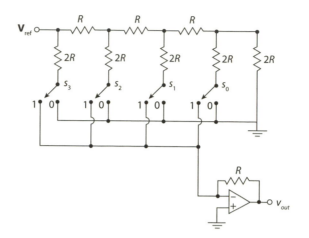

FIGURE 6.8.3. A ladder DAC

1 position), so the current supplied by the V_{ref} source is constant. More important, the current in each of the $2R$ resistors is constant, so all that a switch position does is route its resistor's current to one of two grounds. All of the currents that are directed to the *virtual* ground then flow through the diff-amp's feedback resistor. You may want to review the ladder analysis trick described in note 10, too.

CP 6.6. For the circuit of Figure 6.6.5 find a theoretical expression for the frequency at which the magnitude of the transfer function is maximum and then, using the component values given in the text, numerically evaluate both quantities. Recall that the EWB simulation of the actual circuit gave 0.989 and 111.5 kHz, and the EWB simulation of the GIC version of the circuit gave 0.988 and 111.5 kHz.

CP 6.7. For each of the two circuits in Figure 6.8.4 find an expression for v_{out} as a function of the input(s). Since each circuit has only negative feedback, you can safely make the ideal diff-amp assumption (that is, the voltages on the input terminals of the diff-amp are equal).

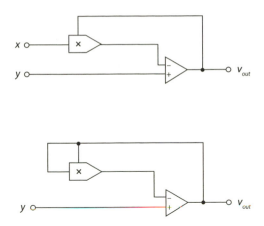

FIGURE 6.8.4. Two circuits with multipliers: what do they do?

6.9 Notes and References

1. After *An Imaginary Tale* appeared I received numerous e-mails from readers—mostly electrical engineers, but also some physicists, and even a couple of mathematicians—telling me of similar experiences. One e-mail in particular was from A. David Wunsch, professor emeritus of electrical engineering at the University of Massachusetts at Lowell, who wrote to tell me that all the Carl and Jerry stories were being reprinted! The original *Popular Electronics* magazines are now mostly decayed piles of pulp paper, long since discarded in abandoned attics or leaky basements or landfills, but the stories have not been forgotten. Each and every one of the 119 Carl and Jerry tales (authored by John T. Frye [1910–1985], who influenced more youngsters than he could possibly have imagined), had been tracked down, computer scanned, and were being reprinted in five volumes, including all the original illustrations. It was a pure labor of love by editor Jeff Duntemann. All five volumes have now been published (by Copperhead Press) and can be purchased on-line at www.lulu.com. I have a complete set of the stories on my study bookshelf and, fifty years after their first appearances, they are still a wonderful treat to read.

2. Electronics kits today are, alas, a fading memory. And even then only in the heads of people my age (sixty nine as I write) and older. In the mid-1950s, however, there were several mail-order outlets selling home electronic kits, with the Heath Company in Benton Harbor, Michigan, taking the number-one spot. Each year when I was in high school I would send away for the latest Heathkit catalog, which I would pounce on when it arrived and then spend the next several days drooling over. Inside its pages were wondrous photos

and descriptions of electronic test equipment like oscilloscopes (one of which I built in my junior year) and VTVMs (vacuum tube voltmeters), televisions and AM/FM radios, and amateur radio receivers and transmitters. The back cover of each year's catalog always had an overhead aerial color photograph of the company, located on the southeastern shore of Lake Michigan. Taken in the fall, with the leaves in full color, it was simply beautiful, especially to a kid living in a hot, dusty, *boring* (to me, anyway) nowhere little burg in Southern California. "Boy," I remember thinking to myself, "could there be a neater place to live and work, *in the entire world,* than the Heath Company in Benton Harbor, designing new, incredible kits?" The answer was an obvious "No, there couldn't be anything better." Ironically, I never made it back to the Heath Company, but my oldest granddaughter is now in school at Lake Michigan College in... Benton Harbor, Michigan! The Heath Company is still there, too, in nearby St. Joseph, but with the relentless development of integrated circuitry, the discrete component kit become obsolete. Heath went out of the electronic kit business twenty years ago, a sad event that made the front page of the *New York Times.*

3. In the case of a capacitor, just remember that $i = C\dfrac{dv}{dt}$, and so if v changed *instantly*, we would have an infinite derivative (that is, an infinite current). Similarly for an inductor, where $v = L\dfrac{di}{dt}$. In both cases, v and i are the voltage drop across, and the current in, the component. In mks units, i, v, L, and C are measured in amperes (A), volts (V), henrys (H), and farads (F), respectively.

4. When I took the electronics sequence at Stanford the text was Jacob Millman's *Vacuum-tube and Semiconductor Electronics* (New York: McGraw-Hill, 1958). (I kept my copy for decades, until it literally fell apart.) That book had a very nice, extensive (fifty-five-page!) presentation on gas discharge tubes in its Chapter 12. It is, of course, now very much out-of-date in general, but in the particular case of its discussion of gas discharge tubes it would be difficult to find the topic treated any better in a modern text. Very little in engineering lasts forever, however. Even as Millman wrote, the glow discharge tube's use in voltage regulation was being phased out by the solid state Zener diode.

5. Another not quite so important refinement of my 1954 understanding of the NE-2 is that the bulb glows *all the while* there is current through it, not just at the instant the current stops. I knew in 1954 that the emitted light was due to free electrons recombining with the positive ions from which they had been ripped, and in so doing giving up energy in the form of light, but I visualized the recombination as occurring only at the end of the current flow. In fact, both ionization and recombination *occur continually* during current flow, and so light is *continually* emitted.

6. I encountered the mysterious (to me then) free-running astable multivibrator circuit of the twinkling Christmas tree one more time while in high school, in another *Popular Electronics* construction project. It appeared in the January 1956 issue ("Electronic Coin Tossing") and was dramatically pitched as a device to help make decisions. (Not to be confused with the October 1955 *Popular Electronics* construction project, "Build a 'Decision Meter,'" which was a weighted sum of the outputs of several independently set potentiometers feeding into a meter. *That* circuit I understood!) The author did include a little tutorial on how the coin tossing circuit worked (writing "it is one which is usually not explained in standard textbooks"), and he included the correct generic voltage waveforms across the neon bulbs, as well as the statement that the voltage drop across a capacitor cannot change instantly. But it was all still just a bit beyond me. I didn't take my first course in calculus until the summer before my senior year (the summer of 1957), and I was still far from being able to derive the blink period equation of (6.2.9). I am, in fact, not at all sure now that I recognized then that it *was* the twinkling Christmas tree circuit I had constructed thirteen months earlier. All it was, though, was the circuit of Figure 6.2.3 (now with just two blinking bulbs instead of fourteen), with values of R and C such that the blinking rate was very high. So high, in fact, that *both* bulbs appeared to always be on. The two bulbs were mounted on an aluminum chassis, along with a push button. Depressing the button simply shorted the C, with the bulb that was on the instant the button was pushed staying on. It gave the appearance of 'freezing' one bulb on and the other off, *at random*, and was pitched to readers as simulating the flip of a fair coin. I remember building the circuit on my workbench and taking the gadget to school, where it was a big hit. I don't think anybody asked me how it worked, however—which was good, because I still didn't know! Magazine writers apparently got a lot of mileage out of that circuit; it appeared yet again (in slightly different form) in the April 1967 issue of *Popular Mechanics* under the title "Electronic 'Coin Flipper.'"

7. See Chapter 17 of my *Mrs. Perkins's Electric Quilt*, (N.J.: Princeton University Press, Princeton 2009), and in particular the MATLAB code **mccube.m**. The reasons I took the wire list approach there, even though I am not a big fan of it, are two: first, it *is* easy to do, and second, the intent there wasn't to do circuit *analysis* at all but rather to demonstrate how to "solve" resistor circuits with a *mathematical* Monte Carlo approach in which the physics of Ohm's law is never used. In a true circuit analysis code, the software implements the actual *physics* (that is, the defining mathematical equations) of each component that appears in the circuit.

8. Another *practical* reason for why I use EWB, v.5, is that I got it for free. That's because it was once used in the electrical engineering department at

the University of New Hampshire, but the department long ago moved on to newer codes (the twenty-first-century version of EWB is called MULTISIM). So there it was when I wandered by one day, loaded on a CD tossed into the back desk drawer of the fellow who runs the EE computer labs. All alone it was, forgotten, unwanted, lost beneath an old, half-used roll of duct tape. Spotting it, I simply asked (perhaps I begged a bit, too) if I could have it. A good reference, if you want to learn more, is John Adams, *Mastering Electronics Workbench: Version 5 & Multisim Version 6*, (New york: McGraw-Hill, 2001). The latest up-date of Multisim is version 11.

9. Infinite ladder circuits, and this trick for solving them, are discussed at length in my *Mrs. Perkins's Electric Quilt*, in that book's Chapter 3. There can be some subtle, delicate complexities associated with such circuits, and with the trick, too, but they are *not* a concern when the ladder is made *only* of resistors as is the infinite ladder of Figure 6.3.1.

10. You could verify this answer by a direct application of the well-known parallel/series resistor formulas, applied from right to left. But that's a bit of work. There is a far easier method that electrical engineers routinely use for ladders. With it you can almost do all the calculations in your head. Start by grounding the bottom "bus," that is, the bottom horizontal wire to which all the vertical resistors connect. That means the bus is at zero volts, and it will serve as the voltage reference for all other points in the circuit. Then *assume* a 1 A current, downward, in the rightmost resistor (16 Ω). That means the node voltage where the two 8 Ω resistors connect must be 24 V. That means the current flowing in the vertical 8 Ω resistor must be 3 A. That means the current in the horizontal 4 Ω resistor must be 4 A. That means that the node voltage where the two 4 Ω resistors connect must be 40 V. That means the current flowing in the vertical 4 Ω resistor must be 10 A. And so on. When you get to the far left node, you'll find the current into node is 48 A and the node voltage is 116 V. So, $R = \frac{116}{48} = \frac{29}{12}$ Ω, as stated.

11. Named after the American mathematical physicist-turned-electrical engineer Hendrik Wade Bode (1905–1982). Bode held BA and MA degrees in mathematics and a PhD in physics, and spent his career in the Mathematical Research Group of the Bell Telephone Laboratories. In the late 1930s he developed the mathematical rules for drawing his famous *Bode plots* from a circuit's transfer function, that is, the amplitude and phase plots that EWB's bode plotter generates directly and almost instantly from a circuit's schematic.

12. EWB allows the user to set the rms (root mean square) value of an ac voltage source, which is the peak voltage divided by $\sqrt{2}$. So, for an ac voltage

with a 29 V peak (that is, 58 V peak to peak), the rms voltage should be set to 20.5 V. If this terminology needs some review, see my *The Science of Radio*, 2nd ed. (New York: Springer, 2001, pp. 354–355). In that same book you can find two more EWB computer simulations involving the Bode plotter. On pp. 382–384 a two-resistor/two-capacitor circuit with a voltage gain greater than one over an *interval* of frequency (not just at a single, isolated frequency) is studied. Another circuit (p. 275), dating from the mid-1950s, is an almost unbelievable one; it produces a very nearly constant 90° phase shift over a 10-to-1 frequency interval (using only a handful of resistors and capacitors) and it plays a crucial role in the construction of single-sideband radio transmitters. It was invented by the American electrical engineer Donald K. Weaver, Jr. (1924–1998). You can find more on the mathematics of Weaver's amazing circuit and single-sideband transmitters in my *Dr. Euler's Fabulous Formula* (Princeton, N.J.: Princeton University Press, 2006, pp. 319–323, 371).

13. The high-gain amplifiers themselves use only resistors, capacitors, and transistors, and so probably it is more logical to call *those* the basic components. But of course all *they* are, fundamentally, is just silicon (and some artfully dispersed impurities to form the doped pn-junctions that are transistors) and so the fundamental building block in modern electronics is nothing but silicon. It's an industry literally built on sand.

14. The differential amplifier might appear to be an awfully simple device, but there is actually a lot of stuff in a diff-amp. The $\mu A741$, for example, one of the earliest of the integrated circuit diff-amps and still available today a half century after its introduction, is built from twenty-four transistors, eleven resistors, and a capacitor.

15. The equations in (6.5.6) are named after the American biologist and actuary Alfred Lotka (1880–1949) and the Italian mathematician Vito Volterra (1860–1940), who independently and nearly simultaneously proposed them in the mid-1920s.

16. These voltages are far larger than one would see in an actual hardware integrated circuit. For EWB, of course, they are just numbers and there will be no sparks, smoke, and blown circuit breakers! The diff-amp model used in the EWB simulation of Figure 6.5.5 has default limits of ±20 V for the output voltage, but I just edited the parameters of the model to allow the diff-amp outputs to be in the interval ±500, 000 V! That's cheating, of course, and if I were to build the Lotka-Volterra analog computer circuit in real hardware I couldn't get away with that. What one would have to do is *scale* the voltages. That is, define (for example) $u(t) = \frac{1}{100}x(t)$ and $v(t) = \frac{1}{100}y(t)$. Then

the equations of (6.5.6) become

$$\frac{d}{dt}\{100u(t)\} = a\{100u(t)\} - b\{100u(t)100v(t)\}$$

$$\frac{d}{dt}\{100v(t)\} = -c\{100v(t)\} + d\{100u(t)100v(t)\}$$

or

$$\frac{du}{dt} = au - 100buv$$

$$\frac{dv}{dt} = -cv + 100duv$$

with the now very reasonable initial conditions of $u(0) = 2$ V and $v(0) = 0.5$ V.

17. Another quite interesting pair of coupled, first-order, nonlinear differential equations appear in the classic problem of a raindrop falling through a uniform mist in a constant, uniform gravitational field. As it falls, the drop continually gains mass according to some rule. For example, the mass may increase in direct proportion to the drop's instantaneous surface area, or perhaps the mass increases in direct proportion to the rate of volume swept out by the falling drop—that is, in proportion to the product of its instantaneous cross-sectional area and its instantaneous speed. I well remember a *very* long night in the old physics library at Stanford when I solved a version of this problem as a homework assignment in Physics 87, during Winter Quarter 1960–61. (See Problem 6–21 in Robert A. Becker, *Introduction to Theoretical Mechanics* [New York: McGraw-Hill, 1954, p. 134].) This can be a quite challenging analytical problem, depending on the mass accretion rule, but it's duck soup to design an EWB simulation circuit for just about any given rule. (For an example of this, see the end of the solution for **CP 6.7.**) For the most recent mathematical details, see Alan B. Sokal, "The Falling Raindrop, Revisited" (*American Journal of Physics*, June 2010, pp. 643–645).

18. Or perhaps I just didn't wait long enough. At the gain of -29.85 it took over one minute of simulated time to see the oscillations. If I had been willing to wait even longer, perhaps I would have seen oscillations with a gain closer to -29. The oscillations I did observe grew in amplitude on the simulated oscilloscope screen as I watched until they reached an amplitude of ±20 V (40 V, peak to peak), the maximum default output of EWB's diff-amp model. This "growing" of the oscillation amplitude is not an artifact of the sequential nature of the EWB circuit simulation code. When I taught the undergraduate electronics lab at the University of New Hampshire, building a phase-shift oscillator out of actual hardware was a standard project, and students could

actually see (on real oscilloscopes) precisely the same oscillation amplitude growth process.

19. How fast a diff-amp's output voltage can change is called the *slew rate*. The ideal diff-amp model in EWB has a default slew rate of 10^{10} V/S, and so to transition from -20 V to $+20$ V (and vice-versa) requires just four nanoseconds.

20. Be very careful here to note that, since our positive feedback circuit is one in which the differential input voltage to the diff-amp is *not* zero, then we can't use the argument (valid in a pure *negative* feedback circuit) that the diff-amp input current is zero, independent of R_{in}. *Now* we have to have a diff-amp that *actually* has a huge input resistance. And, in fact, the default resistance to EWB's diff-amp simulation model is huge: 10^{10} Ω. As I write this I realize I haven't told you what EWB's diff-amp model has for its *output* resistance: it is 1Ω (ideally, it would be zero).

7
THE LEAPFROG PROBLEM

Question: "What's worse than not being able to solve the equations to a problem involving a physical process?"
Answer: "Not even having any equations that you can't solve."
Question: "What do you do then?"
Answer: "Simulate the process."
—Train of thought in the author's mind after getting stuck on the leapfrog problem

7.1 The Origin of the Leapfrog Problem

As the final number-crunching analysis of this book, consider the following pretty little problem that I came across when reading a recent "Ask Marilyn" column, a problem posed to her by a reader (identified only as M. Schwartz, in Ventura, California):

> A friend and I once went from his house to mine with one bike. I started walking and he rode the bike. When he got a couple of blocks ahead, he left the bike on the sidewalk and started walking. When I got to the bike, I started riding, passing him, and then left the bike a couple of blocks ahead. When he got to the bike, he started riding. We did this the whole way. At least one of us was always walking. At times, one was riding; at other times, we were both walking. I'm sure this was faster than if we had no bike. But some people insist that it was no faster because somebody was always walking. Who's right?

The correct reply was:

The reader is right. It's true that someone was always walking. But neither friend walked the whole distance. Both biked part of the way. This increased their average speed, so they saved time.[1]

To a mathematical mind the following question then immediately presents itself (which was not addressed in the "Ask Marilyn" column): What's so special about "a couple of blocks ahead"? Would some other separation distance be "better"?

It seems physically obvious that the "best" (I'll sharpen what *that* means soon) separation between the friends, before the bike-riding one gets off and starts walking, will be a function of all the parameters of the problem: the distance between the two houses (d), the walking speed (w), and the riding speed (r). I also think it reasonable to assume that $r > w$. To be specific, suppose that the two friends (for convenience in writing, I'll present from here on in terms of two boys making the journey along a straight line from one house to the other) each walk at a speed of 2 mph and each ride at a speed of 6 mph. If the houses are one mile apart, then to walk the whole way from one house to the other would take one-half hour ($1,800$ seconds), while to ride the whole way would take one-sixth hour (600 seconds). We therefore know that each boy will thus finish the trip in less than $1,800$ seconds but require more than 600 seconds. Let's call the actual trip times for the boys t_1 and t_2; it should be clear that the two boys will, in general, *not* finish the trip simultaneously. That is, in general, $t_1 \neq t_2$. If we call the longer of these two times the *duration* of the trip, then the criterion for what *best* means is the value of the separation distance that gives the smallest value for the duration. That is, if we denote the separation distance by h, then the optimal h is the value of h that *minimizes the maximum* of t_1 and t_2.

Notice that, to keep things as simple as possible, I'm assuming that the two boys walk at the same speed and ride at the same speed, neither of which is probably true in real life. But matters are already bad enough without introducing those complications! I write this because, even after some effort, I failed to arrive at any analytical development of this process that I thought had even a remote chance of giving me a formula for the optimal separation distance. This could be, of course, because—I'll admit it—I'm not good enough an analyst. Hey, we all get old! (If you think *you* can do better, see Challenge Problem 7.1.)

But, because I have a computer sitting on my desk, I didn't let the lack of formal equations stop me. I decided to write a simulation code that would mimic the defining *physics* of the leapfrog problem (the origin of the name should now be obvious) in software. Here's how I did that.

7.2 Simulating the Leapfrog Problem

To simulate the motion of the boys and of their bike it is crucial to have a clear understanding of all the configurations of the three that could possibly occur. I must admit that that was harder for me to arrive at than I originally thought it would be—Hey, eventually we all get *really* old!—but then I realized the process breaks, in a natural way, into two separate and distinct stages. This may seem ridiculously obvious in hindsight, but it took me a while to appreciate how recognizing these two stages greatly reduces the potential confusion. Stage 1 is where the boys and bike are continually leapfrogging but neither boy has yet traveled distance d; stage 2 begins when one of the boys is the first to travel distance d.

Suppose we denote the locations of the two boys and of the bike, at time t, by the variables $x1$, $x2$, and xb, respectively. At the start of the simulation we'll have $x1 = 0$, $x2 = 0$, $xb = 0$, and $t = 0$. Let's agree to arbitrarily put the boy whose location is designated by $x1$ on the bike at $t = 0$, and start the boy whose location is designated by $x2$ as the initial walker. Their respective speeds will be denoted by $s1$ and $s2$, and so at $t = 0$ we'll have $s1 = r$ and $s2 = w$. To simulate the passage of time, the code will periodically increment t by dt (which in the code will be 0.01 seconds), and the values of $x1$, $x2$, and xb will be updated. Let the variable bs denote the bike's *state*, where $bs = 0$ means the bike is down (not being ridden by one of the boys) and $bs = 1$ means the bike is up and being ridden. The bike's position will change only when $bs = 1$. Finally, let the variable N denote the number of boys who have traveled the entire distance d. Thus, $N = 0$ at $t = 0$, and remains at that value all during stage 1. The code enters stage 2 when N increases from 0 to 1.

Once the code enters stage 2, all that is left to do is to take the location of the boy who has not yet finished and *calculate* (not simulate) how much longer it will take him to finish. There are three possibilities to consider: (1) the first boy to finish did so on the bike, so the second boy must complete his journey by walking all the rest of the way; (2) the first boy to finish did so while walking and therefore left the bike behind him somewhere, which means either (2a) the second boy has not yet reached the bike and so walks the remaining distance to the bike at speed w, then rides all the rest of the way at speed r, or (2b) the second boy had earlier (before the first boy finished) reached the bike and so is already riding and will finish his journey by *continuing* to ride the rest of the way at speed r.

In Figure 7.2.1 I've drawn a very simple flow diagram to show the evolution of the simulation from stage 1 to stage 2.[2] With that diagram, the MATLAB code **leapfrog.m** should now be understandable. In fact, the code will simulate the journey a total of 1, 500 times, once each for values of the separation distance h starting at one foot and continuing, in one-foot increments, up to 1, 500 feet. At the end of the code listing I'll walk you through it, associating each section of the code with the corresponding part of Figure 7.2.1.

leapfrog.m

```
01    d=1;w=2;r=6;sep=linspace(1,1500,1500);
02    d=d*5280;w=w*5280/3600;r=r*5280/3600;dt=.01;
03    for loop=1:1500
04        h=sep(loop);x1=0;x2=0;xb=0;bs=1;N=0;s1=r;
          s2=w;t=0;
05        while N==0
06            x1=x1+s1*dt;x2=x2+s2*dt;xb=xb+r*dt*bs;
              t=t+dt;
07            if max(x1,x2)>=d
08                N=1;
09            end
10            if N==0
11                if bs==0
12                    if min(x1,x2)>=xb
```

(Continued)

(Continued)

```
13                              if x1<x2
14                                      xb=x1;s1=r;
15                              else
16                                      xb=x2;s2=r;
17                                  end
18                                  bs=1;
19                          end
20                      else
21                          if max(x1,x2)-min(x1,x2)>=h
22                              bs=0;
23                                if s1==r
24                                  s1=w;
25                                else
26                                      s2=w;
27                                  end
28                              end
29                      end
30              end
31          end
32          if x1==max(x1,x2)
33              if s1==r
34                      bf=1;
35              else
36                      bf=0;
37                  end
38          else
39              if s2==r
40                      bf=1;
41              else
42                      bf=0;
43                  end
44          end
45          x=min(x1,x2);
46          if bf==1
```

(Continued)

(Continued)

```
47                    T=t+(d-x)/w;
48          else
49              if x<xb
50                      T=t+((xb-x)/w)+((d-xb)/r);
51              else
52                      T=t+((d-x)/r);
53              end
54          end
55          time(loop)=T;
56      end
57      plot(sep,time,'k-')
58      xlabel('separation (feet)','FontSize',16)
59      ylabel('time for second boy to finish (seconds)',
        'FontSize',16)
60      title('FIGURE 7.2.2 - leapfrog.m number-crunches!',
        'FontSize',16)
```

Lines 01 and 02 of the code define the basic parameters of a particular problem: the values of d, w, and r (in units of miles, mph, and mph, respectively), and the values of the separation distance h to be used. In particular, in line 01 the vector *sep* is created with 1,500 elements, starting at one foot and increasing in one-foot steps to 1,500 feet. In line 02 the units of d, w, and r are converted to feet, feet/second, and feet/second, respectively, and the time increment dt is set to 0.01 seconds. Lines 03 and 56 define the main *for/end* loop that runs the code through a complete simulation for a given h (the value of which is set during the initialization process performed in line 04).

Now, with reference to the flow diagram in Figure 7.2.1, lines 05 through 09 are A, line 10 is B, line 11 is C, line 12 is D, lines 13 through 20 are E, line 21 is F, lines 22 through 31 are G, lines 32 through 44 are H (the variable $bf = 1$ means the *bike finished* the journey with the first boy on it, while $bf = 0$ means the first boy to finish did so while walking), and lines 45 through 54 (code block I) calculate the journey time (T) for the second boy (that is, the *duration*). Line 55 stores the

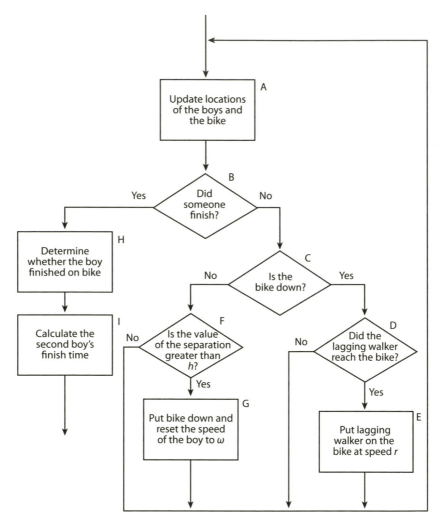

FIGURE 7.2.1. Simulating the leapfrog problem

duration in the vector *time*, and then the whole process is done over again for the next value of *h* stored in *sep*. The final four lines of the code generate Figure 7.2.2, shown below, which is a plot of *time* versus *sep* (*T* versus *h*).

There are a number of interesting observations we can make from Figure 7.2.2; the most striking is, I think, the numerous values of *h* that lead to the same minimum value for the duration (1, 200 seconds). The very small optimal values are probably not attractive ones,

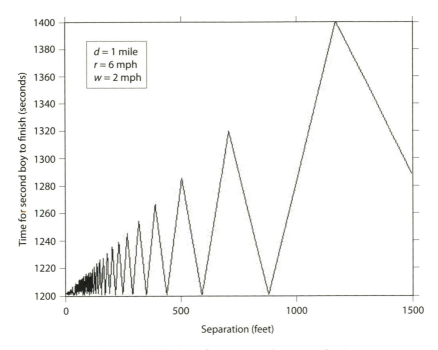

FIGURE 7.2.2. **leapfrog.m** number-crunches!

as they would have the boys continually getting off and on the bike. Notice, however, that there are two optimal values between 500 feet and 1,000 feet (a direct examination of the contents of the *time* vector gives those two optimal h values as 587 feet and 880 feet). This second value is particularly interesting because, if you recall the old grammar school conversion of "twelve city blocks to a mile"—which I hope no reader thinks is a scientific standard!—then the separation of "a couple of blocks" in the original problem statement is ... 880 feet! A mere coincidence, I'm sure, but a curious one.

I'll end here with just a couple more comments. First, it's a lot of fun to play with the code by trying different values of d, w, and r. If you numerically play with the problem by hand it quickly becomes clear that there is a huge amount of number crunching going on (and yet the code is remarkably fast: to generate Figure 7.2.2 took less than seven seconds). And second, if we allow the boys to have different walking or bike-riding speeds, an analytical solution would be hard to come by. But to modify **leapfrog.m** to take such a complication into account would not be very hard to do (see Challenge Problem 7.2).

7.3 Challenge Problems

CP 7.1. Solve the leapfrog problem analytically. If you succeed, send me your analysis and I'll put it in the corrected, paperback edition of this book.

CP 7.2. Repeat **CP 7.1** for the more general case of each boy walking or riding at different speeds. If you can't do it analytically, then generalize **leapfrog.m**.

7.4 Notes and References

1. Marilyn vos Savant, "Ask Marilyn" (*Parade Magazine*, May 2, 2010, pp. 24–26).

2. Some computer scientists I know have argued that drawing flow diagrams is akin to counting on your toes. Rather, they say, one should start writing code right from the get-go. I think that wrongheaded, and I hope present-day students aren't being taught that approach. Without Figure 7.2.1 as a start to my thinking, writing **leapfrog.m** would have been a nightmare task.

8

SCIENCE FICTION: WHEN COMPUTERS BECOME LIKE US

I believe that computing is changing the world more than any other factor today.
— Bill Gates, Moscow, 1997

[He] ceremoniously soldered the final connection with gold ... then moved to a position beside the switch that would ... connect, all at once, all the monster computing machines of all the populated planets in the universe—ninety-six billion planets—into the supercircuit that would connect them all into one supercalculator ... that would combine all the knowledge of all the galaxies.... [He] threw the switch. There was a mighty hum, the surge of power from ninety-six billion planets.... [He] turned to face the machine [and asked][1] "Is there a God?" The mighty voice answered.... "Yes, *now* there is a God." ... He leaped to grab the switch.

A bolt of lightning from the cloudless sky struck him down and fused the switch shut.
— From Fredric Brown's somewhat gloomy 1954 story, "Answer"

8.1 The Literature of the Imagination

I think it is pretty close to the mark to say that the large majority of physicists, mathematicians, and engineers, in their youth, read at least a few science fiction (SF) stories or novels. Certainly the "scientific romances" of H. G. Wells, such as *The Time Machine* and *War of the Worlds*, or the "advanced engineering" works of Jules Verne, such as his *Journey to the Center of the Earth*, *From the Earth to the Moon*, and *20,000*

Leagues Beneath the Sea, would be on the reading list of any youngster with the sort of imaginative mind that would find appeal in living a technical life. Many, however, do drift away from SF as they get older, with the common reason being that an awful lot of the genre is undeniably awful. Still, the fact remains that it is in SF that we find so many wonderful ideas that the often poor quality of the writing can be tolerated much longer than poor writing can be in, say, a badly written mystery or western. If an alien in an SF story "talks funny," well, after all, he/she/it *is* an alien, but if a detective or cowboy doesn't "sound right," that generates an instant toss into the discard pile. In any case, introverted teenage boys—the large majority of regular SF readers—aren't particularly noted for their literary sophistication. I write that from a personal, introspective viewpoint.

It is sometimes said that SF, if successful, predicts *the* future, that is, it foreshadows what actually comes to pass. This isn't strictly true, although successful prediction isn't anything to be ashamed of. To be honest, however, a successful SF prediction is mostly luck. What any good SF writer is actually trying to achieve—beyond the primary goal of writing a tale for which some editor will pay cash—is to entertainingly illustrate a *possible* future. For example, after numerous lander missions to the red planet we now know Earth has not, and will not, battle Martian invaders in a "war of the worlds." When it first appeared in 1898, however, the concept was a possibility. Wells's genius was such that his nineteenth-century novel is still a great read in the twenty-first century.[1]

An imaginative concept in physics that takes this idea of possibility to the limit is that of splitting universes—the so-called "many-worlds" interpretation of reality from quantum mechanics—with each new decision by every sentient being causing all possibilities to be realized. SF writers have had lots of fun with this idea, with Robert Heinlein writing of $(6^6)^6 \approx 10^{28}$ worlds in his novel *The Number of the Beast*, and H. Beam Piper beating that with $10^{100,000}$ worlds in his "paratime" stories. (Both these numbers seem to me, however, to be far too *small* for the splitting universes concept. And with splitting universes the number of worlds would always be increasing anyway.) This view of reality makes it very unlikely that any SF story of the far future will be close to the (eventual) mark. Richard Feynman, for one, didn't put much stock in the many-worlds idea when he wrote a 1982 paper titled "Simulating Physics with Computers,"[2] but it is nonetheless a fascinating speculation.

It is very hard to predict even just a short time into the future. A classic example of this difficulty comes from some final words in a terrific book by John R. Pierce.[3] Pierce (1910–2002) was an extraordinarily creative electrical engineer whose career cut across both industry and academia. He was executive director of Bell Laboratories Communications Sciences Division and was present when the transistor was invented—indeed, the name is due to Pierce. He was chief technologist at Caltech's Jet Propulsion Laboratory and professor of electrical engineering at Caltech where, decades before, he had earned his PhD. He ended his career on the music faculty at Stanford, exploring his interest in computer-generated music. He was also a longtime SF writer (his first published story appeared in a 1930 issue of *Science Wonder Stories*), and no one would ever have accused Pierce of having no imagination.

Pierce's speculations start on a positive note:

> Why is there not more data transmission over existing facilities? Why, for example, do we not send "electronic mail" from home to home or from office to office? [This is remarkable; until I read this I had no idea that *electronic mail* was a term Pierce had used thirty years ago!]... Someday the telephone network itself will provide two-way digital transmission right to telephone sets at a rate of 56,000 bits or more per second.... [This] bit rate is phenomenally high by present standards. [This prediction by Pierce has come and gone, and most users found 56K modem connections—think AOL!—frustratingly slow!]

So far, so good. But then comes the sentence that Pierce lived long enough to almost surely have wished he could take back:

> Two-way digital video communications via optical cable fibers ... would provide digital transmission rates of tens of millions of bits per second, *rates that are far beyond any data-transmission needs that one can foresee* [my emphasis].

Well, actually not, not in today's world of YouTube and streaming video. Even the great can stumble over predicting the *near* future!

With that said, there are several notable examples in SF of stories that did come remarkably close to predicting *the* future; a few came so close to "revealing" supersecret scientific weaponry in World War II, in fact, as to get their authors into trouble with the security authorities. The most famous example is the short story by Clive Cartmill, "Deadline," which appeared in the March 1944 issue of the pulp magazine *Astounding Science Fiction*. "Deadline" contains an amazingly detailed description of the atomic bomb as a U-235 device triggered by a neutron detonator. The hero of the story, an alien commando (all the action takes place on a planet that is not Earth) who has been sent to destroy the bomb of an enemy force, explains that "U-235 has been separated in quantity easily sufficient for preliminary atomic-power research ... but they have not brought the whole amount together ... because they are not at all sure that, once started, it [the resultant explosion] would stop before all of it had been consumed—in something like one micromicrosecond of time." All that was shocking to those in Washington, D.C., who were involved in the security apparatus surrounding the Manhattan Project. The threat of a security leak was enough to bring both the FBI and the Army's Counter-Intelligence Corps down on the heads of both the author and the editor of the magazine.[4]

During my own SF writing career I had (if I may be permitted some self-puffery) two notable brushes with success at predicting the future. One story anticipated the present-day development of *Star Trek*-like "invisibility cloaks," along with how they might work, with a gadget I called a *photon duct*.[5] And in another, cold-fusion-in-water claims similar to the infamous 1989 ones of Pons and Fleischmann make an appearance.[6] But I would be the first to admit that any resemblance between the predicted futures in those stories and what actually came to pass was sheer luck!

8.2 Science Fiction "Spoofs"

There is an important distinction between SF stories that attempt to describe a realistic possible future world and those that, while also serious, are written in an intentional attempt to strain the limits of credibility to illustrate a ludicrous trend in some aspect of current events. Let me give you two examples of such tales of the second kind ("spoofs"), one from physics and one from mathematics, before I turn to the first kind.

Some years ago, at the height of the nuclear arms race during the cold war between the United States and the old Soviet Union, there was much talk of something called the *neutron bomb*. This was an atomic weapon that would produce an immediate huge release of energy but with most of it *not* appearing as blast energy. There would *not* be a big, smoking, glowing hole in the ground after its use; there would *not* be many thousands of people blown to bits; there would *not* be many thousands more cremated in a massive firestorm; but there *would* be an enormous flood of neutron *particle* radiation at the instant of detonation. These particles would pass harmlessly through inanimate matter, like buildings, but would instantly shred human DNA. So, the enemy's infrastructure would remain largely intact while the enemy was massively killed. Unlike the problems that come from the gamma rays of "traditional" atomic bombs, there would be no long-term radiation hazard. There was even some silly talk about the "humanity" of such a weapon!

All that talk of the neutron bomb prompted a young nuclear scientist, Ralph S. Cooper, then on the staff of the Los Alamos Scientific Laboratory, to publish a short (less than 450 words) essay called "The Neutrino Bomb" in the July 13, 1961, issue of the *Los Alamos Scientific Laboratory News*. Written in a serious tone, Cooper's essay took the "humanity" issue to its logical extreme. It was suggested that a neutrino bomb was then under development by the Soviets as a "blastless nuclear weapon" whose energy was almost totally in the form of neutrinos, real particles that have almost zero probability of interacting with anything. "Almost all the energy goes into neutrino production and nothing into blast waves, gamma rays, fission fragments, or fallout," wrote Cooper. The tactical size of the bomb was estimated to be "one microton of TNT"—the muzzle energy of a single bullet from a high-power rifle—using only four pounds of hydrogen that, on detonation, "leaves [only] a temporary vacuum into which the surrounding air rushes with a loud bang, informing the victims in the target area that they have been had." A lot of technical people thought Cooper was actually serious!

For a second example of an SF spoof, this time from mathematics (with just a touch of computers), I'll use one of my own efforts. Written over thirty years ago, and published in a technical journal, it is reproduced below.[7]

Mathematical R&D

The lecture hall was packed, and the air buzzed with whispered excitement. Wild speculations on what was about to happen flew from mouth to ear. The great, Full Professor of Mathematics, Dr. Oliver Osgood, head researcher on five simultaneous government grants, was to speak that day at the graduate seminar on his newest, most astounding discovery.

Not since von Neumann proved the Fundamental Theorem of Games, not even since Fermat scribbled his Last Theorem in a margin, nay, not since Archimedes bounded pi, had such a penetrating intellect laid bare the most intimate details of the Queen of the Sciences. Professor Osgood's insight had once again cast bright light into a hitherto dark corner of mathematics. They were all here to learn what new, incredible secret was now in humanity's possession.

The door burst open (one could almost hear trumpets), and the Great Man strode to the podium. Tall, thin, head awash in a mass of white, unruly hair, eyebrows thickly bushy, he placed a single sheet of paper on the wooden surface before him. He stared at it intently. Impressed, the audience exploded into applause. He acknowledged their acclaim with a boyish (yet mature) smile, then held his arms high to hush the excited crowd.

"Thank you, thank you. You are generous, most kind, indeed. It gives me enormous pleasure to be able to share with you the latest fruits of my research. The Government, which sponsors my work, has agreed to the seminar being unclassified, and I wish to thank those enlightened people for the wisdom of their decision."

"Now, let me get right to it, with no further delay. I have found a new number!"

The audience stared, dumbfounded, at Professor Osgood. This was just incredible.

"Yes, between two and three, somewhat closer to the two

than the three, is a previously unsuspected number. Like a lost penny it has remained hidden through the centuries among the rationals, secreted away beneath the irrationals, but most deeply covered by the transcendentals. Yet, I have found it! With the help, of course, of forty-three graduate students, and hundreds of hours on the university computer system. I thank them all."

A mere associate professor jumped to his feet. "But Professor, this is absolutely marvelous! What *is* this number? Will you write it on the blackboard for us?" A hundred pencils leaped into the hands of the audience, quivering in the excited anticipation of recording the new number.

Professor Osgood shook his head in a charmingly rueful way. "No, I'm sorry, I can't. As you all know, there are two kinds of mathematical proof. The first kind, a constructive proof, would give us the actual value of the number. But my proof is of the second kind, an existence proof. It shows the number exists, but it doesn't tell us *where*, exactly, it is. But we are working on that! The Government has graciously allocated ten million dollars more for my research in the coming fiscal year."

The audience thundered its approval. Those were big time bucks! There would be summer research money for all!

"Well then, Professor, can you tell us more about your proof?"

"Sorry, that's classified DEEP SECRET. Besides myself, only the Secretary of Defense, the Chairman of the Joint Chiefs of Staff, and the President have seen it. I assure you, the secret is safe with *them*. I *can* tell you it is based on the fact that three is odd, and two is even, but beyond that, I'm sure you understand my reluctance to say more."

The mere associate professor nodded his head to show his understanding, and asked, "Tell us, Professor, does that mean this new number has *military* significance?"

Professor Osgood smiled good naturedly, and replied, "We're sure of it. The integrity of our NATO forces in Europe, of our nuclear submarine fleet, of our SAC bases, of our ICBMs, all will be immensely improved by knowledge of this number." He leaned forward toward the spellbound audience.

"I *will* tell you this much. If the new number is squared,

then added to itself, and the result finally multiplied by the cube root of twenty-nine, the theory indicates we *should* have the absolute, total power to . . ." Before Professor Osgood could finish, a short, fat, heavily sweating man pushed forward from the rear of the room. "No, no, Professor, that's classified DEEP SECRET, too! You must say no more!"

Professor Osgood smiled gratefully at the excited man. "Ladies and gentlemen, my apologies for this lapse. This is Colonel Stanley, CIA, and I thank him for setting me straight." The sweaty, almost dripping agent sat down with a plop and a wet squish, relieved at having held back a major leak.

The Professor continued. "Our work for the coming year has the highest urgency. Naturally, we *must* beat the Russians to this new number. The whole strategic balance of technological parity could be unbalanced if we lose this race. There *must* not be a Number Gap!" He paused for a dramatic moment as his eyes blazed, his brow wrinkled, and his fists clenched. He cut a mighty impressive figure, he sure did. The room burst into light from the flash camera of the local news photographer. What a picture it would make on the front page of the late edition!

"And finally, ladies and gentlemen, knowledge of this new number will give us an additional bargaining chip at the upcoming arms control talks! The very future of East-West detente may hinge on our work over the next twelve months!" Professor Osgood then picked up the single sheet of paper before him (having by now decided to go along with his broker's recommendation to dump soybean futures and to buy Acme Gaskets), and strode from the room.

The lecture hall rocked with the thunderous applause of the audience. They knew the worth of what they had heard. With scientists like Professor Osgood, the Free World would never have to worry.

Because, as everybody knew, somewhere in the Soviet Union the Russians probably had a Professor Osgood, too.

A couple of decades after "Mathematical R&D" appeared it became an entry (along with three other of my "mathematical" SF stories) on a website maintained by Professor Alex Kasman, of the mathematics department at the College of Charleston, South Carolina (http://math.cofc.edu/Kasman/MATHFICT/). This site is devoted to the appearances of mathematics in all fictional genres, not just SF. Professor Kasman added the personal comment that the story was making fun of mathematical research, which in fact was not my goal at all. (That's a problem with spoofs, of course—they can be read, because by their very nature they are so unhinged from the real world of today, differently by different people. Recall how Cooper's spoof had similar misreadings.) I wrote to Professor Kasman about this, and he graciously added my response to his website: "This story was a spoof on the Cold War (which in 1979 was roaring hot and heavy). The worry (in the story) about a 'numbers gap' was a take-off on the then much talked-about 'missile gap.' I simply used the unsuspected number between 2 and 3, found via a computer search, as a device that would be immediately understandable by all of *any* technical persuasion. I wasn't attacking mathematics research, but rather those who would profit from dubious 'military paper-pushing' (whatever its nature might be)." And, as Professor Kasman himself wrote elsewhere in the entry, "The audience is thrilled by the discovery both because they know that they will all benefit from the tremendous amount of grant money the government will be contributing to this research and … because they know that if both the US and Soviet Union (this was 1979, remember) continue to waste their efforts on this sort of research, then they have nothing to fear."

8.3 What If Newton Had Owned a Calculator?

Back in Chapter 2 I told you of Newton's fascination with calculating the value of pi out to a large number of digits. He had to do all the grubby arithmetic work by hand, of course. But what if he had possessed an ordinary calculator? Wouldn't that have been great? "What if" questions are the starting point for most SF stories, and this question occurred to me more than thirty years ago. The result was the time travel tale "Newton's Gift."[8] There is a time traveler in it simply because I couldn't think of any other way to get a calculator into

Newton's hands! What happens next, however, may not be quite the happy outcome you expect.

Newton's Gift

Wallace John Steinhope was a sensitive human being, a person deeply concerned about the welfare of his fellow creatures. Any act of injustice, however slight, made his breast pound with righteous indignation. He was a champion of fair play, and his motto in life was taken from the ancient English rule of law— "Let right be done!"

Even while still a lonely, reclusive child, Wallace's heart ached mightily when he read of the laborious, boring, mind-dulling calculations endured by the great mathematicians of old. Just knowing, *thinking*, of Gauss's marvelous mind wasting literally *months* of its precious existence grinding out tedious mathematics that even a dullard could do in a minute, on a home computer, was sheer agony for Wallace. Contemplation of the God-like Newton suffering endless delays in his gravity research, all because of a simple miscalculation of the length of a degree of longitude, was almost unbearable.

Indeed, Newton played a special role in Wallace's life (and he in Newton's, as we shall soon see). While the other great mathematical physicists had merely been hindered in their work by the lack of modern computational aids, Newton had squandered so much valuable time in other, nonscientific pursuits! His quasireligious writings alone, over half a million words, exceeded his scientific writings. What a waste! Wallace wondered endlessly over the reason for this strange misdirection of talent and bored his friends to the edge of endurance with his constant brooding on the mystery. Still, they all liked and admired Wallace enormously and so put up with it. But more than one of them had sworn to throw up the next time Wallace mentioned Newton during a wedding (but that's another story).

So deep was Wallace's anguish for his predecessors that even as he grew older and his own tremendous talents as a mathematical physicist (the result of a lucky genetic mutation induced in a male ancestor some centuries earlier) gained him an international reputation, thoughts of the unmeasurable misery of his scientific ancestors were never far from his mind. It was most appropriate, then, that his greatest discovery gave him an opportunity to *do* something! And Wallace John Steinhope vowed to *help*. He became convinced that it was his purpose on earth—he could not, he *would* not hesitate. As he strapped the knapsack-size time machine onto his chest, his excitement was, therefore, easy to understand.

"It is done! And I am ready. I will travel back and bestow this gift of appreciation, this key to mental relief, on the great Newton himself!" Wallace cradled a small, yet powerful hand-calculator in his palm. It was a marvel of modern electronics. Incorporating large-scale integrated circuitry and a Z-8000 microprocessor solid-state chip, the calculator required only a small, self-contained nuclear battery for its power. It could add, subtract, multiply, divide, do square and cubic roots, trig and hyperbolic functions, take powers, find logarithms, all in mere microseconds. It was programmable, too, able to store up to 500 instructions in its micromemory. The answers it displayed on its red, light-emitting diode readouts would liberate young Isaac from the chains of his impoverished heritage of mathematical calculation. No more Napier's bones for Newton!

But Wallace John Steinhope was no fool. He understood, indeed feared, time paradoxes. He knew Newton could be trusted with the secret, but it wouldn't do for the calculator to survive Newton's time. So Wallace had incorporated a small, self-destructing heat mechanism into it. After five years of use, it would automatically melt itself into an unrecognizable, charred slag mass. But that would be enough time for its task to be completed. The emancipation of Newton's mighty brain from tedium!

Pleased enormously at the thought of the great good he was about to confer, Wallace set the time and space coordinates for merry old England, flipped the power switch on, and vanished.

Materializing in the Lincolnshire countryside in the spring of 1666, he began his rendezvous with destiny. It was the second and final year of the great bubonic plague, and Newton, seeking refuge from the agony and death plundering London and threatening his college of Trinity at Cambridge, had returned home to work in seclusion. The years of the Black Death were Newton's golden years, when the essentials of calculus would be worked out, when the colored spectrum of white light would be explained, and when the principle of the law of gravitation would be grasped. But how much easier it would be if Newton were released from the binding chains of dreary calculation. Wallace's gift would slip the lock on those chains! Accelerate genius! It was early evening when, guided by a map of the area prepared by a friend who was both a cartographer and amateur historian, Wallace reached the quiet little town of Woolsthrope-by-Colsterworth. It was here, in a small farmhouse, that Wallace would meet his hero of the ages. A cold, gentle rain was falling as he approached the door. The soft, hazy light of an oil lamp glowed inside, revealing through the translucent glass the form of a man bent over a table. The fragrant smoke of well-dried wood curled from the chimney, announcing a warm fire within.

With his heart about to burst from excitement, Wallace rapped upon the door. After a pause, the shadow rose and moved away from the window. The door opened, and there stood Isaac Newton, a young man of 23 with an intellect that Hume and Voltaire considered "the greatest and rarest genius that ever rose for adornment and instruction of the species." But for the importance of his self-appointed mission, Wallace would have fainted dead away from the thrill of it all.

"Is this the home of Isaac Newton?" he asked in a voice quavering with the trembling tones normally used by lovers about to reveal their deepest feelings.

The young man, of medium height and thick hair already showing signs of gray, swung open the door and replied, "My home it is, indeed, stranger. Come into the parlor, please, before the wetness takes you ill."

Isaac followed Wallace into the room and stood quietly watching as his visitor removed his soaked coat and hat. The portable time machine was gently placed on the floor next to a wall. The calculator was snug and safe in its plastic case in Wallace's shirt pocket. "Thank you, Master Newton. May we sit while we talk? I am afraid you may wish to take some time to consider my words."

Motioning to a chair near the table, Isaac pulled a second chair from a darkened corner and joined Wallace. "You have a strange sound to your speech, stranger. Are you from here-abouts, or have you traveled far? Please commence slowly your tale."

Wallace laughed aloud at this question, a response prompted by his nervous excitement, and it quite surprised him. "Please forgive me. It is just that I *have* traveled so very, *very* far to see you. You see, I am from the future." Wallace was not one to play his cards close to his chest.

Now it was Isaac's turn to laugh. "Oh, this is most ridiculous. Are you a friend of Barrow's at Trinity? It would be so like him to play such a trick. From the future, indeed!"

Wallace's eyes ached at the sight of the papers on the table where Isaac had been working. What wonders must be there about to be born! In any other situation, Wallace would have asked about their contents, but the die had been cast. He had to convince Isaac of the truth of his tale.

But he had to walk a tight line, too. It just wouldn't do to mis-direct Isaac's interest away from the calculator and toward the time machine itself! He must do something dramatic, something that would rivet his idol's atten-tion and hold it.

"Yes, yes, I understand your reluctance to believe me. But, look here. This will convince you of the honesty of my words." Wallace pulled the shiny black plastic-cased calculator from his shirt pocket and flipped the power switch on. The array of LEDs glowed bright in the gloomy room as they flashed on in a random, sparkling red burst. Isaac's eyes widened, and he pushed his chair back. Was he frightened?

"As the Lord is my Savior, is it a creation of Lucifer? The eyes of it shine with the color of his domain. Are you one of his earthly agents?"

"Oh my, no! Look here, Mas-ter Newton, let me show you that there is no black magic or chicanery involved. It is all per-fectly understandable in terms of the laws of Nature. What I have here is an automatic cal-culator, a device to perform all of your laborious mathematical labors."

So saying, Wallace squeezed the sides of the calculator case together, releasing pressure snap-fittings, and flipped the case open on a hinge at the top. Revealed to Isaac were the innards of the electronic marvel—a tightly packed interior of printed circuit boards, a mass of integrated circuitry, the small LED display, and the sealed nuclear battery. Isaac stared intently at the sight, and Wallace could see the natural curiosity of Newton's great mind begin to drive away the initial apprehension.

"But where are the gears, levers, springs, and ratchets to carry out the calculations? All I see is a black box with lights that glow red—how is *that* done, where is the lamp or candle to provide the light!—and many little isolated fragments of strange shapes. There is clearly nothing in your box that moves!"

"Oh, it is all done with electronics, Master Newton! The central processing unit has access to a solid-state memory that contains the decoding logic necessary to implement the appropriate algorithmic processes to provide the answers to the specific requests entered through these buttons. The actual performance of the box is achieved by the controlled motion of electrons and holes in suitably doped semiconductor material under the influence of electric fields induced—" Wallace, still overcome by his excitement, had rambled on wildly without thought of the essentially infinite technological gap that separated himself from Newton.

"Stop, stop," cried Isaac. "I understand only a few of the words you use and nothing at all of their meaning! But it is obvious that for calculations to be performed, mechanical work must be done, and that implies motion. Pascal's adding machine has shown the veracity of that. I say again, nothing moves in the box. How *can* it work?"

Wallace was embarrassed. The mistake of overlooking the hundreds of years of progress after Newton's time was one a child might make. "I am sorry, Master Newton. I'm going too fast for you." Isaac looked at Wallace with a frown, but Wallace failed to see the pricked vanity of the proud Newton. Going too fast, indeed!

Wallace prepared to lay a firmer technological foundation for Newton, but then he froze.

It couldn't be done! Newton was a genius, certainly, but the task was still impossible. Wallace would have to tell him all about Maxwell's equations, Boolean algebra and computer structure, electronics, and solid-state device fabrication technology. It was just too much, and besides, there was the danger! The potential time paradoxes of all that knowledge out of its proper time sequence! Could Newton, in innocence, reveal some critical bit of knowledge out of its natural place in history! Wallace hesitated, and seeing the suspicion grow again in Isaac's eyes, he realized he had to do something, *anything*, immediately.

"You cannot deny your own eyes," answered Wallace. "Let me *show* you how it works. I'll divide two numbers for you with just the punch of a few buttons. Watch this." And, at random, he entered 81,918 divided by 123. Poor Wallace, of all the numbers to use, they were the worst.

Within milliseconds the answer glowed brightly in fiery red characters. Wallace looked with pride at the result and then, already enjoying in his mind what he knew would be Isaac's amazement, he turned his eyes to the great man. What he saw made his spine tingle and the gooseflesh stand high on his neck. Newton had fallen to his knees, with eyes bulging and hands raised as if in prayer.

"The mark of the Beast, it is the mark of the Beast! It is so written in the Book of Revelation—Here is wisdom. Let him that hath understanding count the number of the beast; for it is the number of man; and his number is six hundred threescore and six!"

Rising to his feet, Newton fell back into his chair. "Your cursed box bears the brand of its master. There can be no doubt now, it *is* the creation of the fallen archangel!"

Wallace was aghast at Isaac's violent reaction. The 17th century genius had now stumbled backward from his chair and had grasped a poker from the hot coals of the fireplace. "Wait, please wait! Watch this, I'll multiply two other numbers together for you, watch!" Wallace quickly punched in the data, and then the answer gleamed steadily in burning red characters on the LEDs. Isaac's eyes first went wide with fear as he again saw the wizard electronics do their marvelous assignment, and then he shut them tight.

Wallace was becoming desperate—this wasn't the way it was supposed to be! "Don't you see—imagine the tedious work, the mind-deadening labor this machine will save you from. And it is yours."

"Yes? But only for the exchange of my soul! That is always the Devil's price for his seductive gifts from Hell!"

As Isaac shrieked these last words at Wallace, he raised the poker over his head. "Begone, you emissary of the Dark World! I know now you must be in the employ of the Father of the Antichrist, but the Lord God Almighty will protect me if I do not waver in my resolve. Begone, or I'll strike your brains out on the floor where you stand!"

Isaac's eyes were wide with fear, nearly rolling back to show all white. Spittle sprayed from his mouth as he yelled at Wallace, who stared in shock at the wild man who threatened him with death.

"Please, please, *listen* to me, please! I beg you to understand—I am a scientist, just like you. The concept of the devil, and all it stands for, is contrary to everything I believe. How *could* I be in the devil's employ, when I don't even accept his existence? You *must* believe me!"

"Blasphemy!" screamed Isaac. "Your own words condemn you. To deny the reality of Satan in a sinful world is to deny that of God, too. Now leave my home, you dark beast from hell, or by the heavens above, *I shall destroy you!*"

As he shrilled these words, Isaac brought the poker down in a wild swing that barely missed Wallace's head.

Struck dumb with confusion at the uncontrolled outburst, Wallace stuffed the calculator into his shirt, grabbed his hat, coat, and time machine and rushed from the house. As he hurried into the cold, wet night, he turned back, just once, to see Isaac Newton framed in the light of the open door. "Go, go, you foul messenger from the Lord of Evil! Back to your stinking pit of burning hell-fire! This is a house that honors the Divine Trinity and is no haven for the likes of you!" Wallace rushed away into the blackness, the time machine bouncing unheeded upon his chest.

He ran, for how long he couldn't recall, until he fell exhausted next to a stream running heavy with the rain. Tears of rage, frustration, and shock streamed from his eyes. Rejected by the great Newton! Well, damn him! Wallace flung the calculator into the stream in his terrible anger and activated the return coordinates. He faded from Newton's world as quickly and as quietly as he had come.

As for Isaac Newton, after having chased the Devil's messenger from his house, he returned on shaking legs to his desk. Pushing aside his rough calculations on the orbit of the moon around the earth, he swore to redeem himself in the eyes of the Savior. Somehow, he had been found lacking and had been tested. And the test was surely not over! He began to reapply his marvelous mind to determine the origin of his failure before the Lord God Jehovah. Taking quill in hand, he wrote the first of the many hundreds of thousands of words that his religious tracts would devour from his allotted time.

Five years later, long after Newton had returned to Cambridge, a group of picnicking children were frightened when a nearby stream suddenly erupted into a geyser of steam. Moments later, as the eruption subsided, the bravest (or most foolhardy) of the boys—who, by an astonishing coincidence that befits any good time travel paradox, would be Wallace's great-grandfather nine times removed—cautiously examined the streambed. All he found were some twisted, hot pieces of what he thought was a hard, black rock, and he tossed them back. They were all that was left of the calculator's nuclear battery. He did receive a tiny radiation dose from them, which caused a recessive genetic mutation that centuries later would suddenly appear as the cause of Wallace's genius, but otherwise the lad was unaffected. The incident was soon forgotten.

Well over 300 years later, Wallace John Steinhope reappeared in his own time. He was essentially the same man as before he left—kind, generous, and sensitive. Ready to come to the aid of any man or beast that might need help, he was giving of himself to a fault. As far as his friends were concerned, in fact, he had even improved (naturally, they didn't know what had brought about the welcome

change, but if they had, they
would have applauded it).

Wallace John Steinhope, you
see, never again had another

kind word for Newton, or for
that matter, any words for him at
all.

Just two final comments about "Newton's Gift." First, notice that
the story does not involve changing the past, a once common story
gimmick (usually involving the infamous "time police") that I take as
being illogical. I make that point, strongly, in my book, *Time Machines*.[9]
Second, I hope you caught the two causal loops in the story. The
first being, of course, that the time traveler makes his trip backward
through time because of a condition in the past (Newton's sudden shift
from science to religion) that his trip, *starting in the future*, causes. And
second, the time traveler's journey itself into the past is the *cause* of the
genetic mutation that results in the time traveler being smart enough
to build the time machine that allows the trip to even occur. Time
travel to the past *is* an odd business, indeed, but both of these causal
loops are logically self-consistent and, for physics, that's all (assuming
no conservation laws are broken) that matters.

8.4 A Final Tale: The Artificially Intelligent Computer

Now, to finally finish the penultimate chapter, I'll complete the circle I started back in the Introduction (and continued throughout the
book), where I wrote of the usefulness of computers to mathematical
physicists. This development is, of course, even more recent than is
the invention of the electronic computer itself, with the first real application to physics—beyond that of "mere" power number crunching in
Monte Carlo and differential equation–solving codes—not occurring
until the early 1950s, as discussed in Chapter 3 on MANIAC-I's discovery of the nonuniversality of the equipartition of energy in nonlinear
multimode systems. There we saw how such twentieth-century giants
as Fermi, Gamow, Von Neumann, and Ulam enthusiastically embraced
the new tool. Still, it was impossible for even those brilliant people
to appreciate how computers would eventually spread throughout our
society to the point where today, almost every teenager in America (if
not the entire developed world) has a computer in the bedroom. These

are machines that make MANIAC-generation computers look much as does a firecracker compared to an atomic bomb.

When I was in high school in the mid-1950s, the prevailing view was that perhaps a handful of huge, MANIAC-like supercomputers, located at top-secret government weapons labs and maybe a few elite schools, would take care of all the computational needs that would ever occur. And as for something like the Internet, well, *that* was simply beyond imagination in those days (except by "Murray Leinster"—see note 4 again—and most thought he was just a crazy SF writer, so what did *he* know?). This narrow view of the use of computers was accepted by even as visionary a thinker as Isaac Asimov, in his various SF stories featuring the fictional Multivac. Today, of course, we have people who suffer real emotional pain if they can't check their e-mail at least once an hour (perhaps you're one of them!).

Later, in the mid-1970s, then a young professor of electrical engineering who was teaching a computer engineering course involving both hardware design and software development, my imagination turned to wondering just what computers might someday do for humans. My thinking was not so much on the pure technical side, however, but more along possible societal impacts, with a target audience much broader than just the close-knit world of mathematical physicists. Science fiction, it seemed to me, was the perfect venue. In one story I imagined a possible future interface between politicians and computers.[10] That story, written in 1979, proposed that candidates for elective office have to demonstrate, on live television, their problem-solving, analytical skills (if any) by displaying a minimal understanding of computer programming, at the level of elementary BASIC. A precursor to such an interface between computers and politics had already actually occurred, more than twenty-five years earlier, when *CBS News* delayed announcing the prediction by the UNIVAC I (UNIVersal Automatic Computer) of Dwight Eisenhower's 1952 presidential win. Built by Remington-Rand, a company founded by the same people who built the ENIAC, the declaration by UNIVAC (the world's first commercial computer) that Adlai Stevenson would lose was considered just too unreasonable to be taken seriously by the political pundits, much to their later considerable embarrassment.

If you look back at the 2008 American presidential campaign, how-ever, you can see that my 1979 tale missed the mark by a lot in predicting technical sophistication in our elected officials (with one modern candidate admitting he didn't even know how to send an e-mail!). I had somewhat better luck, that same year, in predicting today's intimate, symbiotic relationship between warrior and computer in a tale of the near erotic connection between a combat pilot and his aircraft's flight control and weapons computer.[11]

Now, to finish things off, here is what I think my best attempt at SF foreshadowing, with a story that appeared more than thirty years ago in *Omni* magazine.[12] "The Box" deals with a *possible* blurring of the distinction between humans and computers that still lies in the future—but perhaps not *that* far in the future.[13]

The Box

Ambrose P. Talaylee was bad news. At least that's what every-body in the electrical engi-neering department of East Pascatawa University thought. Why, the man was a disheveled monster, with unwashed hair awry, filthy jeans, straggly beard strung with fragments of yester-day's lunch, and his argumen-tative manner. He showed no reverential fear in the presence of his betters, what with his leap-ing at every opportunity to argue with one professor or another. He told vulgar jokes and laughed too much. It didn't help that he was brilliant. And he smelled bad, too.

Ambrose had appeared with a master's degree, three months before, from a small college in Ohio. He'd passed the entrance exams at EPU, and so they'd had to let him in. But right from the start the faculty had the jit-ters about him. Scoring a perfect set of examination papers upsets the evaluation process, you see. Of course all the other grad stu-dents hated his guts since he'd made them look like klutzes. The next-highest man got a sixty-two.

At first they thought they could get rid of Ambrose by stalling him. As old Wingate, the chairman, had roared in a department meeting, "Let the guy take courses until hell freezes over. He'll get no degree without a research dissertation project! Anybody here want to be his advisor?" They had all chuckled at the silliness of such a thing.

But Ambrose zapped them by finding a long-forgotten rule in the Graduate School Catalog. All he needed were the signatures of seven assistant deans on a petition and he could go ahead with doctoral research on his own. The petition, dirty and torn at the edges, and covered with hideous-looking stains, made the rounds. Each A.D. had signed it just to get the strange, smelly person holding it out of his office.

No one really knew what Ambrose was working on. The petition just read "Implementation of a Synthetic Neural Network." As that cocksure young assistant professor Simpson said in another department meeting, "Maybe he's making himself a new brain!" That brought the laughs, and Simpson glowed with joy at having pleased his betters.

For the next eighteen months Ambrose labored in the former double-sized closet (lit by a single, naked, flickering sixty-watt bulb) that they had assigned to him. Occasionally he'd be seen in the stockroom, filling old paper bags with integrated circuit chips, capacitors, wire, and assorted nuts and bolts. He actually seemed to *live* in the closet,

as some said they had heard strange noises coming from it late at night. The area around the closet door was constantly littered with paper trash like greasy pizza boxes, napkins smeared with catsup, and candy wrappers covered with chocolate stains. The man was a disgrace!

No one ever figured out where his money came from. Unlike the other graduate students, Ambrose had no teaching appointment in the sophomore labs.

And then one day Chairman Wingate called a special department meeting. "I have here a petition from Ambrose P. Taleylee for the setting of a date for his oral doctoral examination." He held with obvious disgust, between two fingers, a ragged and soiled piece of paper.

Associate Professor James, the department wise guy, quipped, "Christ, Wingate, it looks like something died on that petition. If you've got any cuts on your hand, you'd better get a tetanus booster shot."

Assistant Professor Simpson got his two cents in, too. "You mean he really got something done? Why, he hasn't talked to anyone in the department for advice or anything. All he's done is swipe the stockroom blind and smell it up something terrible!"

The chairman dropped the petition into a folder and gratefully pushed it aside. "Yes, but by the by-laws of EPU we must honor the request. If his work is a bust, we can deny his degree and run his butt out of here. So who wants to be on his committee? There must be six people, you know. Two of us and four from other departments."

In the end it was decided that Wingate and Simpson would be the electrical engineering representatives, while Professors Lardo, Hogson, Sweeney, and Vreeble, from physics, biology, political science, and mathematics would be the others.

On May 17, at 1:30 P.M., they all gathered in Memorial Hall, except for Ambrose, who came forty minutes later. When he finally swept into the room, the committee members nearest the door reeled in his wake. Wingate and Simpson, almost presciently, had sat as far away from the door as possible. Even so, they still experienced some momentary nausea.

No one knew what to make of it. Ambrose was pushing a cubical metal box mounted on casters to the front of the room. He spoke just two sentences and then sat down. "This is my research project, a synthetic neural network endowed with all of Man's knowledge, and the rules of logic to allow it to synthesize new knowledge. You may address all your questions to the box."

The committee members sat stunned. A mockery was being made of the time-honored tradition of oral doctoral defense. Just what kind of nonsense was this?

But cagey old Wingate knew better than to make a scene. Otherwise Ambrose could file legal complaints, and then all hell would be on their necks if the press got hold of it. Better to humor this obviously deranged person, let him make a fool out of himself, and then have a unanimous vote denying the degree. Wingate stood and walked up to the box. About two feet on an edge, its outer shell was black, crinkly aluminum. A combination loudspeaker-microphone was visible on the front panel.

"Tell me," Wingate began, "what are the limits on the defining integral of the Laplace transform?"

With barely a pause, the box answered in a bell-like voice,

"Lower limit is zero; upper limit is infinity."

Wingate appeared unshaken by the answer he heard. That was just a routine question. Any sophomore would have had that stashed away in his or her minuscule brain. A more impressive test would be to ask a partially ambiguous question to see whether the box could make plausible interpretations, as a human might do. "Correct. Now, what is the color of an electron?"

A slightly longer pause. "Color is the physiological sensation arising from a correlation between radiation wavelength and retinal response. Planck's law gives the relation between energy and frequency. To compute the color of an electron, it is necessary to know its total energy. Please input the physical environment of the electron, and then its kinetic and potential energies may be computed. This will define the frequency and wavelength and, hence, the color."

Wingate stared at the box. "Umph! Well, I pass to one of the other members of the committee."

Simpson leaped to his feet, shouting at Ambrose. "What kind of stunt is this, anyway? Have you got a smart midget inside that thing, Talaylee?" He forgot himself and stalked over to Ambrose.

Ambrose laughed in Simpson's face. Simpson staggered backward as the blast of rancid fumes engulfed him. "Just ask it *anything* you want," Ambrose challenged.

Recovering, Simpson snickered to himself and decided to end this farce. "All right, box, try this on for size. Is there life after death?"

Quick as a wink came the answer: "Yes."

This disconcerted Simpson, but not for long. "Prove it," he yelled.

Back came the machine with a response showing it was indeed Ambrose's creation. "Drop dead and see for yourself!"

Simpson was shocked. "Why, why..." he sputtered. "Such arrogance, such ungracious behavior, is just intolerable."

Ambrose laughed again. "I didn't say it; the *box* did!" The truth of that left Simpson speechless, and he sat down in defeat. The rest of the committee was reeling, too, but they retained enough of the committee instinct to go automatically into a huddle. They reached a unanimous judgement quickly and gave Wingate the pleasure of announcing it.

"Mr. Talaylee, we have concluded that this box of yours is a complete fake. Its premise is absurd, and its answers are too rapid and lacking in sufficient scholarship. We have no wish to embarrass you further. So we will not insist you show us the interior, where we are sure there is a radio transceiver linked to an accomplice elsewhere on campus. Therefore, it is our decision that, as regards your final oral doctoral exam, you fail! Good day, sir." With that, the committee marched out of the room in a column, with Wingate in the lead, like baby ducks following their mother.

Ambrose stood glaring after them, yelling. "You bastards! I'll sue you. I'll sue!" Now that kind of response impressed the committee, even though it did lack a certain polish.

That night there was a secret meeting of the committee. At first, the professors thought they could still dump on Ambrose by awarding the doctorate *to the box* (after all, it, not Ambrose, had answered all the questions). Too risky. Instead, they threatened Ambrose with being blackballed if he persisted in his threat. "Look here, Talaylee," Wingate counseled, from a distance, "if you'll just quietly leave the university, we'll call it quits. We'll even write you a good recommendation, one good enough to get you a teaching job at some reasonably decent high school. Get tough with us, though, and we'll see you never work in academia *anywhere*! Now, what do you say?"

"You stinking bastards! You're scared, aren't you? Well, you should be. I'm suing!"

For their defense, the committee hired Sylvester Shylock Shyster, a highly respected criminal lawyer. Ambrose elected to serve as his own lawyer. This is, of course, most unusual, but the judge relented after listening to Ambrose's arguments in closed chambers. Once the judge had caught the drift of Ambrose's arguments (to say nothing of catching the drift of Ambrose himself), he quickly agreed, and just as quickly he rushed from the room.

Since all the physical evidence was circumstantial, Shyster was confident of victory. But then the prosecution called its surprise witness—the box! Yes, the box had been in the laboratory/closet with Ambrose when the committee made its threats. The box had heard it all, but, after all, which of the professors cared what they said in front of a mere box?

Naturally, S. S. Shyster stood in court to protest.

"Your Honor, this is most unusual. This box is only a conglomeration of inanimate wires, circuit chips, and solder. It has no soul, no moral code of behavior, no knowledge of right from wrong. Its metabolism is that of amperes and volts, not that of a human's beating heart and the pulse of warm blood. It has no sensitivity, no ability to give voice to the feelings of humanity. Therefore, I move the testimony of this box not be allowed."

"Do you feel the machine is capable of lying?" the judge asked.

"No, Your Honor. It has been shown to be a machine of computational infallibility. But the intricate complexities, the subtle nuances, of this incredibly convoluted case are just too involved for its mechanistic principles to grasp. It cannot, to make an analogy, appreciate the infinite variations that separate the pure white of innocence from the deep black of guilt."

"So," said the judge, "your objection really comes down to a feeling that the box may not be able to express itself in a sufficiently human manner. Well, let me ask it the crucial question, and we'll see whether you're correct."

Turning to the cubical box (positioned comfortably on a purple velvet cushion in the witness chair), the judge asked, "Tell me, box, did the committee, in fact threaten to do damage to the reputation of your creator, one Ambrose P. Talaylee? And further, did the committee deny him the royal annointment of the doctorate?"

"Does a bear crap in the woods, Your Honor?" the box said. "Is a hog's fanny pork?"

"I object," yelled S. S. Shyster.

"We confess," cried the committee in unison.

"Guilty as charged," the judge roared.

"Damn right," chuckled the box.

"Double damn right!" snorted Ambrose, who in his ecstasy failed to notice that said snort caused all those nearby to turn green. But he wouldn't have cared even if he had noticed.

Thus did the committee learn, as its members were led

away in shame and humiliation, of the error in treating an artificially intelligent box lightly.

It was later, as the committee sat glumly in jail while they waited for Shyster to come see them, that they learned things were even worse than they had suspected. Wingate had just read a note passed to him by the jailer. It was from his secretary. He hurled it to the floor as he cried, "God damn it! Is there *no* justice left in the world?"

Professor Hogson, from biology, was stunned at the outburst. "My heavens, Wingate, what's wrong now?"

Wingate stared at the floor, his jaw muscles working. "That bastard Talaylee! Not only does he get his Ph.D, but now he's putting us out of business. Do you know what that guy has gone and done? *Do you know?*"

Wingate could see from the puzzled expressions on his colleagues' faces that they didn't know.

"I'll tell you what that creep has done! He's taken that miserable box of his, put a coin slot on it, and stuck it in the Student Union Building! He's got kids lined up getting their homework done for a quarter. It's charging fifty cents for an English essay and two bucks for a master's thesis. A five spot gets you a doctoral dissertation!"

"My Lord, Wingate," gasped Hogson, who was near fainting, "academia as we know it will utterly collapse. What'll we do?"

"He's laid down terms. If we meet his terms, he'll stop." Wingate looked as if he were about to cry.

"Well, come on, man," croaked Sweeney of political science, "what are they? Whatever they are, we'll have to meet them!"

"He wants a faculty appointment at EPU, in my department," said Wingate, bursting into tears. "Oh, mother, I can't bear the thought!"

Sweeney was the first to regain his composure. "Be tough, Wingate, be tough. So he gets an appointment. That doesn't mean he'll get tenure. Then, later, out he goes on his butt!"

Wingate slowly raised his head. "No, no, *no*, you idiot. You don't understand. Talaylee doesn't want the appointment for *himself*. He's already got a research job with the CIA. He can work alone for them breaking codes, and so they don't care *how* bad he smells. It's that damned *box* of his. He wants the *box* to be given a faculty spot! Christ, a bloody damn computer as an assistant professor!" He lowered his head again, and began to weep quietly.

The professors sat motionless for several minutes, immobilized with despair. Then Vreeble, from mathematics, looked as if he'd just had an idea. "I say, Wingate, look at the *bright* side of it all." Wingate stared at him with a stunned expression on his face.

"Just what in hell is *bright*, Vreeble? It sure as heck isn't you!"

"Come, come, don't be bitter. See here, you'll be in excellent shape with the affirmative action, equal opportunity people. Gad, can you think of a better minority faculty hiring decision? Yours will be the first academic department in the entire country to hire a box! You'll be—"

Vreeble couldn't continue, for by now an enraged Wingate was attempting to throttle the unfortunate mathematics professor. Which is why, while the rest of the committee were released, Wingate had to stay another six months pending litigation of an attempted murder charge. But it all worked out okay, since it was Wingate's sabbatical year and he didn't have anything else to do anyway. It gave him plenty of time to figure out where things had started to go to hell and to ponder the evil of all prejudice, be it racial or nasal.

Everything worked out rather well for the box, too. It proved to have all kinds of good points; it was a great joke teller and was able to establish friendly relationships with students, and it didn't smell nearly as bad as Talaylee, either (unless the oil on its hinges got too hot). It was marvelous fun at department parties, and it never drank too much. One year later, when Wingate stepped down from the department chairmanship, the box was elected almost unanimously. Only Simpson and Wingate dissented. It's done just a great job since, too. There's even talk that when the dean retires, then—well, maybe *that's* premature right now.

But still, who can tell? Who can tell?

"The Box" was written as a bit of light-hearted fun. Other writers, however, have had a much less enthusiastic view at the coming of artificial intelligence. Consider, for example, the following excerpt from Lord Dunsany's 1951 novel *The Last Revolution*, in which an inventor has built a mobile machine with a synthetic metallic brain that can reproduce itself:

> "It's been making several of those things lately," said Pender.
> "Good lord!" I exclaimed. "What things?"
> "Things the same as itself," he said.
> "But can it possibly do that?" I asked.
> "I made it," said Pender. "Get it into your head that it is cleverer than I."
> "Then it can," I said.
> "What I can do," said Pender, "it can do much better."

The Last Revolution was clearly inspired by Wells's *War of the Worlds*. While it was Earth's bacteria that defeated Wells's Martian invaders, it is only after the marauding "thinking beasts" accidentally get wet and their brains start to rust that Dunsany's evil calculating machines are reduced to scrap metal.

Pretty gloomy!

I'll end here by remembering an observation the British biologist J. B. S. Haldane is famous for making in his 1927 book *Possible Worlds*: "The universe is not only queerer than we suppose, but queerer than we *can* suppose." Haldane was absolutely correct, but I do think there is one last comment that can be added. Even when humankind someday runs into a physical phenomenon that represents something *so* wonderfully queer that *nobody* has ever before imagined it, then, when the initial awe wears off a bit (as it inevitably will), there will surely be some analyst, somewhere, hunched over his or her desk on which sits a quantum logic computer (humming along at atomic speed while sitting next to a steaming coffee cup and a half-eaten donut) who will be able to explain it all with mathematical physics and massive number crunching.

And that ability will be, I think, the queerest and most wonderful thing of all in the universe.

8.5 Notes and References

1. Outside of SF, too, a well-written story of pure imagination set intentionally in the past can still hook readers despite knowledge of the historical facts;

an example is the 1971 novel by Frederick Forsyth, *The Day of the Jackal.* It enthralled millions around the world (it was made into an equally exciting 1973 film), and still does, even though every one of its readers knows, right from the start, that the late General Charles De Gaulle, president of France, was not assassinated, and so the novel's mysterious, relentless assassin-for-hire must ultimately fail.

2. Richard Feynman, "Simulating Physics with Computers" (*International Journal of Theoretical Physics*, June 1982, pp. 467–488; reprinted in *Feynman and Computation*, ed. Anthony J. G. Hey [New York: Perseus Books, 1999]).

3. J. R. Pierce, *SIGNALS: The Telephone and Beyond* (New York: W. H. Freeman, 1981). The same year his book was published I spent an afternoon with Pierce at his home in Pasadena, California, and we discussed a wide range of topics, including science fiction. I later wrote of that afternoon in the Caltech alumni publication *Engineering & Science* (November 1981, pp. 22–25) in an essay titled "An Interview with John R. Pierce." You can also find it on the Web at http://calteches.library.caltech.edu/527/2/Nahin.pdf.

4. You can read what happened next in an essay ("Let's Call It a Hobby") written by Murray Leinster, the pen name of William F. Jenkins (1896–1975), in a collection of SF stories he edited, *Great Stories of Science Fiction* (New York: Random House, 1951). Murray Leinster is famous among physicists for his story "Sidewise in Time" (*Astounding Science Fiction*, June 1934), which anticipated the quantum mechanical "many-worlds" concept, and among computer scientists for "A Logic Named Joe" (*Astounding Science Fiction*, March 1946), which has a very nice, astonishingly accurate description of today's Internet. More discussion on "Deadline" can be found in Albert I. Berger's *The Magic That Works: John W. Campbell and the American Response to Technology* (San Bernardino, CA: Borgo Press, 1993, pp. 60–69) (the editor of *Astounding Science Fiction* was John W. Campbell [1910–1971]). Cartmill wasn't the only SF writer to get into trouble by writing about an atomic bomb before Hiroshima and Nagasaki. Philip Wylie, the co-author of the SF classic *When Worlds Collide*, was placed under house arrest for attempting to publish his novella, *The Paradise Crater*. Set in 1965, the story has a group of surviving Nazis making an atomic bomb for revenge at losing the war. Pounced on by Army security people when *American Magazine* first rejected the work for being "too absurd and unrealistic," but whose editors then nevertheless reported him to Washington censors, Wylie was actually threatened with death by his guard if the tale proved to indeed be a security leak! The story eventually did appear after the war in *Blue Book* (October 1945), when it suddenly didn't look "too absurd and unrealistic" anymore. It didn't take long for SF writers to see what might lie ahead for the atomic age world after the war, either. The well-known mathematician Chandler Davis (born 1926), for example, published a prophetic story—written while he was still a teenager—on the development of

nuclear-armed ICBMs and of the transition of the control of warmaking powers in America from Congress to the military ("To Still the Drums," [*Astounding Science Fiction*, October 1946]). The hero of the story ends up in prison for trying to reveal the attempt by an early version of a military-industrial complex to undermine democracy. Davis later (1960) spent six months in a federal penitentiary for having refused to testify (several years earlier) before the infamous House Un-American Activities Committee. Soon after his release Davis moved to Canada, obtained a faculty position in the mathematics department at the University of Toronto, and today is the editor of the prestigious journal *The Mathematical Intelligencer*. As a mathematician familiar with the concepts of recursion and self-reference, I suspect the imagined image of the Committee investigating itself (where some *real* un-Americans could actually have been found) would have given him at least some pleasure while he sat in prison. The irony of ending up in the same spot as did the hero of his 1946 story probably didn't escape him, either.

5. Paul J. Nahin, "Archival Voice" (*Analog Science Fiction/Science Fact*, February 1984). *Analog* was the new name *Astounding*'s editor John W. Campbell (see the previous note) gave the magazine in 1960.

6. Paul J. Nahin, "Publish and Perish" (*Analog Science Fiction/Science Fact*, April 1978).

7. Paul J. Nahin, "Mathematical R&D" (*IEEE Transactions on Aerospace and Electronic Systems*, January 1979, pp. 179–180).

8. Paul J. Nahin, "Newton's Gift" (*Omni*, January 1979).

9. Paul J. Nahin, *Time Machines: Time Travel in Physics, Metaphysics, and Science Fiction*, 2nd ed. (New York: Springer, 1999).

10. Paul J. Nahin, "Qualification Test" (*Analog Science Fiction/Science Fact*, December 1979). This story has at least two clear examples of failures to "see" the future. It is set some years in the future of 1979, when I still had television sets with huge glass picture tubes (instead of flat screens) and cars still had carburetors (instead of fuel injection). Glass picture tubes and carburetors, of course, sit today between the typewriter and the slide rule in the museum of obsolete stuff.

11. Paul J. Nahin, "The Man in the Gray Weapons Suit," in *The Future at War*, vol. 1, ed. Reginald Bretnor (New York: Ace Books, 1979; repr. Baen Books, 1988).

12. Paul J. Nahin, "The Box" (*Omni*, January 1980).

13. Computer pioneer John von Neumann, who appeared earlier in this book, would disagree with me. As the mathematical physicist Jeremy Bernstein once wrote in one of his many erudite *New Yorker* magazine essays on science, "As a graduate student, I once asked the late Professor John von Neumann whether he thought computers would replace mathematicians. His answer was 'Sonny, don't worry about it.'" You can find that essay reprinted in Bernstein's book, *A Comprehensible World: On Modern Science and Its Origins* (New York: Random House, 1967, p. 191).

9
A CAUTIONARY EPILOGUE

This statement is a lie.
—Liar's paradox (fourth century BC)

9.1 The Limits of Computation

The speed of any computer is fundamentally limited by how fast its various component parts can send and receive signals among themselves—that is, by the speed of light and by how far those signals have to travel. We can't do anything about the speed of light, but one way to increase the speed of a computer is simply to make it smaller. That means the computer's volume decreases. Suppose, just to be specific, a computer has the shape of a sphere with radius r. As $r \to 0$, the volume decreases as r^3. If the energy required to power our shrinking computer doesn't decrease at least as fast, then the dissipated heat energy density in the computer will increase and the temperature of the computer will rise. Eventually, the computer will melt. That's one sort of limitation on computers, of a *physical* nature, one that can be gotten around—at least for a while—by various means (the most obvious being to add a cooling system to the computer). There are other limitations, however, of a far more profound nature.

Today's computers are of such massive capability that it is easy to suppose, as many actually do, that there is no number-crunching problem that they couldn't, if given sufficient time, grind their way through by, if nothing else, brute force. Unfortunately that's just not true, and it has been known to be false since the English mathematician Alan Turing (1912–1954) proved it in 1936. Turing arrived at his astonishing result by the direct approach of describing, in great detail, particular computational problems that he showed are inherently unsolvable.

Mathematicians had, in fact, seen unsolvability demonstrations before, as in the 1836 proof by the French mathematician Pierre Wentzel (1814–1848) on the unsolvable nature of a classic geometry construction problem from ancient times, that of trisecting an arbitrary angle with compass and ruler alone. Turing's examples, however, were the first demonstrations of specific computational problems that no machine, no matter how "powerful," could complete. And that will still be true a million years from now, too, no matter how much technology may advance over the next ten thousand centuries. Turing's unsolvability proofs constrain what computers will ever be able to do.

This book has been an unabashed ode to the wonder of the modern electronic computer, and so the admission of limitations may seem to be a step backward—particularly so after the up beat ending to the last chapter. Well, honesty compels the inclusion of this final chapter, but even so, I think you'll find the arguments I'll show you next to be beautiful despite their forcing limits on what computers can do.

I'll start by observing that there are problems that have not been solved, even after many decades of intense efforts by the world's greatest mathematicians (and so they are at the least very difficult problems), and yet it *can* be proved that it is impossible to prove them unsolvable. If one could show a perplexing problem to be unsolvable, then mathematicians could just file the vexing thing away forever and move on to something else with a sigh of relief (*Thank God, it isn't our fault and we aren't just stupid!*). Without a proof of unsolvability, however, they have to keep trying, perhaps dooming themselves to yet more endless years of frustration.

The most important unsolved problem in mathematics today—a status it has enjoyed since its original statement in 1859—has that frustrating nature. (It is interesting to note that this problem, the Riemann hypothesis that we encountered back in Chapter 1, fascinated Turing all his professional life.[1]) After calculating only just a few of the first complex zeros of the zeta function, and finding that the real part of each was 1/2, you'll recall that Riemann then made a huge conceptual leap and conjectured that every last one of the infinity of complex zeros would have a real part of 1/2. Trillions of complex zeros have been calculated since 1859, and every last one does indeed have a real part of 1/2. But nobody has been able to *prove* that *all* the complex zeros, with not even a single exception, have a real part of 1/2. Proving or disproving the Riemann hypothesis has stumped the world's best mathematicians for a century and a half. It might continue to stump for another ten million years. Or even forever. It is undeniably a very tough problem. But nobody will ever be able to *prove* it is an unsolvable problem.

Why not? Well, suppose one did have an unsolvability proof. That would mean no one, even by sheer accident, or even if the computation of complex zeros continued at the fantastic rate of $10^{10^{10^{10^{\cdot^{\cdot^{\cdot}}}}}}$ zeros per nanosecond (go as high up as you wish in that stack of exponents), on into the infinite forever, could ever find a complex zero with a real part $\neq 1/2$. (After all, if you did find such a zero you *would* have resolved the Riemann hypothesis, in favor of it being false). But such a conclusion means you *have* resolved the Riemann hypothesis (in favor of it being true), in contradiction to the assumed unsolvability proof. So we are forced—it's the only way to escape the paradox—to conclude that *there just is no such proof*.[2]

9.2 The Halting Problem

Now, let's turn to what Turing did, establishing that there are impossible computational tasks that are beyond any computer in existence, as well as beyond any computer of the future yet to be built.[3] Imagine that there exists a computer program that accepts, as part of its input data, the code for any computer program **P**. This isn't as odd as it might initially seem, as that's just what a compiler does. In the case of FORTRAN, to be specific, one writes a program in the form of high-level English-like statements (that is, in FORTRAN), and then inputs those program statements into a computer running the FORTRAN compiler program. The compiler then outputs the so-called *machine language* version of the original FORTRAN program (the 1s and 0s that are the only things the electronics of a computer understands).

Our supposed program doesn't do anything as exotic as a compiler does, however. All it does—or at least we *imagine* it does—is determine if the input program **P** will, for a given input to that program, eventually halt when run on a computer. A lot of programs don't halt, of course, even if their creators intended them to, because they get trapped in what frustrated programmers call an "infinite loop." (Some of the programs given earlier in this book, in fact, did just that before I figured out what was wrong, and they finally ran correctly!) Figure 9.2.1 shows what our imagined program—I'll call it **H**—would do when presented with the input program **P** and the input **I** for **P**. What we'll now prove is that it is *impossible* to write the program **H**. **H** is like a unicorn—we can *imagine* it, but nevertheless it simply doesn't exist.

What we'll do is demonstrate that there exists a specific program **P**, with a specific input **I**, for which **H** *must* make the *wrong* decision! The proof of this is beautifully simple, indeed, astonishingly simple. In his 1967 book *Computation: Finite and Infinite Machines*, MIT professor of electrical engineering Marvin Minsky writes of this demonstration, "The result of the simple but delicate argument [in the proof of the unsolvability of the halting problem] is perhaps the most important conclusion in this book." The proof takes the form of three steps.

Step 1: As already mentioned, any program **P**, in machine language form, is nothing but a string of 1s and 0s, and as such, we can consider that string of binary symbols as indistinguishable in nature (if not in detail) from any other string of 1s and 0s. So, we are perfectly free to let **P** use *itself* as its input data **I**. Professor Minsky writes of this perhaps curious choice for **I**, "We need not concern ourselves with the question of why anyone would be interested in such introverted calculations; still, there is nothing absurd about the notion of a man contemplating [with his brain] a description of his own brain." Actually, one reason for using **P** as its own data is that this *automatically* specifies the **I** we are going to use for any given **P**.

Step 2: Once we have **P** as the program input to **H**, along with **P** itself as **P**'s input **I**, we'll have the situation shown in Figure 9.2.2, which also

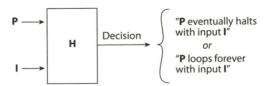

FIGURE 9.2.1. **H** decides if **P** eventually halts, or if **P** loops forever

includes one additional enhancement. With **P** as the "double" input to **H**, **H** will decide whether **P** (with input **P**) either eventually halts or loops forever. The enhancement is a bit of additional code that, after the halt/loop decision has been made, forces the final operation to be a halt if the decision was "loop" or to be an infinite loop if the decision was "halt."[4] As shown in Figure 9.2.2, the resulting enhanced program (**H**, plus the extra code) is a new program called **X** (everything inside the dashed box).

Step 3: This last step is, as Professor Minsky says, the "killer." Let **P** = **X**, that is, apply **X** to itself! Then, Figure 9.2.2 becomes Figure 9.2.3. We see that if **H** decides that **X** halts, then **X** loops, and if **H** decides that **X** loops, then **X** halts. Either way, **H** makes the *wrong* decision! Is your brain melting yet?

Professor Minsky writes of all this, "We have only the deepest sympathy for those readers who have not encountered this type of simple yet mind-boggling argument before. It resembles the argument in 'Russell's paradox' which forces us to discard arguments concerning the 'class of all classes'—a notion whose absurdity is not otherwise particularly evident." The reference is to the English mathematician Bertrand Russell (1872–1970) and his famous puzzle of the village barber who shaves every man in the village who doesn't shave himself. The puzzle occurs when we ask, "Who shaves the barber?" Russell's paradox dates to 1902, and so Turing was clearly aware of it long before 1936.

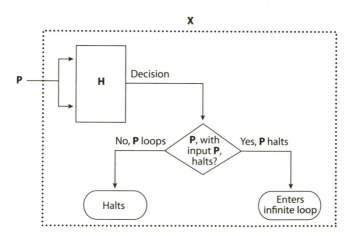

FIGURE 9.2.2. Program **X**

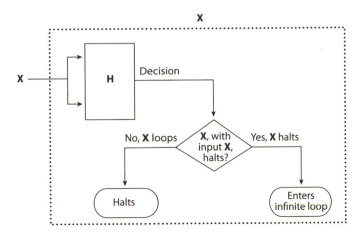

FIGURE 9.2.3. The halting paradox

In fact, Russell himself was not the first to show how a self-referencing condition can carry the seeds of its own destruction in the form of self-contradiction. The earliest example is probably the ancient liar's paradox that opens this chapter. Ask yourself, after reading "This statement is a lie," if it is true or not. It must be one or the other, right? If it is a true statement (that is, it's a lie), then it's *not* true. If it is a lie (that is, it's a true statement), then it's *not* a lie.

Is your brain melting yet?

9.3 Notes and References

1. You can find more on Turing's fascination with the Riemann hypothesis in Andrew Hodges, *Alan Turing: The Enigma* (New York: Simon and Schuster, 1983).

2. In his biography of Turing (*The Man Who Knew Too Much: Alan Turing and the Invention of the Computer* [New York: W. W. Norton, 2006]), David Leavitt writes (p. 130), "So far ... no one [has] been able to prove its [the Riemann hypothesis] unprovability." Well, as you now know, there's a good reason for that!

3. The presentation I am about to show you is not in Turing's 1936 paper, but actually dates from 1952 and is due to the American mathematician Martin Davis (born 1928). It is all in the *spirit* of Turing's paper, however.

4. Here's how that extra code could work. After completing the first part of step 2, we have either a decision that **P** (with input **P**) eventually halts, or that **P** (with input **P**) loops forever. So, suppose we have the variable *decision* equal to 0 if the decision is "halt," and equal to 1 if the decision is "loop." The extra code could then be, in MATLAB:

```
if decision==0
    a=1;
    while a>0
    end
end
```

This code generates Figure 3.8.3 (see note 27 in Chapter 3)

```
fpu.m
N=32;alpha=0.25;TMAX=10000;DT=20;tspan=[0:DT:TMAX];
options=odeset('Reltol', 1e-4, 'OutputFcn', 'odeplot',
    'OutputSel', [1,2,N]);
for l=1:N
    a=1;b(l)=a*sin(pi*l/(N+1));b(1+N)=0;
    omegak2(l)=4*(sin(pi*l/2/N))^2;
end
[T,Y]=ode45('fput1',tspan,b,options,N);
for IT=1:(TMAX/DT)
    TIME(IT)=IT*DT*sqrt(omegak2(1))/2/pi;
    YX(IT,1:N+1)=[0 Y(IT,1:N)]; YV(IT,1:N+1)=[0 Y(IT,N+1:2*N)]
    sXF(IT,:)=imag(fft([YX(IT,1:N+1) 0-YX(IT,N+1:-1:2)]))/
        sqrt(2*(N+1));
    sVF(IT,:)=imag(fft([YV(IT,1:N+1) 0-YV(IT,N+1:-1:2)]))/
        sqrt(2*(N+1));
    Energ(IT,1:N)=(omegak2(1:N).*(sXF(IT,2:N+1).^2)+sVF
      (IT,2:N+1).^2)/2;
    for J=2:N-1
        DifY(IT,J)=Y(IT,J+1)-Y(IT,J);
    end
end
plot(TIME,Energ(:,1),'k', TIME,Energ(:,2),'k',TIME,Energ(:,3),
    'k', TIME, Energ(:,4),'k');
ylabel('energy','FontSize',16)
```

(Continued)

(Continued)

```
xlabel('time, in units of lowest frequency mode period',
    'FontSize',16)
title('Figure 3.8.3-The 'classic' FPU experiment',
    'FontSize',16)
text(21,.055,'1','FontSize',16)
text(10,.014,'2','FontSize',16)
text(58,.055,'3','FontSize',16)
text(35,.035,'4','FontSize',16)
```

SOLUTIONS TO THE CHALLENGE PROBLEMS

Here's what happened just seconds after the frontispiece photos were taken. Notice that Mr. Norris seems remarkably unconcerned. John got the worst of our little experiment (when I was writing this book, John wrote to tell me that he still wonders at how he avoided getting his eyebrows burned off by the sudden, unexpected mini-explosion), but he *did* survive! And so will you, because even if some of the challenge problems have given you a hard time and have blown up in *your* face, here are all the answers.

Introduction

The Buckled Railroad Track Problem. This problem has been around for a long time, generally posed as I have stated it. That is, one only has to provide a *rough* answer (*inches* versus *feet* versus *yards*). A significantly more sophisticated version asks for the value of the maximum buckle accurate, say, to five decimal places. You can find an analysis to that more difficult question in the terrific book by Forman S. Acton (see note 7 in Chapter 3 again), *Real Computing Made Real: Preventing Errors in Scientific and Engineering Calculations* (N.J.: Princeton University Press, Princeton, 1996, pp. 222–223). (His answer is the astonishing 44.49845 feet!) To answer the question posed here, however, requires only the calculation of $\sqrt{\left(2,640 + \frac{1}{2}\right)^2 - (2,640)^2}$, which is the vertical side—the *maximum* buckle—of a right triangle whose hypotenuse is 2, 640.5 feet (one-half mile of track, plus one-half of the one-foot expansion) and whose horizontal side is the original unexpanded half mile of track. This assumes that the shape of the buckled track is straight, when of course it is really curved, but as I've posed it we won't make much error with the straight assumption. One can calculate this maximum buckle using either a hand calculator or, if you're up to it, some grubby arithmetic. But here's an easier approach:

$$\sqrt{\left(2,640 + \frac{1}{2}\right)^2 - (2,640)^2}$$

$$= \sqrt{(2,640)^2 + 2,640 + \frac{1}{4} - (2,640)^2} = \sqrt{2,640 + \frac{1}{4}},$$

which is, by inspection, greater than fifty feet. This is about 10% larger than Acton's result because he took into account the curvature of the buckled track. In any case, the answer is that the maximum buckle is measured in several *yards*.

CP I.1. What makes matters puzzling is that a formal, symbolic derivation of Feynman's identity doesn't appear to identify any problem.

That is, if we change variable to $u = ax + b(1 - x)$, then $\dfrac{du}{dx} = a - b$ and so $dx = du/(a - b)$. Then,

$$\int_0^1 \frac{dx}{[ax + b(1 - x)]^2} = \int_b^a \frac{du/(a - b)}{u^2} = \frac{1}{(a - b)} \int_b^a \frac{du}{u^2}$$

$$= \frac{1}{(a - b)} \left\{ -\frac{1}{u} \right\}\Big|_b^a$$

$$= \frac{1}{(a - b)} \cdot \frac{a - b}{ab} = \frac{1}{ab},$$

just as Feynman claimed. The source of the puzzle is with the integrand (which is, indeed, never negative). The problem is that if a and b are not picked correctly, the integrand blows up to infinity at some point *inside* the integration interval of 0 to 1. That is, for some $x = \widehat{x}$ where $0 \leq \widehat{x} \leq 1$, we'll have $a\widehat{x} + b(1 - \widehat{x}) = 0$. Integrating across an infinity is the source of the puzzle. With that as a start, you should now be able to easily show that to ensure $a\widehat{x} + b(1 - \widehat{x}) = 0$ occurs only *outside* the integration interval requires that a and b have the same sign.

CP I.2. For us, this is a purely mathematical question. It was first encountered as a *physics* problem, however, by Einstein in his famous 1905 paper on special relativity, and you can read more about that in the two papers that motivated the problem's appearance in this book. Both are by Aleksander Gjurchinovski, "Reflection of Light from a Uniformly Moving Mirror" and "Einstein's Mirror and Fermat's Principle of Least Time," which appeared in the *American Journal of Physics* (October 2004, pp. 1316–1324 and 1325–1327, respectively). A later paper, "Einstein's Mirror," co-authored with Aleksander Skeparovski, appeared in *The Physics Teacher* (October 2008, pp. 416–418). To solve for β, recall the identity $\sin(\alpha + \beta) = \sin(\alpha)\cos(\beta) + \cos(\alpha)\sin(\beta)$. Then, the starting equation becomes

$$\sin(\alpha) - \sin(\beta) = -k\sin(\alpha)\cos(\beta) - k\cos(\alpha)\sin(\beta),$$

or, with some rearranging,

$$\sin(\alpha)[1 + k\cos(\beta)] = \sin(\beta)[1 - k\cos(\alpha)].$$

Squaring both sides and using the identity $\sin^2(\theta) + \cos^2(\theta) = 1$ gives

$$[1 - \cos^2(\alpha)][1 + k\cos(\beta)]^2 = [1 - \cos^2(\beta)][1 - k\cos(\alpha)]^2.$$

If you expand all of this out, and are careful in canceling and combining terms, you'll arrive at the following quadratic equation for $\cos(\beta)$:

$$\cos^2(\beta)[1 - 2k\cos(\alpha) + k^2] + \cos(\beta)2k[1 - \cos^2(\alpha)]$$
$$+ 2k\cos(\alpha) - (1 + k^2)\cos^2(\alpha) = 0.$$

Using the quadratic formula,

$$\cos(\beta) = \frac{-k[1 - \cos^2(\alpha)] \pm \sqrt{\begin{array}{l}k^2[1 - \cos^2(\alpha)]^2 - [1 - 2k\cos(\alpha) \\ + k^2][2k\cos(\alpha) - (1 + k^2)\cos^2(\alpha)]\end{array}}}{1 - 2k\cos(\alpha) + k^2}.$$

Since $\cos(\beta) = \cos(\alpha)$ when $k = 0$, we know that we should use the $+$ sign. That simplifies things a little bit. Still, this last result appears to remain an extremely ugly mess—but looks are deceiving. With just a bit of patience and some more very careful algebra, it all reduces to

$$\cos(\beta) = \frac{-k[1 + \cos^2(\alpha)] + \sqrt{\{-k[1 + \cos^2(\alpha)] + (1 + k^2)\cos(\alpha)\}^2}}{1 - 2k\cos(\alpha) + k^2}$$
$$= \frac{-2k + (1 + k^2)\cos(\alpha)}{1 - 2k\cos(\alpha) + k^2},$$

or, finally,

$$\beta = \cos^{-1}\left\{\frac{-2k + (1 + k^2)\cos(\alpha)}{1 - 2k\cos(\alpha) + k^2}\right\}.$$

Just to tantalize you a bit, the physical significance of this result, a discovery due to Einstein, is that when a light beam bounces off a mirror moving either away or towards a light source then the angle of reflection (β) is *not* equal to the angle of incidence (α). It's an effect that is important only at relativistic speeds, however, as $k = \frac{v}{c}$ where v is the mirror's speed and c is the speed of light.

CP I.3. After I wrote this problem up, I thought, I'll let my Math 53 professor himself explain where I had gone wrong! I could do that because I know where he is today, a half century after both of us left Stanford. By some amazing coincidence, John Lamperti is now professor emeritus of mathematics at Dartmouth College, a mere hundred miles north of the University of New Hampshire where I am emeritus. He is an expert in probability theory, a subject I particularly enjoy, and I had read his book *Probability: A Survey of the Mathematical Theory*, 2nd ed. (New York: Wiley, 1996). So I knew where the years had taken him and that's how I came to write to John, to tell him of this book and to remind him of his long-ago Math 53 homework problem. Here's his gracious answer, exactly as he wrote back to me on September 26, 2008:

> First of all, that big "NO!" and a score of 1 out of 10 seem unreasonably negative; young Paul had a good idea about the problem. The reason his solution isn't right, or at least isn't complete, is that the formula he's using for the length of a curve depends on the curve being defined as $y = f(x)$. When you rotate the axes in Paul's solution, representing the curve as a function of x may no longer work. [To understand what John is getting at here, draw a "typical" single-valued curve in some x,y-coordinate system, and then rotate the axes through some angle. In the new coordinate system, the curve may no longer be single-valued.] Here's one way around the difficulty, which *may* be what I had in mind when I assigned the problem almost 50 years ago! To study geometric properties of a curve such as its length, the $y = f(x)$ representation is not a very natural one even when it is possible. It's much better to use a parametric form such as $x = x(t), y = y(t)$ where (say) $0 \leq t \leq 1$. This is neater and also much more general; for example, it allows the curve to have loops. The formula for the length

now looks like this: $L = \int_0^1 \sqrt{\dot{x}^2 + \dot{y}^2}\, dt$ (where $\dot{x} = \dfrac{dx}{dt}$, etc), and it's not hard to check that this quantity really is invariant if the axes are rotated. (That really should be proved, since this is mathematics and not physics.) Now young Paul's argument really works! Suppose we use axes rotated so that $y(0) = y(1)$.

Then $L = \int_0^1 \sqrt{\dot{x}^2 + \dot{y}^2}\, dt \geq \int_0^1 \sqrt{\dot{x}^2}\, dt = \int_0^1 |\dot{x}|\, dt \geq \int_0^1 \dot{x}\, dt = x(1) -$ $x(0)$, and of course that's the straight-line distance from start to finish. Perhaps we were discussing parametric equations at the time I assigned that problem. [After I read John's letter, I have become convinced that we were, indeed, talking about parametric representation.] If so, there would have been some justification for that "NO" on Paul's homework—with a score of, say, 7 out of 10 for a good but not quite right try.

And to *that*, an increase in my long-ago score, I can only say, "Thank you, Professor Lamperti!"

CP 1.4. Each of the 10^7 pixels in a picture can independently have one of 2^{11} gray levels. So, the number of pictures in TUPA is $(2^{11})^{10^7} = (10^{3.3113299})^{10^7} = 10^{33,113,299}$. That's a one followed by more than thirty-three million zeros! The number of seconds in fifteen billion years is $(15 \cdot 10^9)(365)(24)(60)(60) = 4.73 \cdot 10^{17}$. Generating one hundred trillion pictures per nanosecond is a rate of $(100 \cdot 10^{12})/10^{-9} = 10^{23}$ pictures/s. So, in fifteen billion years a total of $4.73 \cdot 10^{40}$ pictures would be generated. This is an infinitesimally tiny, tiny, *tiny* fraction of the total number of pictures in TUPA. And even if we increased the picture generation rate by a factor of a trillion trillion trillion, that conclusion would be unchanged. TUPA is *enormous*.

CP 1.1. The probability of finding a needle is equal to

$$\frac{\text{Volume of a needle in cubic inches}}{\text{Volume of the Earth in cubic inches}}$$

$$= \frac{\pi(0.01/2)^2 2}{\dfrac{4}{3}\pi(3,960 \cdot 5,280 \cdot 12)^3} = 2.37 \cdot 10^{-30}.$$

The probability of picking an integer somewhere in the interval of length 10^{180} is $\dfrac{10^{180}}{6.69 \cdot 10^{370} - 6.62 \cdot 10^{370}} = 1.43 \cdot 10^{-189}$. This probability is *fantastically smaller* than the first probability.

CP 1.2. Following the hint, write the equation $\cos(\mu) = \cos(2\mu) + \cos(4\mu)$. From the identity $\cos(2\mu) = 2\cos^2(\mu) - 1$ it follows, from applying the identity to itself, that $\cos(4\mu) = 2\cos^2(2\mu) - 1 = 2[2\cos^2(\mu) - 1]^2 - 1 = 2[4\cos^4(\mu) - 4\cos^2(\mu) + 1] - 1$, or $\cos(4\mu) = 8\cos^4(\mu) - 8\cos^2(\mu) + 1$. Inserting these expression for $\cos(2\mu)$ and $\cos(4\mu)$ into our starting equation, we have $\cos(\mu) = [2\cos^2(\mu) - 1] + [8\cos^4(\mu) - 8\cos^2(\mu) + 1]$, or $8\cos^4(\mu) - 6\cos^2(\mu) - \cos(\mu) = 0$, or $[8\cos^3(\mu) - 6\cos(\mu) - 1]\cos(\mu) = 0$. For this to be so, either $\cos(\mu) = 0$ or $8\cos^3(\mu) - 6\cos(\mu) - 1 = 0$. The first possibility gives $\mu = 90°$, but that doesn't give us Olmstead's equality. So, writing the second possibility as $[2\cos(\mu)]^3 - 3[2\cos(\mu)] - 1 = 0$ and then defining $x = 2\cos(\mu)$, we have $x^3 - 3x - 1 = 0$. This is a cubic of the form $x^3 = px + q$, with $p = 3$ and $q = 1$. As shown in my book, *An Imaginary Tale: The Story of $\sqrt{-1}$*, pp. 22–24, this cubic has a *real* solution (and we'd expect $\cos(\mu)$ to be real, of course) if $3\sqrt{3q} \leq 2p^{3/2}$. For our cubic, this becomes the requirement $3\sqrt{3} \leq 2(3)^{3/2} = 2 \cdot 3\sqrt{3} = 6\sqrt{3}$, which is certainly true. In addition, it is shown in *An Imaginary Tale* that this real root is given by $x = 2\sqrt{\dfrac{p}{3}}\cos\left\{\dfrac{1}{3}\cos^{-1}\left(\dfrac{3q\sqrt{3}}{2p\sqrt{p}}\right)\right\}$ and

so $x = 2\sqrt{\dfrac{3}{3}}\cos\left\{\dfrac{1}{3}\cos^{-1}\left(\dfrac{3\sqrt{3}}{6\sqrt{3}}\right)\right\} = 2\cos\left\{\dfrac{1}{3}\cos^{-1}\left(\dfrac{1}{2}\right)\right\}$. Since $\cos^{-1}\left(\dfrac{1}{2}\right) = 60°$, then $x = 2\cos(20°)$. But since $x = 2\cos(\mu)$, then $\mu = 20°$, and so $\cos(20°) = \cos(40°) + \cos(80°)$, which is Olmstead's equality, and we are done with the first part of the problem. Since we have $\mu = 20°$, then $\cos(80°) = \dfrac{e^{i4\mu} + e^{-i4\mu}}{2}$, $\cos(40°) = \dfrac{e^{i2\mu} + e^{-i2\mu}}{2}$, and $\cos(20°) = \dfrac{e^{i\mu} + e^{-i\mu}}{2}$. Thus, $\cos(20°)\cos(40°)\cos(80°) = \dfrac{\left(e^{i4\mu} + e^{-i4\mu}\right)\left(e^{i2\mu} + e^{-i2\mu}\right)\left(e^{i\mu} + e^{-i\mu}\right)}{8}$, which, if you carefully multiply out, gives $\cos(20°)\cos(40°)\cos(80°) =$

$$\frac{e^{i7\mu}+e^{i5\mu}+e^{i3\mu}+e^{i\mu}+e^{-i\mu}+e^{-i3\mu}+e^{-i5\mu}+e^{-i7\mu}}{8}=\frac{1}{4}\left[\frac{e^{i7\mu}+e^{-i7\mu}}{2}+\right.$$

$$\left.\frac{e^{i5\mu}+e^{-i5\mu}}{2}+\frac{e^{i3\mu}+e^{-i3\mu}}{2}+\frac{e^{i\mu}+e^{-i\mu}}{2}\right]=\frac{1}{4}[\cos(140°)+\cos(100°)+$$

$\cos(60°)+\cos(20°)]$ or, as $\cos(140°)=-\cos(40°)$ and $\cos(100°)=$ $-\cos(80°)$, we have $\cos(20°)\cos(40°)\cos(80°)=\frac{1}{4}\big[-\cos(40°)$ $-\cos(80°)+\cos(60°)+\cos(20°)\big]=\frac{1}{4}\cos(60°)$, because of Olmstead's equality. Since $\cos(60°)=\frac{1}{2}$, then $\cos(20°)\cos(40°)\cos(80°)=$ $\frac{1}{4}\cdot\frac{1}{2}=\frac{1}{8}$, and we are done.

CP 1.3. Write the integral as $I=\int_0^{2\pi}\sqrt{a^2\sin^2(t)+b^2\cos^2(t)}\,dt=$ $b\int_0^{2\pi}\sqrt{1-\frac{b^2-a^2}{b^2}\sin^2(t)}\,dt$ using the identity $\sin^2(t)+\cos^2(t)=1$. Without any loss of generality, let's take $b\ge a$ (remember, this is a purely *analytical* derivation, and there is now no association of a and b with the major and minor semiaxes of an ellipse; a and b are now just literals) and so $\frac{b^2-a^2}{b^2}\le 1$. Because of the behavior of $\sin^2(t)$ over the interval 0 to 2π it is obvious that $\int_0^{2\pi}=$ $4\int_0^{\pi/2}$, and so $I=4b\int_0^{\pi/2}\sqrt{1-\frac{b^2-a^2}{b^2}\sin^2(t)}\,dt$. This integral looks precisely like the complete elliptic integral of the second kind, that is, like $\int_0^{\pi/2}\sqrt{1-k^2\sin^2(t)}\,dt$, where $k^2=\frac{b^2-a^2}{b^2}$. Now, using the power series expansion given in the problem statement, we have $I=4b\dfrac{\pi}{2\left(1+\dfrac{b-a}{b+a}\right)}\left\{1+\dfrac{n^2}{4}+\dfrac{n^4}{64}+\dfrac{n^6}{256}+\dfrac{25n^8}{16,384}+\cdots\right\}$,

where $n=\dfrac{1-\sqrt{1-\dfrac{b^2-a^2}{b^2}}}{1+\sqrt{1-\dfrac{b^2-a^2}{b^2}}}=\dfrac{b-a}{b+a}$. Or, after a little simplification,

$I = \pi(a+b)\left\{1 + \dfrac{n^2}{4} + \dfrac{n^4}{64} + \dfrac{n^6}{256} + \dfrac{25n^8}{16,384} + \cdots\right\}$. As all the terms in the power series are positive, then if we keep just the first two we can write $I \geq \pi(a+b)\left\{1 + \dfrac{n^2}{4}\right\}$, which if we square says $I^2 \geq \pi^2(a+b)^2\left\{1 + \dfrac{n^2}{2} + \dfrac{n^2}{16}\right\}$, and so, again keeping just the first two power terms, $I^2 \geq \pi^2(a+b)^2\left\{1 + \dfrac{n^2}{2}\right\}$. Inserting the definition of n, we have $I^2 \geq \pi^2(a+b)^2\left\{1 + \dfrac{(b-a)^2}{2(b+a)^2}\right\} =$

$\dfrac{\pi^2}{2}\left[2(a+b)^2 + (b-a)^2\right] = \dfrac{\pi^2}{2}\left[2(a^2+2ab+b^2) + (b-a)^2\right] = \dfrac{\pi^2}{2}$

$\left[2a^2 + 4ab + 2b^2 + (b-a)^2\right] = \dfrac{\pi^2}{2}\left[2a^2 - 4ab + 2b^2 + 8ab + (b-a)^2\right]$

$= \dfrac{\pi^2}{2}\left[2(a^2 - 2ab + b^2) + 8ab + (b-a)^2\right] = \dfrac{\pi^2}{2}\left[2(a-b)^2 + 8ab + \right.$

$(a-b)^2\Big] = \dfrac{\pi^2}{2}\left[8ab + 3(a-b)^2\right] = 4\pi^2\left[ab + \dfrac{3}{8}(a-b)^2\right] = 4\pi$

$\left[\pi ab + \dfrac{3}{8}\pi(a-b)^2\right]$, or, taking the square root, $I \geq$

$\sqrt{4\pi\left[\pi ab + \dfrac{3}{8}\pi(a-b)^2\right]}$. Since $\pi > 3$, we know $\dfrac{3}{8}\pi > \dfrac{9}{8} > 1$, and so we have an even stronger inequality if we write $I \geq \sqrt{4\pi\left[\pi ab + (a-b)^2\right]}$, and we are done. *Historical note:* The power series expansion crucial to this proof appears not to be commonly known. It is due to the German mathematician Ernst Edward Kimmer (1810–1893), who, building on work done by Gauss in 1813, derived it in 1837. See Carl E. Linderholm and Arthur C. Segal, "An Overlooked Series for the Elliptic Perimeter" (*Mathematics Magazine*, June 1995, pp. 216–220).

CP 2.1. Using the resistance/admittance trick on the first three non-trivial cases of $2, 3,$ and 4 stages of the generalized version of Figure 2.2.7, it is not hard to show that $R_2(\alpha) = 1 + \dfrac{1}{\frac{1}{\alpha} + \frac{1}{1+\alpha^2}}$,

$$R_3(\alpha) = 1 + \cfrac{1}{\dfrac{1}{\alpha} + \cfrac{1}{1 + \cfrac{1}{\frac{1}{\alpha^2} + \frac{1}{1+\alpha^3}}}},$$

and

$$R_4(\alpha) = 1 + \cfrac{1}{\cfrac{1}{\alpha} + \cfrac{1}{1 + \cfrac{1}{\cfrac{1}{\alpha^2} + \cfrac{1}{1 + \cfrac{1}{\cfrac{1}{\alpha^3} + \cfrac{1}{1 + \alpha^4}}}}}}.$$

The general result is clearly that the input resistance $R(\alpha)$ is given by the infinite continued fraction

$$R(\alpha) = 1 + \cfrac{1}{\cfrac{1}{\alpha} + \cfrac{1}{1 + \cfrac{1}{\cfrac{1}{\alpha^2} + \cfrac{1}{1 + \cfrac{1}{\cfrac{1}{\alpha^3} + \cfrac{1}{1 + \cfrac{1}{\cfrac{1}{\alpha^4} + \cdots}}}}}}}.$$

The following MATLAB code calculates $R_k(\alpha)$ for a given α, for any *even* value of $k \geq 2$. Limiting ourselves to even k makes the code easier to write than if we have to allow for the odd k case, too. Since all we care about is $R(\alpha) = \lim_{k \to \infty} R_k(\alpha)$, this is not a limitation:

```
cp21.m
digits(100)
a=2;
k=input('How many stages (even number)?');
for loop=1:k
     alpha(loop)=a^loop;
end
d=(1/alpha(k-1))+(1/(1+alpha(k)));
power=k-2;
while power>0
     d=1+(1/d);
     d=(1/alpha(power))+(1/d);
     power=power-1;
end
1+(1/d)
```

When run the code produces the values, correct to the first fourteen digits, of $R_a(0.5) = 1.35329541982623$ and $R_b(2) = 2.24248109286014$.

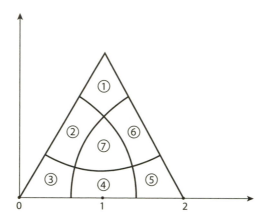

FIGURE S3.1. The seven sub-areas of the triangle

CP 3.1. In Figure S3.1 I've sketched an equilateral triangle with side length 2 showing the seven subareas into which it is divided by three circular arcs of radius $\sqrt{2}$ centered on the vertices of the triangle. In what follows the notation $\langle x \rangle$ means the area of subarea x, $1 \le x \le 7$. The answer to our problem, then, is $\dfrac{\langle 7 \rangle}{\text{Total area of triangle}}$.

Since the base of the triangle is 2 and its height is $\sqrt{3}$, we have the total area of the triangle as

$$\langle 1 \rangle + \langle 2 \rangle + \langle 3 \rangle + \langle 4 \rangle + \langle 5 \rangle + \langle 6 \rangle + \langle 7 \rangle = \frac{1}{2} \cdot 2 \cdot \sqrt{3} = \sqrt{3}. \tag{a}$$

By symmetry we can write $\langle 1 \rangle + \langle 2 \rangle + \langle 6 \rangle + \langle 7 \rangle = \langle 2 \rangle + \langle 3 \rangle + \langle 4 \rangle + \langle 7 \rangle = \langle 4 \rangle + \langle 5 \rangle + \langle 6 \rangle + \langle 7 \rangle$ as the areas of $60°$ circular sectors with radii $\sqrt{2}$, which is one-sixth the area of a circle with radius $\sqrt{2}$, that is, $\dfrac{\pi \left(\sqrt{2}\right)^2}{6} = \dfrac{\pi}{3}$, and so

$$\langle 1 \rangle = \frac{\pi}{3} - \langle 2 \rangle - \langle 6 \rangle - \langle 7 \rangle, \ \langle 3 \rangle = \frac{\pi}{3} - \langle 2 \rangle - \langle 4 \rangle - \langle 7 \rangle, \ \langle 5 \rangle$$
$$= \frac{\pi}{3} - \langle 4 \rangle - \langle 6 \rangle - \langle 7 \rangle.$$

Adding these three results, we get $\langle 1 \rangle + \langle 3 \rangle + \langle 5 \rangle = \pi - 2\langle 2 \rangle - 2\langle 4 \rangle - 2\langle 6 \rangle - 3\langle 7 \rangle$, or

$$\langle 1 \rangle + \langle 3 \rangle + \langle 5 \rangle + 2\left[\langle 2 \rangle + \langle 4 \rangle + \langle 6 \rangle\right] + 3\langle 7 \rangle = \pi. \tag{b}$$

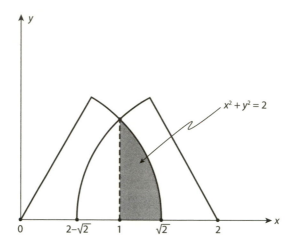

FIGURE S3.2. Forming $\langle 4 \rangle + \langle 7 \rangle$

Now, again by symmetry, $\langle 2 \rangle + \langle 7 \rangle = \langle 4 \rangle + \langle 7 \rangle = \langle 6 \rangle + \langle 7 \rangle$. In particular, Figure S3.2 shows how $\langle 4 \rangle + \langle 7 \rangle$ is generated as the common area in the intersection of the two circular arcs centered on the bottom two vertices of the triangle. By symmetry, these two arcs intersect at a point somewhere on the line normal to the x-axis, midway between $x = 0$ and $x = 2$, that is, at a point directly above $x = 1$. The equation of the circular arc centered on the vertex at the origin is $x^2 + y^2 = 2$ (remember, the radius of each arc is $\sqrt{2}$) and, by symmetry, the area $\langle 4 \rangle + \langle 7 \rangle$ is twice the shaded area in Figure S3.2, that is, twice the area between the circular arc and the x-axis, from $x = 1$ to $x = \sqrt{2}$. So, from freshman calculus we have $\langle 4 \rangle + \langle 7 \rangle = 2 \int\limits_{1}^{\sqrt{2}} y \, dx = 2 \int\limits_{1}^{\sqrt{2}} \sqrt{2 - x^2} \, dx =$

$$2 \left[\frac{x\sqrt{2 - x^2}}{2} + \sin^{-1}\left(\frac{x}{\sqrt{2}} \right) \right]_{1}^{\sqrt{2}} = 2 \left[\sin^{-1}(1) - \frac{1}{2} - \sin^{-1}\left(\frac{1}{\sqrt{2}} \right) \right] =$$

$$2 \left[\frac{\pi}{2} - \frac{1}{2} - \frac{\pi}{4} \right] = 2 \left[\frac{\pi}{4} - \frac{1}{2} \right], \text{ or}$$

$$\langle 4 \rangle + \langle 7 \rangle = \frac{\pi}{2} - 1, \ \langle 2 \rangle + \langle 7 \rangle = \frac{\pi}{2} - 1, \ \langle 6 \rangle + \langle 7 \rangle = \frac{\pi}{2} - 1.$$

(c)

From (c) we can write

$$\langle 2\rangle = \frac{\pi}{2} - 1 - \langle 7\rangle, \quad \langle 4\rangle = \frac{\pi}{2} - 1 - \langle 7\rangle, \quad \langle 6\rangle = \frac{\pi}{2} - 1 - \langle 7\rangle$$

which, when added, results in $\langle 2\rangle + \langle 4\rangle + \langle 6\rangle = \dfrac{3\pi}{2} - 3 - 3\langle 7\rangle$ or,

$$\langle 2\rangle + \langle 4\rangle + \langle 6\rangle + 3\langle 7\rangle = \frac{3\pi}{2} - 3. \tag{d}$$

Subtracting (a) from (b) gives

$$\langle 1\rangle + \langle 3\rangle + \langle 5\rangle + 2\left[\langle 2\rangle + \langle 4\rangle + \langle 6\rangle\right] + 3\langle 7\rangle$$

$$- \left[\langle 1\rangle + \langle 2\rangle + \langle 3\rangle + \langle 4\rangle + \langle 5\rangle + \langle 6\rangle + \langle 7\rangle\right] = \pi - \sqrt{3},$$

or

$$\langle 2\rangle + \langle 4\rangle + \langle 6\rangle + 2\langle 7\rangle = \pi - \sqrt{3}. \tag{e}$$

Finally, subtracting (e) from (d) gives

$$\langle 7\rangle = \frac{3\pi}{2} - 3 - \left[\pi - \sqrt{3}\right] = \frac{\pi}{2} - 3 + \sqrt{3}.$$

So, the probability a point randomly selected from the interior of the triangle is from subarea 7 is $\dfrac{\langle 7\rangle}{\text{Total area of triangle}} = \dfrac{\frac{\pi}{2} - 3 + \sqrt{3}}{\sqrt{3}} = \dfrac{\pi}{2\sqrt{3}} + 1 - \sqrt{3}$, which is (3.2.1), and we are done.

CP 3.2. Write $u(x, t) = X(x)Y(y)$, and Laplace's equation separates to give $\dfrac{1}{X}\dfrac{d^2X}{dx^2} = -\beta^2$ and $-\dfrac{1}{Y}\dfrac{d^2Y}{dy^2} = -\beta^2$, where β is a positive constant. Thus, $\dfrac{d^2X}{dx^2} + \beta^2X = 0$ and $\dfrac{d^2Y}{dy^2} - \beta^2Y = 0$. For $Y(y)$: try $Y(y) = Ae^{cy}$, where c is a constant, which gives $c^2Ae^{cy} - \beta^2Ae^{cy} = 0$, or $c = \pm\beta$. So, $Y(y) = A_1e^{-\beta y} + A_2e^{\beta y}$. Since $\lim_{y\to\infty} u(x, y) = \lim_{y\to\infty} X(x)Y(y)$, then $\lim_{y\to\infty}\left(A_1e^{-\beta y} + A_2e^{\beta y}\right) = 0$, which says that $A_2 = 0$. Thus, $Y(y) = Ae^{-\beta y}$. Also, $u(0, y) = 0$ says that $X(0)Y(y) = 0$, and $u(\pi, y) = 0$ says that $X(\pi)Y(y) = 0$, and these two results

tell us that $X(0) = 0$, $X(\pi) = 0$. For $X(x)$: try $X(x) = Be^{cx}$, where again c is some constant, which gives $c^2 Be^{cx} + \beta^2 Be^{cx} = 0$, or $c = \pm i\beta$. So, $X(x) = B_1 e^{i\beta x} + B_2 e^{-i\beta x}$. Since $X(x) = 0$, we have $0 = B_1 + B_2$ or $B_2 = -B_1$, and therefore $X(x) = Be^{i\beta x} - Be^{-i\beta x} = 2iB \sin(\beta x)$. Since $X(\pi) = 0$, we have $\sin(\beta \pi) = 0$, and so, with n an integer, $\beta \pi = n\pi$, or $\beta = n$. In summary, $X_n(x) = B_n \sin(nx)$ and $Y_n(y) = A_n e^{-ny}$, and so $u_n(x, y) = X_n(x) Y_n(y) = A_n B_n e^{-ny} \sin(nx)$, $n = 1, 2, \ldots$. The most general solution is $u(x, y) = \sum_{n=1}^{\infty} u_n(x, y)$, or $u(x, y) = \sum_{n=1}^{\infty} F_n e^{-ny} \sin(nx)$. (I have written $F_n = A_n B_n$.) Since $u(x, 0) = T$, then $T = \sum_{n=1}^{\infty} F_n \sin(nx)$, and so

$$T \int_0^{\pi} \sin(mx)\, dx = \sum_{n=1}^{\infty} F_n \int_0^{\pi} \sin(mx) \sin(nx)\, dx \quad .$$

Now, recall the orthogonality property of the sine functions, that is, $\int_0^{\pi} \sin(mx) \sin(nx)\, dx = \begin{cases} 0, & m \neq n \\ \frac{\pi}{2}, & m = n \end{cases}$. Then the left-hand side of the expression in the box is $T \int_0^{\pi} \sin(mx)\, dx = T \left[-\frac{\cos(mx)}{m} \right]_0^{\pi} = T \left[-\frac{\cos(m\pi) - 1}{m} \right] = \begin{cases} 0, & m \text{ even} \\ \frac{2T}{m}, & m \text{ odd} \end{cases}$. And for the right-hand side, it is zero if $m \neq n$ and is $\frac{F_m \pi}{2}$ if $m = n$. So, $\frac{F_m \pi}{2} = \frac{2T}{m}$ when m is odd, that is, $F_m = \frac{4T}{m\pi}$ when m is odd, and $F_m = 0$ when m is even. Thus, $u(x, y) = \frac{4T}{\pi} \left\{ \sum_{m=1,3,5,\ldots}^{\infty} \frac{e^{-my}}{m} \sin(mx) \right\} = \frac{4T}{\pi} \left\{ e^{-y} \sin(x) + \frac{1}{3} e^{-3y} \sin(3x) + \frac{1}{5} e^{-5y} \sin(5x) + \ldots \right\}$.

CP 3.3. Following the hint and putting $w = re^{i\theta}$ in the power series expansion for $\frac{1}{2} \ln\left(\frac{1 + re^{i\theta}}{1 - re^{i\theta}} \right)$, we have $\frac{1}{2} \ln\left(\frac{1 + re^{i\theta}}{1 - re^{i\theta}} \right) = re^{i\theta} +$

$\dfrac{r^3 e^{i3\theta}}{3} + \dfrac{r^5 e^{i5\theta}}{5} + \cdots$. On the other hand, if we write the com-

plex quantity $\dfrac{1 + re^{i\theta}}{1 - re^{i\theta}}$ with explicit real and imaginary parts, we

have $\dfrac{1 + re^{i\theta}}{1 - re^{i\theta}} = \dfrac{1 + re^{i\theta}}{1 - re^{i\theta}} \cdot \dfrac{1 - re^{-i\theta}}{1 - re^{-i\theta}} = \dfrac{1 + re^{i\theta} - re^{-i\theta} - r^2}{1 - re^{i\theta} - re^{-i\theta} + r^2} =$

$\dfrac{1 - r^2 + i2r\sin(\theta)}{1 + r^2 - 2r\cos(\theta)}$. If we now write this complex quantity in polar

coordinates, as $\rho e^{i\phi}$, then in particular $\phi = \tan^{-1}\left\{\dfrac{2r\sin(\theta)}{1 - r^2}\right\}$.

Draw a triangle if this isn't obvious. (As you'll see, we won't
need to know ρ in what follows.) What we have at this

point is $\dfrac{1}{2}\ln\left(\dfrac{1 + re^{i\theta}}{1 - re^{i\theta}}\right) = re^{i\theta} + \dfrac{r^3 e^{i3\theta}}{3} + \dfrac{r^5 e^{i5\theta}}{5} + \cdots = \dfrac{1}{2}\ln\left(\rho e^{i\phi}\right) =$

$\dfrac{1}{2}\left\{\ln\left(\rho\right) + \ln\left(e^{i\phi}\right)\right\} = \dfrac{1}{2}\ln\left(\rho\right) + i\dfrac{1}{2}\phi$. So, using Euler's identity,

$\dfrac{1}{2}\ln\left(\rho\right) + i\dfrac{1}{2}\phi = \left[r\cos(\theta) + ir\sin(\theta)\right] + \left[\dfrac{r^3}{3}\cos(3\theta) + i\dfrac{r^3}{3}\sin(3\theta)\right] +$

$\left[\dfrac{r^5}{5}\cos(5\theta) + i\dfrac{r^5}{5}\sin(5\theta)\right] + \cdots$ or, if we equate imaginary parts,

$\dfrac{1}{2}\phi = \dfrac{1}{2}\tan^{-1}\left\{\dfrac{2r\sin(\theta)}{1 - r^2}\right\} = r\sin(\theta) + \dfrac{r^3}{3}\sin(3\theta) + \dfrac{r^5}{5}\sin(5\theta) + \cdots$.

If we now write $\theta = x$ and $r = e^{-y}$, we have $e^{-y}\sin(x) +$

$\dfrac{1}{3}e^{-3y}\sin(3x) + \dfrac{1}{5}e^{-5y}\sin(5x) + \cdots = \dfrac{1}{2}\tan^{-1}\left\{\dfrac{2e^{-y}\sin(x)}{1 - e^{-2y}}\right\} = \dfrac{1}{2}$

$\tan^{-1}\left\{\dfrac{2\sin(x)}{e^{y} - e^{-y}}\right\} = \dfrac{1}{2}\tan^{-1}\left\{\dfrac{\sin(x)}{\sinh(y)}\right\}$. You'll recall that the solution

to the previous problem is $\dfrac{4T}{\pi}$ times the left-hand side of this last

result. So, the steady-state temperature in the infinite strip from the

last problem is, in closed form, $u(x, y) = \dfrac{4T}{\pi} \cdot \dfrac{1}{2}\tan^{-1}\left\{\dfrac{\sin(x)}{\sinh(y)}\right\} =$

$\dfrac{2T}{\pi}\tan^{-1}\left\{\dfrac{\sin(x)}{\sinh(y)}\right\}$. If we assume we are on the isotherm

$u(x, y) = T_0$, we have $\dfrac{T_0}{\dfrac{2T}{\pi}} = \dfrac{\pi T_0}{2T} = \tan^{-1}\left\{\dfrac{\sin(x)}{\sinh(y)}\right\}$, or express-

ing the isotherm curve as $y = y(x)$, $y = \sinh^{-1}\left\{\dfrac{\sin(x)}{\tan\left(\dfrac{\pi}{2} \cdot \dfrac{T_0}{T}\right)}\right\}$.

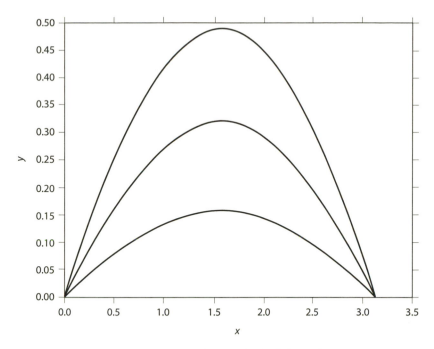

FIGURE S3.3. Isotherms for infinite strip

The MATLAB code **iso.m** does all the grubby arithmetic and plots the three requested isotherms in Figure S3.3; the upper, middle, and lower curves are for $\dfrac{T_0}{T} = 0.7$, 0.8, and 0.9, respectively.

---

iso.m

```
x=linspace(0,pi,1000);
f(1)=0.9;f(2)=0.8;f(3)=0.7;
num=sin(x);
for i=1:3
    den=tan(pi*f(i)/2);
    y(i,:)=asinh(num/den);
end
for i=1:3
    plot(x,y(i,:),'k-')
    hold on
end
```

(Continued)

(Continued)

xlabel('x','FontSize',16)

ylabel('y','FontSize',16)

title('Figure S3.3 - Isotherms for infinite strip',

    'FontSize', 16)

CP 3.4. When in mode 2 for $N = 2$, the total energy is $E_2 = \frac{1}{2}\left(\frac{dx_{12}}{dt}\right)^2 + \frac{1}{2}\left(\frac{dx_{22}}{dt}\right)^2 + \frac{1}{2}x_{12}^2 + \frac{1}{2}(x_{22} - x_{12})^2 + \frac{1}{2}x_{22}^2$. So, $\frac{dE_2}{dt} =$

$\frac{dx_{12}}{dt} \cdot \frac{d^2x_{12}}{dt^2} + \frac{dx_{22}}{dt} \cdot \frac{d^2x_{22}}{dt^2} + x_{12}\frac{dx_{12}}{dt} + (x_{22} - x_{12})\left(\frac{dx_{22}}{dt} - \frac{dx_{12}}{dt}\right) +$

$x_{22}\frac{dx_{22}}{dt} = \frac{dx_{12}}{dt}\left(\frac{d^2x_{12}}{dt^2} + x_{12}\right) + \frac{dx_{22}}{dt}\left(\frac{d^2x_{22}}{dt^2} + x_{22}\right) + x_{22}\frac{dx_{22}}{dt} -$

$x_{12}\frac{dx_{22}}{dt} - x_{22}\frac{dx_{12}}{dt} + x_{12}\frac{dx_{12}}{dt} = \frac{dx_{12}}{dt}\left(\frac{d^2x_{12}}{dt^2} + 2x_{12} - x_{22}\right) + \frac{dx_{22}}{dt}$

$\left(\frac{d^2x_{22}}{dt^2} + 2x_{22} - x_{12}\right)$. In mode 2, the two masses oscillate

out of phase, and so we have $x_{12} = -x_{22}$, $\dfrac{dx_{12}}{dt} = -\dfrac{dx_{22}}{dt}$, and

$\frac{d^2x_{12}}{dt^2} = -\frac{d^2x_{22}}{dt^2}$. Writing $\frac{dE_2}{dt}$ in terms of just mass 1, $\frac{dE_2}{dt} =$

$\frac{dx_{12}}{dt}\left(\frac{d^2x_{12}}{dt^2} + 3x_{12}\right) - \frac{dx_{12}}{dt}\left(-\frac{d^2x_{12}}{dt^2} - 3x_{12}\right) = 2\frac{dx_{12}}{dt}\left(\frac{d^2x_{12}}{dt^2} + 3x_{12}\right)$.

The mode 2 frequency is $\sqrt{3}$, and so $x_{12} = a_{11}\cos(\sqrt{3}t)$ and

$\frac{d^2x_{12}}{dt^2} = -3a_{11}\cos(\sqrt{3}t)$. Thus, $\frac{dE_2}{dt} = 2\frac{dx_{12}}{dt}\left(-3a_{11}\cos(\sqrt{3}t) + 3a_{11}\right.$

$\left.\cos(\sqrt{3}t)\right) = 0$, as was to be shown.

CP 4.1. Starting with $\dfrac{dx}{dt} = ax - by$ and $\dfrac{dy}{dt} = cx - ay$, differentiate the

differential equation for x to get $\dfrac{d^2x}{dt^2} = a\dfrac{dx}{dt} - b\dfrac{dy}{dt} = a\left(ax - by\right) -$

$b\left(cx - ay\right) = \left(a^2 - bc\right)x$, or $\dfrac{d^2x}{dt^2} + \alpha x = 0$, where $\alpha = bc - a^2 > 0$. If

this inequality is satisfied, *then* we have the harmonic differential

equation of (4.1.1), and x is sinusoidal. If, on the other hand, the

inequality is not satisfied, then the differential equation becomes

$\frac{d^2x}{dt^2} - \alpha x = 0$, where $\alpha = a^2 - bc > 0$, in which case the behavior of x is exponential. That is, $x(t) = C_1 e^{t\sqrt{\alpha}} + C_2 e^{-t\sqrt{\alpha}}$, where the values of C_1 and C_2 depend on the initial conditions. The very same argument leads to exactly the same conclusion for y.

CP 4.2. The apparent "closures" of the phase-plane portraits for the $N = 2$ and $N = 5$ hanging masses problems are illusions, caused by plotting on two-dimensional paper surfaces. When there is just one state variable, as in the problem discussed in Section 4.3, portrait closure in 2-space *does* mean periodicity because the "phase space" for that case is itself two-dimensional. But when there is more than one state variable, as in the hanging masses problems, then a true plot of the trajectory of a state variable in phase space would be a plot in a *high-dimensional* space (4-space for $N = 2$ and 10-space for $N = 5$). Plotting on a paper surface *collapses* the state-variable trajectory, making it appear to have crossings (closures) when they don't actually exist.

CP 5.1. The gravity force of attraction on m is $F = m\frac{d^2x}{dt^2} = -\frac{GMm}{x^2}$, or $\frac{d^2x}{dt^2} = -\frac{GM}{x^2}$. From the chain rule $\frac{d^2x}{dt^2} = \frac{dv}{dt} = \frac{dv}{dx} \cdot \frac{dx}{dt} = v\frac{dv}{dx}$, where v is the speed of m, then $v\frac{dv}{dx} = -\frac{GM}{x^2}$, or $v\,dv = -GM\frac{dx}{x^2}$. Integrating indefinitely gives $\frac{1}{2}v^2 + C = \frac{GM}{x}$, where C is a constant. To evaluate C, notice that $v = 0$ when $x = d$, and so $C = \frac{GM}{d}$, and thus $\frac{1}{2}v^2 + \frac{GM}{d} = \frac{GM}{x}$, or $\frac{1}{2}v^2 = GM\left(\frac{1}{x} - \frac{1}{d}\right)$ and so $v = \sqrt{2GM\frac{d-x}{xd}}$. Thus, $\frac{dx}{dt} = \sqrt{\frac{2GM}{d}}\sqrt{\frac{d-x}{x}}$, or $dt = \sqrt{\frac{d}{2GM}}\sqrt{\frac{x}{d-x}}\,dx$, and so we can write $\int_0^T dt = T = \sqrt{\frac{d}{2GM}}\int_d^0 \sqrt{\frac{x}{d-x}}\,dx$. Making the change of variable $u = d - x$ (and so $du = -dx$), we have $T = \sqrt{\frac{d}{2GM}}\int_0^d \sqrt{\frac{d-u}{u}}\,du = d\sqrt{\frac{d}{2GM}}\int_0^d \sqrt{\frac{1 - \left(\frac{u}{d}\right)}{\left(\frac{u}{d}\right)}}\,d\left(\frac{u}{d}\right)$, and so, with

$y = \dfrac{u}{d}$, $T = \sqrt{\dfrac{d^3}{2GM}} \displaystyle\int_0^1 \sqrt{\dfrac{1-y}{y}}\, dy$. Now, with $y = \cos^2(z)$ (and so $dy =$

$-2 \sin(z) \cos(z)\, dz$), we have $T = \sqrt{\dfrac{d^3}{2GM}} \displaystyle\int_{\pi/2}^{0} \sqrt{\dfrac{1 - \cos^2(z)}{\cos^2(z)}} (-2 \sin(z)$

$\cos(z) dz) = 2\sqrt{\dfrac{d^3}{2GM}} \displaystyle\int_0^{\pi/2} \dfrac{\sin(z)}{\cos(z)} \sin(z) \cos(z)\, dz = 2\sqrt{\dfrac{d^3}{2GM}} \displaystyle\int_0^{\pi/2} \sin^2(z)$

dz, and so, using a table of definite integrals, we have $T = 2\sqrt{\dfrac{d^3}{2GM}} \cdot$

$\dfrac{\pi}{4} = \dfrac{\pi}{2}\sqrt{\dfrac{d^3}{2GM}}$, or finally, $T = \pi\sqrt{\dfrac{d^3}{8GM}}$.

CP 5.2. Since $f(x)$ is non-negative, then somewhere in the interval $0 \le x \le d$ the function $f(x)$ will have a smallest value. Call this smallest value s, and then the minimum force on m will be $-GMs$. If we assume that this force acts on m over the entire interval of time 0 to T_m, where $t = T_m$ is when m collides with M, then T_m will certainly be greater than the actual time required for m to hit M. That is, for this minimum force we have $\dfrac{d^2x}{dt^2} = -GMs$, or $\dfrac{dx}{dt} = -GMst + C_1$. Since $\dfrac{dx}{dt} = 0$ when $t = 0$, we have $C_1 = 0$. Thus, $x(t) = -GMs\dfrac{t^2}{2} + C_2$. Since $x = d$ when $t = 0$, we have $C_2 = d$. So, $x(t) = d - GMs\dfrac{t^2}{2}$. Now, $x(T_m) = 0$ by definition, and so $0 = d - GMs\dfrac{T_m^2}{2}$, or $T_m = \sqrt{\dfrac{2d}{GMs}}$, which is finite, and certainly the actual time of collision, T, is less than T_m. For Newtonian gravity specifically we have the smallest value for $f(x)$ as $s = \dfrac{1}{d^2}$, which gives $T_m = \sqrt{\dfrac{2d^3}{GM}} = \sqrt{\dfrac{16d^3}{8GM}} = 4\sqrt{\dfrac{d^3}{8GM}}$, which is indeed greater than the exact value for $T = \pi\sqrt{\dfrac{d^3}{8GM}}$ calculated in the previous solution. But not greater by much; only as 4 exceeds π. That is, T_m is "only" 27.3% greater than the exact T, the time until collision, for Newtonian gravity.

CP 5.3. We have $\mathbf{a} \cdot \mathbf{b} = a_x b_x + a_y b_y + a_z b_z$ and so $(\mathbf{a} \cdot \mathbf{b})^2 = (a_x b_x + a_y b_y + a_z b_z)^2 = a_x^2 b_x^2 + a_y^2 b_y^2 + a_z^2 b_z^2 + 2a_x a_y b_x b_y + 2a_x a_z b_x b_z + 2a_y a_z b_y b_z$. Also,

$$\mathbf{a} \times \mathbf{b} = \begin{vmatrix} \mathbf{i} & \mathbf{j} & \mathbf{k} \\ a_x & a_y & a_z \\ b_x & b_y & b_z \end{vmatrix} = \mathbf{i}\,(a_y b_z - a_z b_y) - \mathbf{j}\,(a_x b_z - a_z b_x) + \mathbf{k}\,(a_x b_y - a_y b_x),$$

and so $|\mathbf{a} \times \mathbf{b}| = \sqrt{(a_y b_z - a_z b_y)^2 + (a_x b_z - a_z b_x)^2 + (a_x b_y - a_y b_x)^2}$, which says that $|\mathbf{a} \times \mathbf{b}|^2 = a_y^2 b_z^2 - 2a_y a_z b_y b_z + a_z^2 b_y^2 + a_x^2 b_z^2 - 2a_x a_z b_x b_z + a_z^2 b_x^2 + a_x^2 b_y^2 - 2a_x a_y b_x b_y + a_y^2 b_x^2$. Thus, $(\mathbf{a} \cdot \mathbf{b})^2 + |\mathbf{a} \times \mathbf{b}|^2 = \left[a_x^2 b_x^2 + a_y^2 b_y^2 + a_z^2 b_z^2 + 2a_x a_y b_x b_y + 2a_x a_z b_x b_z + 2a_y a_z b_y b_z \right] + \left[a_y^2 b_z^2 - 2a_y a_z b_y b_z + a_z^2 b_y^2 + a_x^2 b_z^2 - 2a_x a_z b_x b_z + a_z^2 b_x^2 + a_x^2 b_y^2 - 2a_x a_y b_x b_y + a_y^2 b_x^2 \right] = a_x^2 \left(b_x^2 + b_y^2 + b_z^2 \right) + a_y^2 \left(b_x^2 + b_y^2 + b_z^2 \right) + a_z^2 \left(b_x^2 + b_y^2 + b_z^2 \right) = \left(a_x^2 + a_y^2 + a_z^2 \right) \left(b_x^2 + b_y^2 + b_z^2 \right) = a^2 b^2$, as was to be shown. Now, let $\mathbf{a} = \mathbf{r}$ and $\mathbf{b} = \mathbf{v}$, the position and velocity vectors, respectively, of a mass moving under Newtonian gravity. Then, direct substitution into the above result gives $(\mathbf{r} \cdot \mathbf{v})^2 + |\mathbf{r} \times \mathbf{v}|^2 = r^2 v^2$. Since $\mathbf{v} = \dfrac{d\mathbf{r}}{dt}$, and since $\mathbf{c} = \mathbf{r} \times \mathbf{v}$ as derived in (5.2.7), we have $\left(\mathbf{r} \cdot \dfrac{d\mathbf{r}}{dt} \right)^2 + |\mathbf{c}|^2 = r^2 v^2$. We know from note 4 of Chapter 5 that $\mathbf{r} \cdot \dfrac{d\mathbf{r}}{dt} = r \dfrac{dr}{dt}$, and so $r^2 \left(\dfrac{dr}{dt} \right)^2 + c^2 = r^2 v^2$ or, $v^2 = \left(\dfrac{dr}{dt} \right)^2 + \dfrac{c^2}{r^2}$, as was to be shown.

CP 5.4. The total gravitational force on each of the Earth-like planets (each with mass m) is $\dfrac{GMm}{R^2} + \dfrac{Gm^2}{(2R)^2} = \dfrac{GMm}{R^2} \left(1 + \dfrac{m}{4M} \right)$, where M is the mass of the Sun-like star that the two planets orbit. In the units of the problem, $GM = 4\pi^2$ and $R = 1$, and so this force is $4\pi^2 m \left(1 + \dfrac{m}{4M} \right)$. It is this force that provides the centripetal acceleration force required for each planet's circular orbit, and so if v is the orbital speed of each planet, we have (remember, $R = 1$) $\dfrac{mv^2}{R} = mv^2 = 4\pi^2 m \left(1 + \dfrac{m}{4M} \right)$, or $v = 2\pi \sqrt{1 + \dfrac{m}{4M}}$. The orbital period for each of the two (duplicate) planets is $T_D = \frac{2\pi R}{v} = \frac{2\pi}{v}$, or $T_D = \frac{1}{\sqrt{1 + \frac{m}{4M}}} \approx \frac{1}{1 + \frac{m}{8M}} \approx 1 - \frac{m}{8M}$, since $m \ll$

M. The unit of time is one Earth-year, and so T_D is just a bit less than one Earth-year. Specifically, T_D is $\frac{m}{8M} = \frac{5.98 \cdot 10^{24} \text{ kg}}{8 \left(1.99 \cdot 10^{30} \text{ kg}\right)} = 0.375 \cdot 10^{-6}$ Earth years less than Earth's year. Since one Earth-year is $31.5 \cdot 10^6$ seconds, then a "year" on each of the two duplicate planets is just about 12 seconds 'shorter than an Earth-year. Not too much of a difference, but one easily measurable with modern atomic clocks.

CP 5.5. This is essentially a problem in the geometry of an equilateral triangle. It should be clear, by symmetry, that the net gravitational force on each of the three m's is always radially inward. The common center of mass of the three m's is located at the intersection of the three radii (each of length R) that bisect the vertex angles (each of $60°$, of course). For any pair of the three masses, this forms two $30°$–$60°$–$90°$ triangles, and from them it is easy to calculate that the two masses of a pair are distance $2 \left(\frac{1}{2}R\sqrt{3}\right) = R\sqrt{3}$ apart. The force each of the three masses directly exerts on each of its two companions is, therefore, given by $\frac{Gm^2}{\left(R\sqrt{3}\right)^2} = \frac{Gm^2}{3R^2}$. The component of that force that is inwardly-directed directly along an orbital radius is $\frac{Gm^2}{3R^2} \cos(30°) = \frac{Gm^2}{3R^2} \cdot \frac{\sqrt{3}}{2} = \frac{Gm^2}{2\sqrt{3}R^2}$. Each mass "feels" *twice* this force, since each mass is attracted by its *two* companions, and so the total inwardly directed radial force felt by each m (the force that produces the required centripetal acceleration for a cicular orbit) is $2\frac{Gm^2}{2\sqrt{3}R^2} = \frac{Gm^2}{\sqrt{3}R^2} = \frac{mv^2}{R}$, where v is the orbital speed of each m. Thus, $v^2 = \dfrac{Gm}{R\sqrt{3}}$, or $v = \sqrt{\dfrac{Gm}{R\sqrt{3}}}$. The orbital

period T is given by $vT = 2\pi R$, and so $T = 2\pi\dfrac{R}{v} = 2\pi R\sqrt{\dfrac{R\sqrt{3}}{Gm}}$, or

$$T = 2\pi\sqrt{\dfrac{R^3\sqrt{3}}{Gm}}.$$

CP 5.6. Multiplying through (5.5.24) by 2 and rearranging terms,
we have $2U = 4\pi^2\left(\dfrac{r_1^2}{l^3} + \dfrac{2}{r_1}\right) + 4\pi^2\alpha\left(\dfrac{r_2^2}{l^3} + \dfrac{2}{r_2}\right)$, which is of

the form $2U = 4\pi^2\left[A(r_1) + \alpha B(r_2)\right] = 4\pi^2\left[A(r_1) + \dfrac{1-\mu}{\mu}B(r_2)\right]$, $0 \le \mu \le 1$,

where $A(r_1) = \dfrac{r_1^2}{l^3} + \dfrac{2}{r_1}$ and $B(r_2) = \dfrac{r_2^2}{l^3} + \dfrac{2}{r_2}$. Clearly, $A(r_1) \ge 0$ and
$B(r_2) \ge 0$ because of the physical meaning of l, r_1, and r_2. From
the *claim* (see the last part of this solution) that $A + \dfrac{1-\mu}{\mu}B \ge$

$\dfrac{\min(A, B)}{\mu}$ if $A, B \ge 0$, we have $2U \ge \dfrac{4\pi^2}{\mu}\min\left(\dfrac{r_1^2}{l^3} + \dfrac{2}{r_1}, \dfrac{r_2^2}{l^3} + \dfrac{2}{r_2}\right)$.

Now, the minimum of $f(x) = \dfrac{x^2}{l^3} + \dfrac{2}{x}$ occurs at the solution

to $\dfrac{df}{dx} = 0$, that is, at the solution to $\dfrac{2x}{l^3} - \dfrac{2}{x^2} = 0$. So, $2x^3 =$
$2l^3$, or $f(x)$ is minimum when $x = l$. Since the two argu-
ments to the min function are of the same form as $f(x)$,

we have $\min\left(\dfrac{r_1^2}{l^3} + \dfrac{2}{r_1}, \dfrac{r_2^2}{l^3} + \dfrac{2}{r_2}\right) = \dfrac{1}{l} + \dfrac{2}{l} = \dfrac{3}{l}$. Thus, $2U \ge \dfrac{4\pi^2}{\mu}$.

$\dfrac{3}{l} = \dfrac{4\pi^2}{\mu}l^2 \cdot \dfrac{1}{l^2} \cdot \dfrac{3}{l} = 3l^2 \cdot \dfrac{4\pi^2}{\mu l^3} = 3l^2\omega^2 = 3(wl)^2$, and we are done.
That is, we are done once we show that the above claim
is true. Here's how to do that. It is certainly true, depend-
ing on the values of r_1 and r_2, that either $A > B$ or $A < B$.
Let's consider each possibility in turn (remember, $0 \le \mu \le 1$).
If $A > B$, then $A + \dfrac{1-\mu}{\mu}B = \dfrac{A\mu + (1-\mu)B}{\mu} > \dfrac{B\mu + (1-\mu)B}{\mu} =$

$\dfrac{B\mu + B - \mu B}{\mu} = \dfrac{B}{\mu} = \dfrac{\min(A, B)}{\mu}$. On the other hand, if $A <$

B, then $A + \dfrac{1-\mu}{\mu} B = \dfrac{A\mu + (1-\mu)B}{\mu} > \dfrac{A\mu + (1-\mu)A}{\mu} = \dfrac{A\mu + A - \mu A}{\mu} =$

$\dfrac{A}{\mu} = \dfrac{\min(A, B)}{\mu}$. So, no matter whether $A > B$ or $A < B$, we have

$A + \dfrac{1-\mu}{\mu} B \geq \dfrac{\min(A, B)}{\mu}$, which is the claim, and now we really are done.

CP 5.7. Assume that a stable triple-star system does form. Then, because of the time-reversible nature of Newtonian gravity, if we run time backward the *stable* triple-star system will *remain* a triple-star system. That, after all, is what *stable* means. The original single star and the original binary will *not* suddenly reappear. But since that contradicts time reversibility for the original single star + binary system configuration, we conclude that the assumption a stable triple-star system could form in the first place is false.

CP 5.8. From (5.8.1) we have $\dfrac{d^2 z}{dt^2} = -\dfrac{z}{\left(z^2 + \dfrac{1}{4}\right)^{3/2}}$, with the ini-

tial conditions $z(0) = 0$, $\dfrac{dz}{dt}\big|_{t=0} = V$, and the final condition of $v(z = \infty) = 0$. This final condition says the mass comes to rest after having moved infinitely far up the z-axis; that is, the mass has just escaped the binary pair. By the chain rule we have $v = \dfrac{dz}{dt} = \dfrac{dz}{dv} \cdot \dfrac{dv}{dt}$, and so $\dfrac{dv}{dt} = \dfrac{v}{\dfrac{dz}{dv}} = v \dfrac{dv}{dz}$. Since the accel-

eration of the mass is $\dfrac{dv}{dt} = \dfrac{d^2 z}{dt^2}$, we then have the alter-

native to (5.8.1) of $v \dfrac{dv}{dz} = \dfrac{dv}{dt} = \dfrac{d^2 z}{dt^2}$, or $v \dfrac{dv}{dz} = -\dfrac{z}{\left(z^2 + 1/4\right)^{3/2}}$.

Thus, $v\, dv = -\dfrac{z\, dz}{\left(z^2 + 1/4\right)^{3/2}}$, and so, integrating indefinitely, $\dfrac{1}{2} v^2 +$

$C = -\displaystyle\int \dfrac{z\, dz}{\left(z^2 + 1/4\right)^{3/2}}$. Changing variable to $u = z^2 + \dfrac{1}{4}$ (and so

$du = 2z\, dz$, which says $dz = \dfrac{du}{2z}$), we have $\dfrac{1}{2} v^2 + C = -\dfrac{1}{2} \displaystyle\int \dfrac{du}{u^{3/2}}$, or $\frac{1}{2}v^2 + C = -\frac{1}{2}\left(-2u^{-1/2}\right) = u^{-1/2} = \frac{1}{\sqrt{u}} = \frac{1}{\sqrt{z^2 + 1/4}}$. Since $v =$

V when $z = 0$, we have $\frac{1}{2}V^2 + C = \frac{1}{\sqrt{1/4}} = 2$, or $C = 2 - \frac{1}{2}V^2$.

Thus, $\frac{1}{2}v^2 + 2 - \frac{1}{2}V^2 = \frac{1}{\sqrt{z^2 + 1/4}}$. Since $v(z = \infty) = 0$, we have

$2 - \frac{1}{2}V^2 = 0$, and so the escape speed along the z-axis is $V = 2$.

CP 6.1. With reference to Figure 6.8.1, before the switch is closed, C is charged to V, and there are no currents flowing. In particular, the inductor current $i_1 = 0$. Thus, just *after* the switch is closed the inductor current is still zero. Therefore, at $t = 0+$ the initial capacitor discharge current of $\frac{V}{R}$ is the switch current. That is, $i_S(0+) = \frac{V}{R}$. Then, after a long time has passed (and so the circuit has reached its steady-state condition), the battery current is $\frac{V}{R}$ because then the inductor is a "short" ; also, the zero-resistance switch completely bypasses the now completely discharged C, and so all that current flows through the switch. That is, $i_S(\infty) = \frac{V}{R}$, and so we see that $i_S(0+) = i_S(\infty)$. Once the switch has been closed we can write Kirchhoff's voltage drop law around each of the circuit's two loops as follows:

$$-V + L\frac{di_1}{dt} + i_1 R = 0, \quad i_1(0+) = 0 \qquad (a)$$

$$-V + \frac{1}{C}\int i_2\, dt + i_2 R = 0, \quad i_2(0+) = \frac{V}{R}. \qquad (b)$$

From (a), $\frac{di_1}{dt} + \frac{R}{L}i_1 = \frac{V}{L}$ with $i_1(0+) = 0$, for which (from experience with first-order differential equations), we can immediately write the solution $i_1(t) = \frac{V}{R}\left(1 - e^{-\frac{R}{L}t}\right)$. If we differentiate (b), we have $R\frac{di_2}{dt} + \frac{1}{C}i_2 = 0$ with $i_2(0+) = \frac{V}{R}$ and, again from experience, we immediately have $i_2(t) = \frac{V}{R}e^{-\frac{1}{RC}t}$. Since Kirchhoff's current node law tells us that $i_S(t) = i_1(t) + i_2(t)$, then

$i_S(t) = \dfrac{V}{R}\left(1 - e^{-\frac{R}{L}t}\right) + \dfrac{V}{R}e^{-\frac{1}{RC}t} = \dfrac{V}{R}\left(1 + e^{-\frac{1}{RC}t} - e^{-\frac{R}{L}t}\right)$. Notice that this mathematical result says $i_S(0+) = \dfrac{V}{R} = i_S(\infty)$, just as we argued above from physics. Notice, too, however, that the mathematical result tells us even more: if $\dfrac{R}{L} = \dfrac{1}{RC}$, then the two exponentials cancel and $i_S(t) = \dfrac{V}{R}$ at every instant. To find the extrema of $i_S(t)$ I'll set $\dfrac{di_S}{dt} = 0$. Thus, $\dfrac{di_S}{dt} = \dfrac{V}{R}\left(\dfrac{R}{L}e^{-\frac{R}{L}t} - \dfrac{1}{RC}e^{-\frac{1}{RC}t}\right)$, and this is zero if $\dfrac{R}{Le^{\frac{R}{L}t}} = \dfrac{1}{RCe^{\frac{1}{RC}t}}$ or, $\dfrac{e^{\frac{1}{RC}t}}{e^{\frac{R}{L}t}} = \dfrac{L}{R^2C} = e^{\frac{1}{RC}t - \frac{R}{L}t} = e^{t\left(\frac{1}{RC} - \frac{R}{L}\right)}$. Thus, $t\left(\dfrac{1}{RC} - \dfrac{R}{L}\right) = \ln\left(\dfrac{L}{R^2C}\right)$, and so the extrema of $i_S(t)$ occurs at $t = \dfrac{\ln\left(\dfrac{L}{R^2C}\right)}{\dfrac{1}{RC} - \dfrac{R}{L}}$. For the given component values ($L = 0.1H = 10^{-1}$, $R = 1,000\Omega = 10^3$, and $C = 0.01\mu fd = 10^{-8}$) we have the extrema of $i_S(t)$ occurring when $t = \dfrac{\ln\left(\dfrac{10^{-1}}{10^6 \cdot 10^{-8}}\right)}{\dfrac{1}{10^3 \cdot 10^{-8}} - \dfrac{10^3}{10^{-1}}} = \dfrac{\ln(10)}{10^5 - 10^4} = \dfrac{\ln(10)}{90,000} = 25.58 \ \mu\,\text{sec}$. The switch current at that time is, using $V = 100$ volts, $i_S(t) = \dfrac{100}{1,000}\left(1 + e^{-\frac{25.58 \cdot 10^{-6}}{10^3 \cdot 10^{-8}}} - e^{-\frac{10^3 \cdot 25.58 \cdot 10^{-6}}{10^{-1}}}\right)$ A $= 0.1\left(1 + e^{-2.558} - e^{-0.2558}\right)$ A $= 0.0303$ A. Since $i_S(0+) = i_S(\infty) = \dfrac{V}{R} = 0.1$ A, then $i_S(25.58\mu\,\text{sec}) = 0.0303$ A is the *minimum* switch current.

CP 6.2. In the notation of Figure S6.2 (which is a near reproduction of Figure 6.3.4), we can write the following equations relating the input current I to the two node voltages X and Y: $I = \dfrac{V - X}{1} + \dfrac{V - Y}{1}$, $\dfrac{V - X}{1} = \dfrac{X}{2} + \dfrac{X - Y}{2}$, and $\dfrac{V - Y}{1} + \dfrac{X - Y}{2} = \dfrac{Y}{1}$, and so $I = V - X + V - Y$, $2V - 2X = X + X - Y$, and $2V - 2Y + X - Y = 2Y$, or $I = 2V - X - Y$, $2V = 4X - Y$, and $2V = 5Y - X$. Thus, $X = 5Y - 2V$. Then $2V = 20Y - 8V - Y = 19Y - 8V$, or $19Y = 10V$, and so

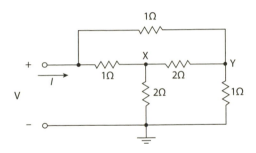

FIGURE S6.2. A bridged-tee

$Y = \dfrac{10}{19}V$. From this we have $X = \dfrac{50}{19}V - 2V = \dfrac{50}{19}V - \dfrac{38}{19}V = \dfrac{12}{19}V$,

and so $I = 2V - \dfrac{12}{19}V - \dfrac{10}{19}V = \dfrac{38}{19}V - \dfrac{22}{19}V = \dfrac{16}{19}V$. This gives $R =$

$\dfrac{V}{I} = \dfrac{19}{16}\ \Omega = 1.1875\Omega$, exactly as computed by EWB.

CP 6.3. To calculate the phase-shift through the second stage, we must account for the "loading" effect of the third stage, as shown in Figure S6.3. In the notation of that figure we

have $\mathbf{V}_x = \dfrac{\mathbf{VZ}}{\mathbf{Z} + \dfrac{1}{i\omega C}}$, where $\mathbf{Z} = \dfrac{R(\dfrac{1}{i\omega C} + R)}{2R + \dfrac{1}{i\omega C}} = \dfrac{R + i\omega R^2 C}{1 + i2\omega RC}$ (as

usual, all complex quantities are in bold). $\mathbf{V}$ and $\mathbf{V}_x$, as complex quantities, each have a phase angle in the complex plane, and in fact, the angle of $\mathbf{V}_x$ minus the angle of $\mathbf{V}$ is the phase shift through the second stage that we are to calculate. If we denote the second-stage phase shift by ϕ_2, then electrical engineers write $\phi_2 = \angle \mathbf{V}_x - \angle \mathbf{V}$, where

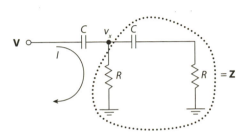

FIGURE S6.3. The "loaded" second stage

the symbol $\angle$ means "angle of." (I offer this explanation because, in one of my previous books, I used this notation without explanation and was rather sharply informed by one academic mathematician that he had no idea, none at all, of what it could possibly mean. I had thought it all rather self-evident in context, but still, if he really was stumped, then perhaps this *is* necessary to say.) Now, since the angle of a quotient of complex quantities is the angle of the numerator minus the angle of the denominator, we have $\phi_2 = \angle \dfrac{\mathbf{V}_x}{\mathbf{V}}$. From above we have $\dfrac{\mathbf{V}_x}{\mathbf{V}} = \dfrac{\mathbf{Z}}{\mathbf{Z} + \dfrac{1}{i\omega C}} =$

$$\frac{i\omega C \mathbf{Z}}{1 + i\omega C \mathbf{Z}} = \frac{i\omega C \dfrac{R + i\omega R^2 C}{1 + i2\omega RC}}{1 + i\omega C \dfrac{R + i\omega R^2 C}{1 + i2\omega RC}} = \frac{-\omega^2 R^2 C^2 + i\omega RC}{1 + i2\omega RC + i\omega RC - \omega^2 R^2 C^2} =$$

$\dfrac{-\omega^2 R^2 C^2 + i\omega RC}{1 - \omega^2 R^2 C^2 + i3\omega RC}$. At the $180°$ phase-shift frequency of

$\omega = \omega_0 = \dfrac{1}{RC\sqrt{6}}$, this becomes $\dfrac{-\dfrac{1}{6} + i\dfrac{1}{\sqrt{6}}}{1 - \dfrac{1}{6} + i\dfrac{3}{\sqrt{6}}} = \dfrac{-1 + i\sqrt{6}}{5 + i3\sqrt{6}}$.

Remember, ϕ_2 is the angle of this complex quantity. Noticing that the numerator is in the second quadrant, we can now immediately write that $\phi_2 = 180° - \tan^{-1}(\sqrt{6}) - \tan^{-1}(\dfrac{3\sqrt{6}}{5})$, and a few button pushes on a hand calculator gives us our answer: $\phi_2 = 56.440°$. Since the total phase shift through all three stages is $180°$, and since we calculated (in the text) that the third-stage phase shift is $\phi_3 = 67.792°$, then the first-stage phase shift must be $\phi_1 = 55.768°$. We could calculate ϕ_1 *directly* by repeating the above calculation for the first stage while accounting for the loading on it by the second *and* the third stages. That is *not* a nice calculation—I know, I've done it!—and if you want to do it, too, then you should eventually arrive at the result that ϕ_1 is the angle of the complex quantity $\dfrac{-3\omega^2 R^2 C^2 + i\omega RC(1 - \omega^2 R^2 C^2)}{1 - 6\omega^2 R^2 C^2 + i\omega RC(5 - \omega^2 R^2 C^2)}$. Further, if you then substitute $\omega = \dfrac{1}{RC\sqrt{6}}$ into this you'll find that its angle is indeed $55.768°$.

CP 6.4. Notice that the circuit of Figure 6.8.2 has positive feedback as well as negative feedback. So, I'll *assume* that the ideal diff-amp assumption still holds. Then, if we write the voltage on both of the diff-amp's inputs as x, and v and y as the input and output voltages (as shown in the figure), respectively, we have $\dfrac{0-x}{R_1} = \dfrac{x-y}{R_2}$, or $-xR_2 = xR_1 - yR_1$, or $yR_1 = x(R_1 + R_2)$, or $x = \dfrac{R_1}{R_1 + R_2}y$. Also, $\dfrac{v-x}{R_3} = C\dfrac{dx}{dt} + \dfrac{x-y}{R_4}$, or $v - x = R_3C\dfrac{dx}{dt} + \dfrac{R_3}{R_4}(x - y)$, or $v = R_3C\dfrac{dx}{dt} + \left(1 + \dfrac{R_3}{R_4}\right)x - \dfrac{R_3}{R_4}y$, or $v = R_3C\dfrac{R_1}{R_1 + R_2}\cdot\dfrac{dy}{dt} + \left(1 + \dfrac{R_3}{R_4}\right)\dfrac{R_1}{R_1 + R_2}y - \dfrac{R_3}{R_4}y$, or $v = \dfrac{R_1R_3C}{R_1 + R_2}\cdot\dfrac{dy}{dt} + \dfrac{R_1(R_3 + R_4)}{R_4(R_1 + R_2)}y - \dfrac{R_3}{R_4}y$. Now, *if* the last two terms on the right *cancel*, then v and $\dfrac{dy}{dt}$ would be directly proportional, and we would have an integrator. This cancelation occurs if $\dfrac{R_1(R_3 + R_4)}{R_4(R_1 + R_2)} - \dfrac{R_3}{R_4} = 0$, which says $R_1(R_3 + R_4) - R_3(R_1 + R_2) = 0$, or $R_1R_3 + R_1R_4 - R_1R_3 - R_3R_2 = 0$, or $R_1R_4 = R_3R_2$, as claimed in the problem statement. Assuming we have cancelation, then $v\dfrac{R_1 + R_2}{R_1R_3C} = \dfrac{dy}{dt}$, or $y(t) = \dfrac{R_1 + R_2}{R_1R_3C}\int v(t)$ dt, which is a *positive* gain integrator. For this circuit to work requires precisely matched resistors, however, which is a significant negative feature.

CP 6.5. As pointed out in the problem statement, each of the $2R$ resistors in the input ladder circuit is connected to a ground, and so we can draw the equivalent ladder circuit of Figure S6.5. There I have assumed that the current in the rightmost $2R$ resistor is 1 A, and then, using the analysis trick of note 10 in Chapter 6, you should be able to quickly see that the other currents are $I_0 = 1$ A, $I_1 = 2$ A, $I_2 = 4$ A, and $I_3 = 8$ A. The voltage at the far left, V_{ref}, is then equal to $16R$ V and so $R = \dfrac{1}{16}V_{ref}$. So, the total current routed to the virtual ground input of the diff-amp (and so into the feedback resistor R) is $S_0I_0 + S_1I_1 + S_2I_2 + S_3I_3 = S_0 + 2S_1 + 4S_2 + 8S_3$, where the S_k are binary variables, that is, equal to either 0 or 1, depending on each switch position. Thus, $\dfrac{0 - v_{out}}{R} = S_0 + 2S_1 + 4S_2 + 8S_3$, or

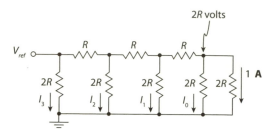

FIGURE S6.5. The $R/2R$ DAC's input ladder circuit

$v_{out} = -(S_0 + 2S_1 + 4S_2 + 8S_3)\,R$. But, since $R = \dfrac{1}{16}V_{ref}$, then $v_{out} = -\left(\dfrac{1}{16}S_0 + \dfrac{1}{8}S_1 + \dfrac{1}{4}S_2 + \dfrac{1}{2}S_3\right)V_{ref}$, as was to be shown.

CP 6.6. From (6.6.10), using $s = i\omega$, we have the transfer function of the circuit as

$$
\mathbf{H} = \frac{\dfrac{R_b\,\dfrac{1}{i\omega C}}{R_b + \dfrac{1}{i\omega C}}}{R_a + i\omega L + \dfrac{R_b\,\dfrac{1}{i\omega C}}{R_b + \dfrac{1}{i\omega C}}} = \frac{\dfrac{R_b}{1 + i\omega R_b C}}{R_a + i\omega L + \dfrac{R_b}{1 + i\omega R_b C}}
$$

$$
= \frac{R_b}{(R_a + i\omega L)(1 + i\omega R_b C) + R_b}
$$

$$
= \frac{R_b}{R_a + i\omega R_a R_b C + i\omega L - \omega^2 R_b L C + R_b}
$$

$$
= \frac{R_b}{R_a + R_b - \omega^2 R_b L C + i\omega(R_a R_b C + L)}
$$

$$
= \frac{1}{1 + \dfrac{R_a}{R_b} - \omega^2 LC + i\omega\left(R_a C + \dfrac{L}{R_b}\right)},
$$

and so

$$
|\mathbf{H}|^2 = \frac{1}{\left[1 + \dfrac{R_a}{R_b} - \omega^2 LC\right]^2 + \omega^2\left[R_a C + \dfrac{L}{R_b}\right]^2}
$$

$$= \cfrac{1}{\left[1 + \dfrac{R_a}{R_b}\right]^2 - 2LC\left[1 + \dfrac{R_a}{R_b}\right]\omega^2 + \omega^4 L^2 C^2 + \omega^2 \left[R_a C + \dfrac{L}{R_b}\right]^2},$$

or

$$|\mathbf{H}|^2 = \cfrac{1}{\omega^4 L^2 C^2 + \omega^2 \left\{\left[R_a C + \dfrac{L}{R_b}\right]^2 - 2LC\left[1 + \dfrac{R_a}{R_b}\right]\right\} + \left[1 + \dfrac{R_a}{R_b}\right]^2}.$$

$|\mathbf{H}|^2$ will be maximum when the denominator is a minimum. So, differentiating the denominator and setting that equal to zero (when $\omega = \omega_m$) gives $4\omega_m^3 L^2 C^2 + 2\omega_m \left\{\left[R_a C + \dfrac{L}{R_b}\right]^2 - 2LC\left[1 + \dfrac{R_a}{R_b}\right]\right\} = 0$, and so $2\omega_m^2 L^2 C^2 + \left\{\left[R_a C + \dfrac{L}{R_b}\right]^2 - 2LC\left[1 + \dfrac{R_a}{R_b}\right]\right\} = 0$, or $\omega_m = \dfrac{1}{LC}\sqrt{\dfrac{2LC\left[1 + \dfrac{R_a}{R_b}\right] - \left[R_a C + \dfrac{L}{R_b}\right]^2}{2}}$.

Notice that if the numerator of the fraction under the radical sign isn't positive, then ω_m is imaginary (!)—which is simply the math telling us that in that case $|\mathbf{H}|^2$ *has no maximum* (as frequency increases the transfer function magnitude will just decrease monotonically). For our problem, as you'll next see, the numerator *is* positive. If $R_a = 10^6 \ \Omega$, $R_b = 8 \cdot 10^6 \ \Omega$, $L = 1$ H, and $C = 10^{-12}$ F, then

$$f_m = \frac{\omega_m}{2\pi} = \frac{1}{2\pi \cdot 1 \cdot 10^{-12}}$$

$$\times \sqrt{\frac{2 \cdot 1 \cdot 10^{-12}\left(1 + \dfrac{10^6}{8 \cdot 10^6}\right) - \left(\dfrac{1}{8 \cdot 10^6} + 10^6 \cdot 10^{-12}\right)^2}{2}}$$

$$= \frac{10^{12}}{2\pi} \sqrt{\frac{2 \cdot 10^{-12} \cdot 1.125 - \left(0.125 \cdot 10^{-6} + 10^{-6}\right)^2}{2}}$$

$$= \frac{10^{12}}{2\pi} \sqrt{\frac{2.25 \cdot 10^{-12} - 1.265625 \cdot 10^{-12}}{2}}$$

$$= \frac{10^{12} \cdot 10^{-6}}{2\pi} \sqrt{\frac{2.25 - 1.265625}{2}}$$

$$= \frac{10^6}{2\pi} (0.70156) = 111.7 \, \text{kHz}.$$

And at this frequency,

$$|\mathbf{H}|^2 = \cfrac{1}{\begin{aligned}&\left(2\pi \cdot 1.117 \cdot 10^5\right)^4 \left(10^{-12}\right)^2 + \left(2\pi \cdot 1.117 \cdot 10^5\right)^2 \\ &\left\{\left(10^6 \cdot 10^{-12} + \frac{1}{8 \cdot 10^6}\right)^2 - 2 \cdot 10^{-12}\left(1 + \frac{10^6}{8 \cdot 10^6}\right)\right\} \\ &\qquad\qquad\qquad\qquad + \left(1 + \frac{10^6}{8 \cdot 10^6}\right)^2\end{aligned}}$$

$$= \cfrac{1}{\begin{aligned}&0.2426 + 49.2568 \cdot 10^{10}(1.265625 \cdot 10^{-12} \\ &\qquad -2.25 \cdot 10^{-12}) + 1.265625\end{aligned}}$$

$$= \frac{1}{0.2426 - 0.4849 + 1.2656} = \frac{1}{1.0233}$$

and so

$$|\mathbf{H}| = \sqrt{\frac{1}{1.0233}} = 0.9885.$$

Both f_m and $|\mathbf{H}|$, as we've just calculated, are in *excellent* agreement with EWB's simulation results.

CP 6.7   In the first circuit, the output voltage of the multiplier is $x v_{out}$. Because of the ideal diff-amp assumption, that voltage equals y. So, $x v_{out} = y$, or $v_{out} = \frac{y}{x}$. The first circuit is a *divider*. In the second circuit, the output voltage of the multiplier is v_{out}^2. So, $v_{out}^2 = y$, or $v_{out} = \sqrt{y}$. For this square-root circuit to work it is required that

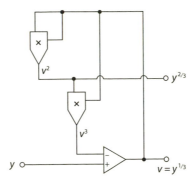

FIGURE S6.7. An EWB circuit for $y^{1/3}$ and $y^{2/3}$

the voltage y never go negative. Can you see what would happen if $y < 0$? Hint: v_{out} would *not* enter the fourth dimension and take on imaginary values (now wouldn't *that* be something!—what, I wonder, would a four-dimensional electrical spark look like?) but rather would simply rail to its most negative possible value (an EWB simulation agrees). See if you can explain *why*, and you might find it helpful to start with the observation that the multiplier output is *always* non-negative.

As a final comment on circuits with multipliers in a negative feedback loop, you should now be able to see how easy it is to design new circuits to calculate almost any fractional power you wish. For example, suppose that given a voltage y you need to create $y^{1/3}$ and/or $y^{2/3}$ for an analog EWB differential equation solution (the $y^{2/3}$ case, in particular, occurs in the falling raindrop problem mentioned in note 17 of Chapter 6 for both of the mass accretion rules given there). The circuit in Figure S6.7 generates both voltages. Since this circuit is generating *cube* roots, it is now perfectly okay for voltage y to go negative without causing any problems.

ACKNOWLEDGMENTS

Readers of my earlier books know that I often mention the companionship I receive, while I write, from my cats Heaviside, Maxwell, and Tigger. Sadly, Heaviside and Maxwell are now both gone, to romp forever with other literary cats among the endless rows of books in the Great Library in the Sky. But Tigger is still with me, and still doing his job (along with his new friend Vixen, just arrived from the local animal shelter). As I write they are snuggled up together in the chair across from my desk, giving me their practiced, pathetic "we are starving" looks while they watch me type. (Don't feel sorry for them—they are both fat!)

If you ask most professors to list their top ten gripes, I suspect somewhere in just about every list would be students complaining about homework or test scores. And so, when you read the answer to Challenge Problem I.3 and learn that, while writing this book, I complained to one of my Stanford University math professors about a homework assignment from a half-century ago, you'll understand why I absolutely must thank John Lamperti (now emeritus professor of mathematics at Dartmouth College) for being a good sport in dealing with my pleading after five decades for more points. I admit it—that *is* pathetic!

There are many people behind the creation of a book in addition to the author. At Princeton University Press I thank Dimitri Karetnikov, Debbie Tegarden, Carmina Alvarez, Stefani Wexler, and (of course!) my wonderful editor Vickie Kearn. In Chicago, I am indebted, once again, to my favorite copyeditor Marjorie Pannell. At the University of New Hampshire (UNH) the talented folks in the Photographic Services Department made jpeg and tiff images for me with, seemingly, the wave of a magic wand. In the UNH Electrical Engineering Department, it was from Frank Hludik's desk drawer that I retrieved (with Frank's blessing) the discarded disk of *Electronics Workbench* that stars in Chapter 6. My high school friend, John Baker, deserves special thanks for Fedexing his 1956 class yearbook to me—and again to Princeton—for production of the frontispiece and solution section photographs.

And, as she has before (and no doubt will again), my talented wife Pat saved me from numerous horrific writing disasters with her phenomenal computer skills.

Paul J. Nahin
Lee, New Hampshire
October 2010

INDEX

ALSO BY PAUL J. NAHIN

Oliver Heaviside (1988, 2002)

Time Machines (1993, 1999)

The Science of Radio (1996, 2001)

An Imaginary Tale (1998, 2007)

Duelling Idiots (2000, 2002)

When Least Is Best (2004, 2007)

Dr. Euler's Fabulous Formula (2006, 2011)

Chases and Escapes (2007)

Digital Dice (2008)

Mrs. Perkins's Electric Quilt (2009)

Time Travel (2011)